Sensitizing Agents for Cancer Resistant to Cell Mediated Immunotherapy

IMMUNOTHERAPY IN RESISTANT CANCER: FROM THE LAB BENCH WORK TO ITS CLINICAL PERSPECTIVES

VOLUME 2

Sensitizing Agents for Cancer Resistant to Cell Mediated Immunotherapy Series

Series Editor: Benjamin Bonavida, PhD

Volume 1: Autophagy in Immune Response: Impact on Cancer Immunotherapy
Edited by Salem Chouaib

Volume 2: Immunotherapy in Resistant Cancer: From the Lab Bench Work to Its Clinical Perspectives
Edited by Jorge Morales-Montor and Mariana Segovia-Mendoza

Upcoming Volumes:

Immunotherapeutic Strategies for the Treatment of Glioma
Edited by Michael Lim and Christopher Jackson

Breaking Tolerance to Unresponsiveness to Immunotherapy by Natural Killer Cells
Edited by Sandro Matosevic

NK cells in cancer immunotherapy: Successes and Challenges
Edited by Anahid Jewett

Breaking Tolerance to Pancreatic Cancer Unresponsive to Immunotherapy
Edited by Kasuya Hideki and Itzel Bustos Villalobos

Sensitizing Agents for Cancer Resistant to Cell Mediated Immunotherapy

IMMUNOTHERAPY IN RESISTANT CANCER: FROM THE LAB BENCH WORK TO ITS CLINICAL PERSPECTIVES

VOLUME 2

Edited by

JORGE MORALES-MONTOR
Departamento de Inmunología, Instituto de Investigaciones Biomédicas, Universidad Nacional Autónoma de México, Mexico City, México

MARIANA SEGOVIA-MENDOZA
Departamento de Farmacología, Facultad de Medicina, Universidad Nacional Autónoma de México, Mexico City, México

ELSEVIER

ACADEMIC PRESS
An imprint of Elsevier

Academic Press is an imprint of Elsevier
125 London Wall, London EC2Y 5AS, United Kingdom
525 B Street, Suite 1650, San Diego, CA 92101, United States
50 Hampshire Street, 5th Floor, Cambridge, MA 02139, United States
The Boulevard, Langford Lane, Kidlington, Oxford OX5 1GB, United Kingdom

Library of Congress Cataloging-in-Publication Data
A catalog record for this book is available from the Library of Congress

British Library Cataloguing-in-Publication Data
A catalogue record for this book is available from the British Library

ISBN: 978-0-12-822028-3

For information on all Academic Press publications
visit our website at https://www.elsevier.com/books-and-journals

Publisher: Stacy Masucci
Acquisitions Editor: Rafael Teixeira
Editorial Project Manager: Samantha Allard
Production Project Manager: Punithavathy Govindaradjane
Cover Designer: Greg Harris

Typeset by SPi Global, India

Cover Figure Insert

The cover figure shows how a tumor cell (gray circle) is being recognized by different immune cells, soluble factors, and antibodies. This recognition is strengthened after pharmacological or biological therapies (white molecules).

The editors would like to thank Lic. Rosa Flores Hernández and Mtro. Andi Espinoza Sánchez, Facultad de Medicina, Universidad Nacional Autónoma de México (UNAM), for the realization of the cover image.

Aims and Scope for Series "Sensitizing Agents for Cancer Resistant to Cell Mediated Immunotherapy"

The role of the immune system in the eradication of cancers has been investigated for several decades with controversial findings. The controversy was the result of a poor understanding of the underlying mechanisms that govern responsiveness and unresponsiveness. Hence, significant advances have been made with respect to the regulation of the host immune response against cancer and several immunotherapeutics have been recently introduced and used clinically. These include both antibody and cell-mediated immunity targeting the cancer cells. Such immunotherapies led to significant clinical responses in various cancer types that were unresponsive to conventional therapies. However, only a subset of cancer patients responds to such immunotherapeutics and also there is a responding subset that develops resistance to further treatment. Various studies have examined potential underlying mechanisms involved in resistance and identified a variety of gene products that play pivotal roles in maintaining the resistant phenotype of the cancer cells to cell-mediated immunotherapy.

The main objective of the proposed series "Sensitizing Agents for Cancer Resistance to Cell Mediated Immunotherapy" is the development of individual volumes that are focused on the application of particular sensitizing agents that, when used in combination with cell-mediated immunotherapy, result in the reversal of resistance.

A variety of different classes of immunosensitizing agents has been reported. Each individual volume will focus on one class of immunosensitizing agents and their effects on the reversal of cell-mediated immune resistance in different cancers. Emphasis will be on biochemical, molecular, and genetic mechanisms by which the sensitizing agents mediate their effects individually and/or in combination with immunotherapy. Each editor will compile non-overlapping review chapters on the therapeutic role of specific sensitizing agents used in combination with conventional immunotherapy and the reversal of resistance. There will also be an emphasis on discrimination of responses obtained in various cancer types.

The scope of the series is to provide updated information to scientists and clinicians that is valuable in their quest to gather information, carry out new investigations, and develop novel immunosensitizing agents that are both more potent and also that might be active whereby the existing ones were not active.

Dr. Benjamin Bonavida PhD
(Series Editor)

About the Series Editor

Dr. Benjamin Bonavida, PhD (Series Editor), is currently distinguished research professor at the University of California, Los Angeles (UCLA). He is affiliated with the Department of Microbiology, Immunology and Molecular Genetics, UCLA David Geffen School of Medicine. His research career, thus far, has focused on investigations in the fields of basic immunochemistry and cancer immunobiology. His research investigations have ranged from the biochemical, molecular, and genetic mechanisms of cell-mediated killing and tumor cell resistance to chemo-immuno cytotoxic drugs. The reversal of tumor cell resistance was investigated by the use of various selected sensitizing agents based on molecular mechanisms of resistance. In these investigations, there was the newly characterized dysregulated NF-κB/Snail/YY1/RKIP/PTEN loop in many cancers that was reported to regulate cell survival, proliferation, invasion, metastasis, and resistance. Emphasis was focused on the roles of the tumor suppressor Raf kinase inhibitor protein (RKIP) and the tumor promoter Yin Yang 1 (YY1) and the role of nitric oxide as a chemo-immuno-sensitizing factor. Many of the aforementioned studies are centered on the clinical challenging features of cancer patients' failure to respond to both conventional and targeted therapies. The editor has been active in the organization of regular sequential international miniconferences that are highly focused on the roles of YY1, RKIP, and nitric oxide in cancer and their potential therapeutic applications.

Several books edited or coedited by the editor have been published. In addition, the editor has been the series editor of books (over 23) published by Springer on *Resistance to Anti-Cancer Targeted Therapeutics*. In addition, the editor is presently the series editor of three series published by Elsevier/Academic Press on *Cancer Sensitizing Agents for Chemotherapy, Sensitizing Agents for Cancer Resist to Cell Mediated Immunotherapy*, and *Breaking Tolerance to Anti-Cancer Immunotherapy*. Lastly the editor is the editor in chief of the Journal *Critical Reviews in Oncogenesis*. The editor has published over 500 research publications and reviews in various scientific journals of high impact.

Acknowledgments: The Editor wishes to acknowledge the excellent editorial assistance of Ms. Inesa Navasardyan, who has worked diligently in the completion of this volume, namely, in both the editing and formatting of the various contributions of this volume.

The Editor acknowledges the Department of Microbiology, Immunology, and Molecular Genetics and the UCLA David Geffen School of Medicine for their continuous support. The Editor also acknowledges the assistance of Mr. Rafael Teixeira, Acquisitions Editor for Elsevier/Academic Press, and the excellent assistance of Ms. Samantha Allard, Editorial Project Manager for Elsevier/Academic Press, for their continuous cooperation throughout the development of this book.

Aims and Scope of the Volume

We edited a book, that is dedicated to the assembly of the latest information of anticancer-mediated immunotherapy and biological factors associated with tumor microenvironment, to break possible resistance to immunotherapies. The role of the immune system in the eradication of cancers has been investigated for several decades with controversial findings. The controversy was the result of a poor understanding of the underlying mechanisms that govern responsiveness and unresponsiveness. Hence, significant advances have been made with respect to the regulation of the host immune response against cancer and several immunotherapeutics have been recently introduced and used clinically. These include both antibody and cell-mediated immunity targeting the cancer cells. Such immunotherapies led to significant clinical responses in various cancer types that were unresponsive to conventional therapies. However, only a subset of cancer patients responds to such immunotherapeutics and also there is a responding subset that develops resistance to further treatment. Various studies have examined the potential underlying mechanisms involved in resistance and identified a variety of gene products that play pivotal roles in maintaining the resistant phenotype of the cancer cells to cell-mediated immunotherapy. A variety of different classes of immunosensitizing agents has been reported. Our book focuses on immunosensitizing agents and their effects on the reversal of cell-mediated immune resistance in different cancers. Emphasis was put on the biochemical, molecular, and genetic mechanisms by which the sensitizing agents mediate their effects individually and/or in combination with immunotherapy. The chapters in this book are related to the therapeutic role of specific sensitizing agents used in combination with conventional immunotherapy and the reversal of resistance. There is also an emphasis on discrimination of responses obtained in various cancer types. The scope of the book is to provide updated information to scientists and clinicians that is valuable in their quest to gather information, carry out new investigations, and develop novel immunosensitizing agents that are both more potent and also might be active whereby the existing ones were not active.

About the Volume Editors

Dr. Jorge Morales-Montor studied biology at the Iztacala UNAM Faculty of Higher Studies, obtaining the title in 1992. He obtained a doctor's degree in October 1997. His doctoral thesis was recognized with the Lola and Igo Flisser-PUIS Award to the best graduate thesis at the national level in the area of parasitology, a recognition that he has also later received as a tutor, since one of his doctorate students won the same award in 2008. In November 1997, he began a postdoctoral stay at the Department of Cellular Biology at the University of Georgia, in the laboratory of Dr. Raymond T. Damián, one of the most recognized parasitologists in the world. Dr. Morales received a grant from the Fogarty Foundation (one of the most prestigious in Ibero-America) to carry out research on schistosomiasis in the mandrel, being a Pan-American Fellow for 4 years. Dr. Ray Damián would write years later, which assured that without a doubt, Jorge Morales-Montor had been the best postdoctoral researcher with whom he collaborated in his nearly 35-year career. He was repatriated to Mexico in 2001 by CONACYT and joined the Department of Immunology of the Institute of Biomedical Research of UNAM as Associate Investigator "C". In just 9 years, he managed to climb the entire ladder of university academic positions, to become a Definitive C Titular Researcher at the Institute of Biomedical Research. The same is reflected in the Level of Premiums for Academic Performance, where it has reached the highest level currently: Level D, for the third consecutive period. Also in the National System of Researchers, he has had the same growth, starting in 1997 as a candidate, and, to date, being promoted to Level III, the highest, for the third consecutive period. Dr. Morales-Montor has been invited to participate in different congresses (more than 100). In addition, he is part of the editorial committee of more than 15 indexed international journals. Some of his most important contributions are partially determining the role of steroid hormones in immunological sexual dimorphism, in the polarization of the immune response, and in the antigenic presentation. He has also made very relevant studies in relation to how different physiological stages, how the estrous cycle, age, sex, or pregnancy affect the functioning of the immune, endocrinological, and nervous system, and what molecules could be the determinants in this context of net. It has been shown that the central nervous system is involved in the regulation of the immune response to parasitic infections, and the effect of this activation on various behaviors of the infected host. But the central nervous system has provided interesting data about its impact in the parasitology approach. For instance, a modern concept is depicted by how the central nervous system modulates the gene and proteomic regulation of the different sex steroids in parasites, which are involved in important functions of parasites such as establishment, growth, and reproduction. Finally, the practical use of the knowledge acquired by the earlier mentioned studies has been applied to a theory that he calls old drugs, new uses: the use of hormones and antihormones as antiparasitic therapy. He has also entered the study of environmental contamination, specifically endocrine disruptors and disease, studying their role in two very important diseases in the country: cancer and obesity, projects with which he has formed two consortiums of investigation. Its results are a very important contribution to the health of both Mexicans and Latin Americans in general, since this is where serious health problems related to parasitic infections, cancer, and obesity are concentrated. His investigations are characterized by an exhaustive and meticulous experimental work, and his scientific production already has 153 articles in international indexed journals, and the majority as the first author or corresponding author. He has more than 3000 citations to his works, and an h-index of 25, one of the highest in the country's scientific community. His articles published in high-impact international journals include Nature, PlosOne, Journal of Immunology, Journal of Infectious Diseases, Journal of Interferon and Cytokine Research, and among others. In fact, recently, his 2015 article, The Role of Cytokines in Breast Cancer Development and Progression, published in the Journal of Interferon and Cytokine Research, was the subject of a press release released by Mary Ann Liebert Publications. This is sent all over the world, to newspapers, Journals, scientists, radio, TV, popular magazines, to what is considered as a very important contribution in a certain area of science. Very few scientific articles are released as "press release." He is also the 4th most cited author in the area of parasitology in the country. He has also edited several books and published more than 55 chapters in books, national and foreign. In this area, recently, the chapter "The Role of Sex Steroids in the Host-Parasite Interaction," published in the international book "Sex Steroids" in 2012, reached the figure of 28,000 downloads, which means the degree of attention that has after receiving his work; the foregoing makes it clear that Dr. Morales-Montor's work is highly relevant and widespread among the national and international academic community, and his brilliant career has earned him more than 30 awards, such as the Miguel Alemán Valdez Award in the area of Health 2006, the Distinction National University for Young Academics in the area of Research in Natural Sciences 2006, the CANIFARMA Veterinary Prize 2007 and 2009 in the area of Basic Research, and the 2009 Heberto Castillo Martínez Capital City Award for Young Latin American Academics in Research Basic, and for the third consecutive congress, in 2011, one of his works was awarded the "Dr. José Eleuterio González" Award, for the best work of and research at the XXVI National Congress of Research in Medicine, to name just a few of its achievements. He has also been awarded many distinctions, such as joining the Mexican Academy of Sciences (2005), and being one of the few Mexican scientists to be inducted to the Latin American Academy of Sciences (2008), The National Academy of Medicine, the New York Academy of Sciences, the American Association of Immunologists are deserved recognitions for his academic quality and career. He has graduated and directs more than 35 bachelor's, 8 master's, and 15 doctorate students; has directed more than 30 social service students; and on research stays in his laboratory. His academic leadership is reflected in the trust and respect that his peers confer on him, having been President of the Mexican Society of Parasitology (one of the oldest and most prestigious scientific societies in the country) and currently being President, and founding member, of the Mexican Society of Neuroimmunoendocrinology, since 2011. Due to its scientific curiosity, it is in the process of founding the Mexican Society for Translational Environmental Biomedicine. He has been invited to edit special volumes in various magazines with international circulation and is a member of the editorial committee of magazines of importance in his area of work, such as Parasite Immunology, The Open Parasitology Journal, and among others. He has been a jury for the Arturo Rosenblueth Awards for the best CINVESTAV Doctoral Thesis, a jury for the Lola and Igo Flisser-PUIS 2010 Awards, and a jury for the Heberto Castillo Award, for the best Latin American Researcher 2012, awarded by the Federal District Government. It is noteworthy that he is an outstanding scientist, who has contributed to the scientific research of Mexico with the generation of new frontier knowledge in the world and with the training of high-level human resources.

Affiliations and expertise

Head of Laboratory of Neuroinmunoendocrinology, Department of Immunology, Institute of Biomedical Research, National Autonomous University of Mexico (UNAM), his expertise is in translational biomedicine, tumor microenvironment, immunology, breast and colorectal cancer, and infectious diseases. All of them related to environment and the neuroimmunoendocrine network.

Dr. Mariana Segovia-Mendoza studied the Bachelor of Biological Pharmaceutical Chemistry at the Universidad Nacional Autónoma Metropolitana-Xochimilco in Mexico City obtaining the title in 2011. Later, she studied a master's degree at the National Institute of Cancerology in Mexico City. In this period, she delved into the study of different endocrine therapies and treatments used in cervical cancer. In July 2012, she carried out her doctoral studies at the "Salvador Zubirán" National Institute of Nutrition in Mexico City, where she specialized in breast cancer, hormones, and targeted therapy against this disease. During her doctoral studies, she did a research stay in the Department of Biochemistry and Molecular Biology with Dr. Mauricio Reginato of the Drexel University College of Medicine in the city of Phil1-adelphia, United States, which led to an international collaboration. At the end of her doctoral studies, she received a postdoctoral grant from the National Institute of Nutrition "Salvador Zubirán" which lasted one year. She also completed a second postdoctoral stay in the Department of Neuroimmunoendocrinology of the Institute of Biomedical Research at UNAM with Dr. Jorge Morales-Montor, a research group with which she currently collaborates researching the infiltrating immune cells and their alterations by endocrine disruptor compounds in patients with breast cancer.

Affiliations and expertise

Dr. Mariana Segovia-Mendoza has extensive expertise in the subject of cancer, especially in breast cancer and hormones. She belongs to the National System of Researchers in Mexico and to date, she has been promoted to Level I. At the moment, she is an Associate Professor of the Endocrinology Laboratory of the Faculty of Medicine from the Universidad Nacional Autónoma de México (UNAM).

Preface

This book is a valuable source for cancer researchers, medical doctors, clinicians, and several members of biomedical field who need to understand the mechanisms to fight resistance to immunotherapy in cancer. This book addresses the functioning of different signaling pathways, the modification of the metabolism of the immune system cells, and the modification of chemical compounds with estrogenic activity involved in the generation of resistance to immune therapies. The discussions are intended to provide an understanding to broaden the landscape for the creation of different therapeutic strategies for cancer.

Immune tolerance is the state in which the immune system does not generate a response toward otherwise immunogenic antigens. Tolerance is crucial for life, and a failure to tolerate leads to autoimmune diseases. However, there are moments when our own cells happen to be the disease, as with cancer. Tolerance may be achieved by preventing autoreactive clones from maturing, early in their development, or by either eliminating or suppressing them if they reach the periphery. Palacios-Arreola and Nava-Castro describe the basic mechanism mediating those processes, namely, clonal deletion, anergy induction, immunomodulatory cytokines, and regulatory cells. They conclude by commenting on the immune recognition of cancer cells and why tolerance is a target for immunotherapy. Segovia-Mendoza et al. deal with the resistance of breast cancer therapy. It is a critical problem that is not fully understood. In this sense, the molecular classification of this pathology determines the treatment that the patient needs. Different immune and nonimmune therapies have been provided for this pathology. In this regard, in their chapter, they compile and describe the use of immune clinical options such as immunoconjugates as an alternative for avoiding breast cancer resistance to conventional therapy. In addition, their work is focused on addressing the in vitro approaches until clinical studies about the effects of these drugs in breast cancer are available. Also, in regard to breast cancer, Gomez-De Leon mention that the majority of immunotherapeutic approaches have focused on the postoperative setting; nevertheless, strategies that utilize autologous tumors are the most promising of the immunotherapeutic approaches, generating a response against a host of relevant tumor antigens. Even more, benefits of systemic intravenous therapies have been observed only in a small population of patients and have been associated with toxicity profiles, which are sometimes very severe. In this chapter, the authors thoroughly resume the intratumoral immunotherapeutic approaches that have been successful in the inhibition of tumoral growth and/or metastasis in different types of cancer, paying special attention to the different delivery systems that have been developed. Another hot topic in cancer immunotherapy is adoptive cell therapy (ACT). Volpedo et al. suggests that it is a promising cancer therapy in terms of safety and efficacy in that the expression of FasL by therapeutic lymphocytes can induce apoptosis in Fas-expressing tumor cells.

However, malignant cells can adapt by decreasing Fas to escape from therapeutic lymphocytes and increasing FasL expression to induce apoptosis of immune cells, thereby mounting resistance to ACT. To overcome this problem, the expression of Fas and FasL can be modulated in the tumor microenvironment to immunosensitize tumor cells for ACT-killing. This combined strategy represents higher cost-effectiveness for cancer patients.

Prado-García et al. mention that T cells are essential for the immune response against tumor cells; in addition, immunotherapies such as immune checkpoint blockade (ICB), adoptive T cell therapy, and CAR (Chimeric antigen receptor) T cells rely on these cells to trigger the elimination of tumor cells. In order to exert their effector functions, T cells depend on metabolism to obtain energy in the form of ATP and metabolites that are used for macromolecular biosynthesis. However, tumor cells reprogram their metabolism to survive and proliferate, which has deleterious consequences on T cell metabolism. The authors discuss some potential strategies that may restore the T cell effector function by enhancing T cell metabolism. On the other hand, Ostoa-Saloma mentions that breast cancer is the most common cancer in women worldwide, and therefore, research aimed at finding ways of early detection of this disease assumes great importance. Tools that contribute to the recognition of cancer warning signs can lead to early diagnosis in a healthy population to identify people with the disease at an initial stage when symptoms are not yet apparent. Early detection will increase the survival rate of patients and, therefore, an effective treatment option to fight the disease; he continues to describe the advantages of IgM, in their innate form called natural IgM, to identify tumor antigens in order to find a candidate as an early diagnostic tool for breast cancer. Furthermore, questions in the precancerous condition should be considered: Is it possible to identify and differentiate the antigens recognized by both IgM and IgG and determine which immunoglobulin is best for identifying antigens in early stages? How would it be the recognition of IgM toward tumor antigens in the middle of a genetic background associated with resistance and/or susceptibility? By measuring the pattern of IgM recognition of tumor antigens, would it be possible to differentiate which individuals are more susceptible and which are less susceptible? And therefore, by means of IgM, could antigens associated with susceptibility and/or resistance to breast cancer be recognized? On another topic, Terrazas et al. term inflammation in colorectal cancer as a double-edged sword. They mention that in the past two decades, a close relationship between inflammation and carcinogenesis has been described. In fact, since 2011, inflammation has been considered as one of the ten hallmarks of cancer. However, natural selection did not evolve inflammation to cause troubles and diseases to our body. Both types of inflammation, acute and chronic, are natural events orchestrated by the immune system to generate protection against infectious and noninfectious insults, including colorectal cancer (CRC). Strong evidence in humans and some animal models of CRC indicates that the absence of molecules associated with inflammation induces an increased number of tumors and a reduction of apoptosis, suggesting that inflammation is involved in protection when the tumor has been established. However, another line of evidence, also well supported, suggests that the inflammatory processes during cancer development, especially in CRC, are highly undesirable because it accelerates in many cases the tumor growth. Therefore, many questions arise about the relationship between inflammation and CRC: Which are the steps and decisions taken by the immune system to guide the inflammation to favor CRC development? Or maybe, is inflammation a process involved in protection against CRC establishment? The balance between inflammation, its causes and checkpoints are extremely important during CRC development; research is underway striving to answer these questions. In this chapter, we try to establish some evidence that suggest a major role for inflammation in protection during early CRC development. The understanding of the complex relationship between inflammation and cancer may help to develop combinatory therapies (drugs and immunomodulators) to blockade the possible collateral damage induced by chronic inflammation that favors CRC elimination. Also, in regard to CRC, it has been postulated that environmental pollution is nowadays the first risk factor associated with the development of lung cancer, stroke, and heart disease. Within the universe of a complex mix of pollutants that shape the environment, the importance of endocrine-disrupting compounds (EDCs), such as bisphenols, phthalates, aromatic polycyclic hydrocarbons (AHPs) and pesticides, has been highlighted in recent years because of our chronic and ubiquitous exposition, particularly by ingestion of food and water, and their interference in hormone and immune functions. EDCs are associated principally with breast cancer development, but there is a poor understanding of the relationship with other neoplasms such as colorectal cancer (CRC). The potential carcinogenic role of EDC in the development of CRC, focusing on inflammation and neuroendocrine-immune regulation, is discussed by Rodríguez-Santiago et al. They point out that the interaction between inflammatory bowel disease and CRC in terms of disruption of intestinal permeability, dysbiosis, and immune response induced by EDCs is a crucial factor in CRC. Moreover, the authors delve into the neuroendocrine axes involved in the progression of this pathology. Also regarding CRC, ongoing evidence suggest that the signal transducer and activator of transcription 6 (STAT6) play a pivotal role in CRC development. Compelling evidence from both human and experimental models shows that STAT6 not only contributes to mediating immune response, but is also involved in the pathology associated with disease by altering the epithelial barrier function, promoting proliferation of intestinal epithelial cells and regulating the expression of prosurvival and prometastatic proteins. These studies have led to the approach of targeting STAT6 as an effective treatment strategy in alleviating CRC. Thus, silencing or hindering STAT6 signaling may strengthen the chemotherapy response with positive effects in the current treatments. Therefore, the role of STAT6 in colorectal cancer biology and its potential as a new therapeutic target for the prevention and treatment of this disease is suggested by Leon-Cabrera et al. Moreover, macrophage migration inhibitory factor (MIF) is a pleiotropic protein with cytokine and chemokine properties that regulates a diverse range of physiological functions related to innate immunity, inflammation, and glucocorticoid-mediated immunosuppression and is highly expressed by cancer cells, through which it affects angiogenesis, tumor growth, and metastasis. The role of MIF in colorectal cancer (CRC) is underscored by data showing that its overexpression in the chronic stages of CRC is associated with clinical severity. The specific functions of MIF are now being defined in CRC, and MIF-targeted biologic therapeutics are in early-stage trials. Rodríguez-Sosa et al. summarize the current knowledge about the role of MIF in cancer with special emphasis on CRC.

Antibody therapeutics have been at the heart of transformative cancer treatment for more than 20 years, first with antibodies directly targeting the tumor and more recently with immune checkpoint blockade. Despite their impact and widespread utilization, antibody therapy mechanisms of action and factors governing response or resistance in patients are still poorly understood. One aspect that has emerged as important for all clinically developed antibodies is antibody Fc interactions with Fcγ receptors. Antibody Fc acts to connect the specificity of antibody to the power of the innate immune system and ensuing adaptive immunity. What has become clear is that in the context of both direct tumor targeting and ICB mAb, these interactions can be pivotal to therapeutic activity and survival. Through improved understanding and evolving strategies of Fc-engineering, FcγR-blockade, and pharmacological modulation of immune effector cell FcγR expression, we are at the dawn of harnessing the power of already clinically validated and new classes of antibody-based cancer immunotherapeutics, as commented by Frendeus et al. They point out that the current impact and consequence of the proper interaction of the Fc:Fcγ receptor and its relation to the antibody efficacy and resistance. In addition, they discuss the development of novel classes of antibodies as great strategies to enhance the patient responses and overcome the drug resistance. Finally, Garay-Canales et al. mention that the common cancer therapy includes surgery, radiotherapy, and chemotherapy; these techniques have some disadvantages such as lack of specificity, causing side effects, and the possibility of relapse. The novel design of specific, nontoxic therapies for cancer require an in-depth knowledge of the biology and the molecules involved, which would result in a better outcome for patients. The matrix metalloproteinases (MMPs) are a family of zinc-dependent endopeptidases; some of the major substrates are the components of the extracellular matrix (ECM). MMPs are responsible for regulating numerous physiological and pathological events including bone development, wound repair, and different stages in carcinogenesis. Therefore, these enzymes represent highly relevant targets for cancer therapy. Overexpression of MMPs is well documented in most types of cancer; upregulated expression is present not only in cancer cells, but also in stroma cells, which modify and regulate the tumor microenvironment. To break the cancer tolerance, they discuss the evolution of some strategies: First efforts using MMPs were addressed to inhibit the catalytic site or the binding zinc domain; although several chemical inhibitors were synthesized, poor specificity to target only cancer cells was achieved, and none of them passed human clinical trials. Then, studies of crystal structure, phage display, and production of monoclonal antibodies revealed specific epitopes or domains within the MMPs, these cryptic sites are helpful to design new targets. One novel strategy is the drug delivery systems (DDS); this strategy uses the upregulated expression of MMPs in cancer cells and is based on the specificity of sensitive MMP substrates or monoclonal antibodies attached to a highly toxic drug that will be "activated or released" only in the presence of the selected MMP. Altogether, knowing the features of the MMPs, it is possible to design more specific and less toxic therapies for several types of cancer and combining these strategies with standard therapies, we could achieve the goal of better survival rate for cancer patients.

Thus, the book it is an actualized, up-to-date compilation that intends to be of use to all people involved in the world of cancer.

Jorge Morales-Montor
Mariana Segovia-Mendoza

Contents

7. The inflammation during colorectal cancer: A friend or a foe?

Itzel Medina-Andrade, Jonadab E. Olguín, Tonathiu Rodríguez, and Luis I. Terrazas

8. Environmental pollution as a risk factor to develop colorectal cancer: The role of endocrine-disrupting chemicals in the inflammatory process as a risk factor to develop colorectal cancer

Yair Rodriguez-Santiago, Karen Elizabeth Nava-Castro, and Jorge Morales-Montor

9. Targeting the STAT6 signaling pathway as a therapy against colon cancer

Ana Catalina Rivera Rugeles, Yael Delgado-Ramirez, Luis I. Terrazas, and Sonia Leon-Cabrera

10. Macrophage migration inhibitory factor (MIF): Its role in the genesis and progression of colorectal cancer

Imelda Juárez-Avelar, Tonathiu Rodríguez, Ana P. García-García, and Miriam Rodríguez-Sosa

11. Fcγ receptors—Master regulators of antibody therapy

Stephen A. Beers and Björn Frendéus

Contributors

Stephen A. Beers Antibody and Vaccine Group, Centre for Cancer Immunology, University of Southampton Faculty of Medicine, Southampton, United Kingdom

Daniela Alejandra Castro-Flores Departamento de Enfermedades Cronico-Degenerativas, Instituto Nacional de Enfermedades Respiratorias, "Ismael Cosio Villegas", Ciudad de México, México

Marianna de Carvalho Clímaco Instituto de Ciências Biológicas, Departamento de Parasitologia, Universidade Federal de Minas Gerais, Belo Horizonte, Minas Gerais, Brazil

Yael Delgado-Ramirez Unidad de Investigación Biomédica, Universidad Nacional Autónoma de México, Tlalnepantla, México

Laura Díaz-Alvarez Posgrado en Ciencias Biológicas, Universidad Nacional Autónoma de México, Ciudad Universitaria, Ciudad de México, México

Björn Frendéus BioInvent International AB, Lund, Sweden

Claudia A. Garay-Canales Departamento de Inmunología, Instituto de Investigaciones Biomédicas, Universidad Nacional Autónoma de México, Ciudad de México, México

Rocio García-Becerra Departamento de Biología Molecular y Biotecnología, Instituto de Investigaciones Biomédicas, Universidad Nacional Autónoma de México, Ciudad de México, México

Ana P. García-García Unidad de Biomedicina, Facultad de Estudios Superiores Iztacala, Universidad Nacional Autónoma de México, Tlalnepantla, México

Carmen T. Gomez de Leon Departamento de Inmunología, Instituto de Investigaciones Biomédicas, Universidad Nacional Autónoma de México, Ciudad de México, México

Imelda Juárez-Avelar Unidad de Biomedicina, Facultad de Estudios Superiores Iztacala, Universidad Nacional Autónoma de México, Tlalnepantla, México

Cristina Lemini Departamento de Farmacología, Facultad de Medicina, Universidad Nacional Autónoma de México, Ciudad de México, México

Sonia Leon-Cabrera Unidad de Investigación Biomédica; Escuela de Medicina, Facultad de Estudios Superiores Iztacala, Universidad Nacional Autónoma de México, Tlalnepantla, México

Georgina I. Lopez-Cortes Departamento de Inmunología, Instituto de Investigaciones Biomédicas, Universidad Nacional Autónoma de México, Ciudad de México, México

Itzel Medina-Andrade Laboratorio Nacional en Salud: Diagnóstico Molecular y Efecto Ambiental en Enfermedades Crónico-degenerativas; Unidad de Biomedicina, Facultad de Estudios Superiores Iztacala, Universidad Nacional Autónoma de México, Tlalnepantla, México

Jorge Morales-Montor Departamento de Inmunología, Instituto de Investigaciones Biomédicas, Universidad Nacional Autónoma de México, Ciudad de México, México

Karen Elizabeth Nava-Castro Departamento de Mutagénesis y Genotoxicidad Ambientales, Centro de Ciencias de la Atmósfera, Universidad Nacional Autónoma de México, Ciudad de México, México

Jonadab E. Olguín Laboratorio Nacional en Salud: Diagnóstico Molecular y Efecto Ambiental en Enfermedades Crónico-degenerativas; Unidad de Biomedicina, Facultad de Estudios Superiores Iztacala, Universidad Nacional Autónoma de México, Tlalnepantla, México

Pedro Ostoa-Saloma Departamento de Inmunología, Instituto de Investigaciones Biomédicas, Universidad Nacional Autónoma de México, Ciudad de México, México

Thalia Pacheco-Fernández Unidad de Biomedicina, Facultad de Estudios Superiores Iztacala, Universidad Nacional Autónoma de México, Tlalnepantla, México

M. Isabel Palacios-Arreola Departamento de Mutagénesis y Genotoxicidad Ambientales, Centro de Ciencias de la Atmósfera, Universidad Nacional Autónoma de México, Ciudad de México, México

Heriberto Prado-Garcia Departamento de Enfermedades Cronico-Degenerativas, Instituto Nacional de Enfermedades Respiratorias, "Ismael Cosio Villegas", Ciudad de México, México

Tonathiu Rodríguez Unidad de Biomedicina, Facultad de Estudios Superiores Iztacala, Universidad Nacional Autónoma de México, Tlalnepantla, México

Yair Rodriguez-Santiago Departamento de Inmunología, Instituto de Investigaciones Biomédicas, Universidad Nacional Autónoma de México, Ciudad de México, México

Miriam Rodríguez-Sosa Unidad de Biomedicina, Facultad de Estudios Superiores Iztacala, Universidad Nacional Autónoma de México, Tlalnepantla, México

Susana Romero-Garcia Departamento de Enfermedades Cronico-Degenerativas, Instituto Nacional de Enfermedades Respiratorias, "Ismael Cosio Villegas", Ciudad de México, México

Ana Catalina Rivera Rugeles Unidad de Investigación Biomédica, Universidad Nacional Autónoma de México, Tlalnepantla, México

Abhay R. Satoskar Department of Pathology, The Ohio State University Medical Center; Department of Microbiology, The Ohio State University, Columbus, OH, United States

Mariana Segovia-Mendoza Departamento de Farmacología, Facultad de Medicina, Universidad Nacional Autónoma de México, Ciudad de México, México

Luis I. Terrazas Laboratorio Nacional en Salud: Diagnóstico Molecular y Efecto Ambiental en Enfermedades Crónico-degenerativas; Unidad de Biomedicina, Facultad de Estudios Superiores Iztacala, Universidad Nacional Autónoma de México, Tlalnepantla, México

Greta Volpedo Department of Pathology, The Ohio State University Medical Center; Department of Microbiology, The Ohio State University, Columbus, OH, United States

CHAPTER

1

Cancer vs immune tolerance—The challenge of fighting "self"

M. Isabel Palacios-Arreola and Karen Elizabeth Nava-Castro

Departamento de Mutagénesis y Genotoxicidad Ambientales, Centro de Ciencias de la Atmósfera, Universidad Nacional Autónoma de México, Ciudad de México, México

Abstract

Immune tolerance is the state in which the immune system does not generate a response towards otherwise immunogenic antigens. Tolerance is crucial for life and its failures lead to autoimmune diseases. Tolerance may be achieved by preventing autoreactive clones from maturing, early in their development, or either eliminating or suppressing them if they reach periphery. However, there are moments when our own cells happen to be the disease, like with cancer. Herein, we describe the basic mechanism mediating those processes, namely clonal deletion, anergy induction, immunomodulatory cytokines, and regulatory cells. Finally, we comment on the immune recognition of cancer cells and why tolerance is a target for immunotherapy.

Abbreviations

AIRE	autoimmune regulator
APCs	antigen-presenting cells
BCR	B cell receptor
cAMP	cyclic adenosine monophosphate
CTLA-4	cytotoxic T lymphocyte antigen-4
DCs	dendritic cells
IFN-γ	interferon gamma
IL	interleukin
iTreg	induced regulatory T cell
MHC	major histocompatibility complex
nTreg	natural regulatory T cell
PD-1	programmed death 1 receptor
STAT	signal transducer and activator of transcription
TCR	T cell receptor
TFG-β	transforming growth factor beta

Immunotherapy in Resistant Cancer:
From the Lab Bench Work to Its Clinical Perspectives
https://doi.org/10.1016/B978-0-12-822028-3.00018-2

TGF-βRI	TFG-β receptor I
TGF-βRII	TFG-β receptor II
Th	T helper lymphocyte
TNF-α	tumor necrosis factor alpha

Conflict of interest

No potential conflicts of interest were disclosed by the authors.

Introduction

Immune tolerance

Since the early 1900s, it was observed that the immune system was able to distinguish self from non-self. Ehrlich was one of the first in noticing that immune system would respond to everything that was injected to an individual, including cells from another individual from the same species, but not to cells of their own body [1]. Before this, it was considered obvious that the immune system does not attack its own cells and tissues, but it has been demonstrated that this is a complex well-orchestrated phenomenon that is actively pursued to maintain homeostasis.

Immune tolerance is a state in which the immune system does not generate a response towards otherwise immunogenic antigens. But how does the immune system get to know and recognize itself? There is no general rule as to which antigens the immune system is tolerant, immune tolerance is shaped throughout development and during the adult life and relies on multiple mechanisms.

Tolerance is indispensable for life and failures in it lead to autoimmune disease. However, as this book will discuss, there are moments when our own cells happen to be the disease. This is the case with cancer, and tolerance implies that the immune system is conditioned not to attack those own cells, leading to a poor immune antitumoral response. This is the issue that immunotherapy tries to overcome.

Mechanisms of immune tolerance

When talking about the immune system, there are two basic branches: innate and adaptive immunity. Innate immune activation is based on the recognition of pathogen-associated molecules which are evolutively absent in our tissues or danger-associated molecules, which are common to all individuals of a species, genre, and even taxonomical families. On the contrary, adaptive immune activation is much more diverse, based on the recognition of an infinite number of possible antigens. Among all those possible antigens that the adaptive immune system may recognize, there may be antigens present in the own tissues, that is, self-antigens or autoantigens. For this reason, immune tolerance is generally focused on T and B cells, although some mechanisms involve also cells from the innate immune system.

There are two main ways to achieve immune tolerance, one is to avoid the generation or maturation of autoreactive T or B cells, which is deemed as central tolerance, and the other is

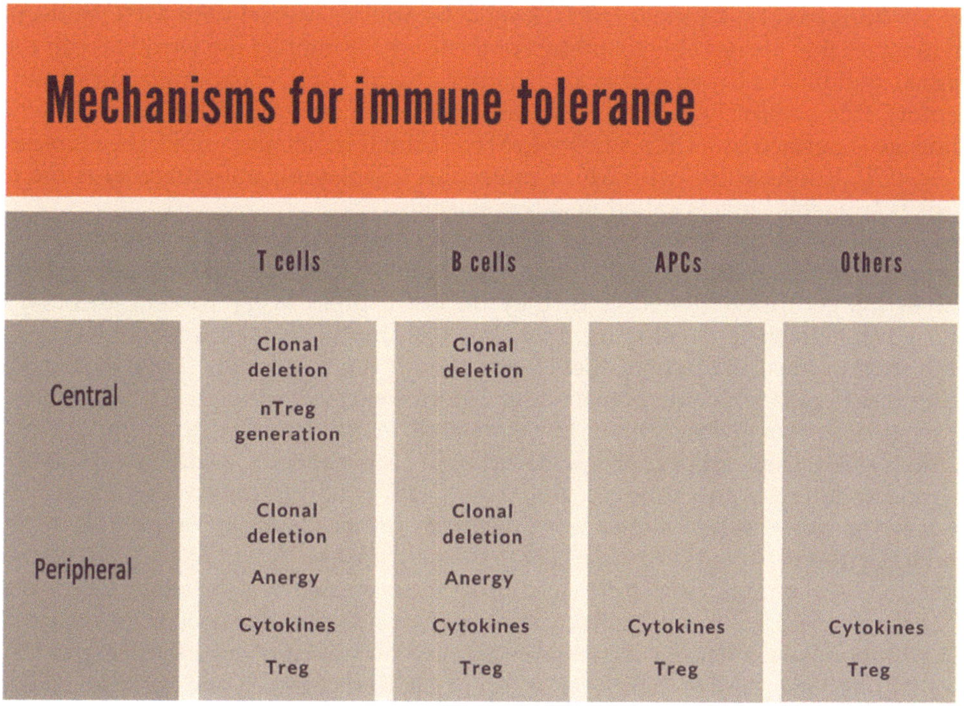

FIG. 1 Mechanisms for immune tolerance. Central immune tolerance occurs in primary lymphoid organs, before the cells enter the periphery. T and B lymphoid lineages undergo clonal deletion of autoreactive cells during their development; additionally, some surviving autoreactive T cells may develop into nTregs. Peripheral immune tolerance allows for modulation and/or inhibition of immune cells once in the periphery. This is achieved through clonal deletion and anergy of T and B cells, the secretion of inhibitory cytokines, and action of Tregs.

to eliminate or suppress those autoreactive cells that might have scaped the first mechanisms, which is termed peripheral tolerance (see Fig. 1).

Central tolerance

Central tolerance is based on the prevention and depletion of autoreactive T and B cells. The development and maturation of those cell lineages are different and so are the mechanisms used to confer tolerance.

Immature T cell progenitors emerge from the bone marrow and migrate to the thymus where they arrive in a so-called double-negative phenotype, lacking both CD4 and CD8 co-receptors and T cell receptor (TCR). As thymocytes (developing T lymphocytes) mature, they acquire both CD4 and CD8 molecules and a preliminary version of the TCR and then they undergo a selection process in which TCR's ability to bind MHC molecules is tested [2]. It is important to note that MHC molecules expressed on cortical thymic epithelial cells display self-antigens [3]. A functional TCR must be able to interact with MHC molecules, so that thymocytes that are not able to interact or those which interact too weakly receive no survival signal and die. If a thymocyte is able to interact with MHC class I molecules (using CD8 co-receptor), it is positively selected and commits to the cytotoxic $CD8^+$ lineage, while the $CD4^+$ lineage is defined by TCR's affinity to MHC class II molecules with CD4 as co-receptor. However, if TCR interaction with MCH molecules (displaying self-antigens) is too strong, apoptosis is triggered, this is called negative selection [2].

It would appear as this first negative selection managed to eliminate autoreactive T cells, but cortical thymic epithelial cells are only able to display self-antigens present in themselves [3]. As differentiated cells, they do not express every protein encoded in the genome. Further in their developing path, lineage-committed, single positive thymocytes enter the thymus medulla, where they interact with dendritic cells (DCs) and medullary thymic epithelial cells which display a wider repertoire of self-antigens [2]. What allows this wider repertoire is a transcription factor named AIRE (from autoimmune regulator), which induces the transcription of a wide array of tissue-specific antigens [4]. Once again, if a thymocyte interacts too strongly with any MHC molecule displaying an autoantigen, it will be negatively selected and conducted to either death or a natural regulatory T phenotype (nTreg).

Unlike T cells, B lymphocytes fully develop in the bone marrow. Whether B cells undergo positive selection process is still unclear, but negative selection does occur. Immature B cells interact with bone marrow cells expressing so-called housekeeping molecules (proteins, carbohydrates, or glycolipids); if their B cell receptor (BCR) recognizes one of those molecules, it receives a signal to hold maturation and undergoes a "second chance" BCR recombination [5]. In the case that this second rearranged BCR is still autoreactive, the cell receives a pro-apoptotic signal [6].

Peripheral tolerance

Despite the negative selection processes, some autoreactive clones may still enter the periphery; however, there are ways to eliminate them or suppress their activity. Peripheral tolerance is achieved by multiple mechanisms, including peripheral clonal deletion, anergy induction, immunomodulatory cytokines, and regulatory cells [1,7].

The first three mechanisms for peripheral tolerance rely on the three-signal paradigm for lymphocyte activation. Peripheral T and B lymphocytes do not acquire effector functions by just encountering their antigen, but require different signals to fully activate and proliferate. Those requisites are referred to as the three-signal paradigm [8]. The first signal is the presentation of the antigen in the context of MHC molecules, which interact with the TCR. The second signal is provided by co-stimulatory molecules expressed by antigen-presenting cells (APCs) or helper T cells. The main co-stimulatory molecule for T cells is CD28 whose APC counterparts are B7-1 (CD80) and B7-2 (CD86), while B cell primary co-stimulatory molecule is CD40 which binds to CD40L expressed by helper T cells (Th) [9,10]. The expression of co-stimulatory molecules in APCs is context-dependent and its upregulation often depends on the recognition of pathogen-associated or danger-associated molecules [11,12]. Signals derived from co-stimulation prevent apoptosis (inducing the antiapoptotic Bcl-xl), thus increasing activation pathways and promoting cytokine production of activated cells [9,10]. The third signal is provided by cytokines produced by innate immune cells upon activation [8]. Besides aiding in the activation, cytokines also shape the kind of adaptive immune response generated.

Peripheral clonal deletion

Peripheral clonal deletion of T cells occurs in the lymph nodes, where T cells interact with immature DCs that routinely take up and process apoptotic cells which leads to autoantigen presentation. Since there is no pathogen infection or danger signals, those DCs express very low levels of co-stimulatory molecules, so that the second signal is deficient [1]. This results in the activation of apoptotic Fas and Bim pathways [13] that lead to apoptosis in the absence of antiapoptotic Bcl-xl signaling [9].

B cell peripheral clonal deletion also occurs, but seems to be a more plastic phenomenon. It has been demonstrated that autoreactive B clones that persist in the periphery may be deleted upon encounter of highly multivalent membrane-bound antigens [14], but recent evidence suggests that multiple antigen encounters may be required for efficient deletion of those autoreactive clones; otherwise, those clones may persist, although functionally anergic, with a short life span [15].

Anergy induction

Anergy is a mechanism in which lymphocytes are functionally inactivated following suboptimal antigen encounter, but remain alive for a certain period of time in a hyporesponsive state [16], with defective phospholipase activation and intracellular calcium mobilization [17]. As described earlier, co-stimulation provides survival signals and activates multiple pathways for proper activation. Nevertheless, suboptimal co-stimulation may provide sufficient antiapoptotic signals, but not enough activation ones.

Interestingly, co-stimulatory molecules not only provide positive signaling, some provide negative second signals that inhibit T cell responses, mediating tolerance [13]. One of the most important negative co-stimulatory molecules is the programmed death 1 receptor (PD-1), which is expressed on T, B, and myeloid cells [7] and binds to PD-L1 and PD-L2 ligands. Upon ligand binding, PD-1 mediates the inhibition of PI3K and Akt pathways, preventing T cell proper activation [18].

Another key negative co-stimulatory molecule known to induce anergy is CTLA-4, which also binds to B7 molecules but induces a completely different signaling that prevents cycle cell progression [13]. CTLA-4 inhibits CD28-dependent T cell activation, cell cycle progression, and IL-2 production [19,20]. The role of CTLA-4 in anergy is also related to the expression of genes like *Cbl-b*, *p25Kipl*, *Dgkz*, *Itch*, *NFAT1*, and *Grail* [21].

Regarding autoreactive B cells, if avidity to self-antigens is significant but too low to induce clonal deletion, those cells may be maintained in the periphery in an anergic state, unable to mobilize calcium, upregulate activation markers, or proliferate [22]. Anergy is achieved by chronic binding of antigen in the absence of secondary signals (co-stimulation by Th), pathogen- or danger-associated signals [22]. However, anergy is not permanent and dependents on continued occupancy of antigen receptors and removal of antigen results in the restoration of responsiveness [23].

Inhibitory cytokines

Cytokines are key molecules that shape immune response, either by stimulating or by inhibiting certain types of response. Some cytokines are considered particularly inhibitory: IL-10 and TFG-β. Those cytokines are produced primarily by regulatory T cells, but may also be secreted by anergic T cells [1].

Upon binding to its receptor, IL-10 signaling activates the Jak-STAT pathway, particularly STAT3 [24]. One of the main effects of IL-10 is the suppression of TLR signaling, which is crucial for APC activation [25]. This cytokine also downregulates the expression of pro-inflammatory cytokines like TNF-α, which exerts stimulatory effects on APCs [25], as well as IL-1α, IL-1β, IL-6, and IL-12 [26]. IL-10 also interferes with IFN-γ signaling, which is necessary for the expression of MHC class II and co-stimulatory molecules, pro-inflammatory cytokine production, and proper activation of monocyte-macrophages [27].

TGF-β is a pleiotropic family of cytokines expressed by many immune and nonimmune cells [28]. It regulates multiple processes, from embriogenesis to carcinogenesis [29] and of course immune tolerance. TGF-β binds to a tetrameric receptor composed of two TGF-βRI and two TGF-βRII molecules, which results in the activation of Smad transcription factors [29]. One of the tolerogenic mechanisms of this cytokine is the inhibition of both helper and cytotoxic T lymphocytes, which is mediated by the suppression or IL-2 production, downregulation of c-myc, and upregulation of cyclin-dependent kinase inhibitors [28]. Regarding cytotoxic T lymphocytes, they are further regulated by TGF-β through the inhibition of perforin and IFN-γ production and may even be subject to apoptosis, since TGF-β upregulates the pro-apoptotic protein Bim [28]. Another mechanism by which TGF-β promotes immune tolerance is the induction of regulatory T cells (Tregs). Early during negative selection in the thymus, TGF-β is able to induce Foxp3 expression [29], favoring the nTreg phenotype. Later, in the periphery, TGF-β promotes the conversion of naïve $CD4^+$ T cells into induced Tregs (iTregs), upregulating Foxp3 expression [28].

Regulatory cells

Regulatory T cells (Tregs) account for approximately 5%–10% of peripheral $CD4^+$ T cells and modulate immune response through suppressive mechanisms [30]. Nowadays, it is recognized that there are different types of Tregs, but they are typically characterized by the expression of Foxp3 and/or high expression of the IL-2 receptor, CD25 [31]. Tregs have been reported to exert inhibitory effects on T cells, B cells, APCs, and other immune cell types, such as NK, NKT, mast cells, and osteoblasts [32,33]. Tregs suppress or modulate immune response through several mechanisms, including cytolysis, modulation of DC maturation or function, metabolic disruption, and the production of inhibitory cytokines [34].

Cytolysis is a contact-dependent mechanism of Treg suppression and is based on the secretion of perforin and granzyme A/B, leading to apoptosis of target cells [33].

Tregs inhibit APC maturation and function through contact-dependent mechanisms that involve CTLA-4, and LAG-3 [30,32]. This results in APC downregulation of co-stimulatory molecules, mainly CD80/CD86, which diminishes the ability of APCs to properly activate other cells.

The most studied mechanism for metabolic disruption is IL-2 deprivation. Since Tregs express high levels of CD25, those lymphocytes consume great amounts of IL-2, which leads to the depletion of this cytokine, being unavailable for other T subpopulations like cytotoxic and helper T cells, reducing their activation and proliferation potential. However, transfer of cAMP and secretion of adenosine have been recently recognized as metabolic disruption mechanisms, particularly for effector T cells [35,36].

The Tregs are one of the most important sources of IL-10, TGF-β, IL-27, and IL-35 [30], key inhibitory cytokines. The secretion of those cytokines expands Treg effector functions, since many different cells are modulated by them, from both innate and adaptive branches.

Immune recognition of cancer. When only self is left

Despite being self, cancerous cells are potentially immunogenic, at least up to a certain degree. Tumors arising from viral infection are theoretically the most immunogenic, since they express foreign viral antigens, which are non-self and may elicit the immune response. However, virus-derived tumors are a minory. Most tumors arise from transformed self-tissues.

However, genetic instability, which is a hallmark of cancer [37], creates a source for neo-antigens [38], i.e., mutated proteins which may lead to distinct, previously nonexistent self-antigens. Immune response towards those mutation-derived neo-antigens has been documented [39]. Nevertheless, not only the immune response identifies specific antigens but also innate immune system is able to recognize damage- and stress-associated molecules which are the first alert signs in cancer development.

In the 1950s, Burnett proposed the immunovigilance theory [40], which poses that innate immune cells detect transformed cells, attack them, and elicit further response from adaptive immune system in order to eliminate those transformed cells. If this process fails, transformed cells proliferate, thus indicating cancer. However, this process is much more complex than originally thought and comprises not only a lineal event series, but alternative processes and stages, such as immunoedition and immunosubversion [41,42].

Even when transformed cells are detected during immunovigilance and a robust immune response is generated, some transformed cells may escape immune detection, due to genetic instability, which may lead to antigen loss. Those cells are then able to proliferate and may be subject to immune recognition, posing a new selection pressure. During this process, called immunoedition, the immune system can recognize and destroy the immunogenic cancer cells expressing tumor antigens, but new, less immunogenic clones arise. After this process, it is probable that the neo-antigens that were initially recognized are lost [43].

In this scenario, only self-antigens are left, which are poorly immunogenic due to the central and peripheral tolerance mechanisms. Moreover, cancer cells induce their own tolerogenic mechanisms, such as the secretion of immunomodulatory cytokines and induction of multiple lineage suppressor cells.

Conclusion

Immune tolerance is indispensable for homeostasis. This basic principle of auto-preservation drives the multiple mechanisms through which tolerance to self-antigens is achieved, from early developmental stages and throughout life, from hematopoietic sites to the periphery and both before and after antigen encounter.

However, cancer is a particular circumstance in which danger arises from self-tissues, which are inherently protected from immune response. Even though cancer cells may initially express immunogenic neo-antigens, selection process by *immunoedition* eliminates those cancer clones. When only self is left, tolerance becomes a challenge.

This is what immunotherapy tries to overcome: the natural and acquired immune tolerance of cancer cells, our dangerous, damaged self.

References

[1] Mak TW, Saunders ME. Immune tolerance in the periphery. In: The immune response. Elsevier; 2006. p. 433–62. https://doi.org/10.1016/b978-012088451-3.50018-1.

[2] Germain RN. t-cell development and the CD4-CD8 lineage decision. Nat Rev Immunol 2002;2:309–22. https://doi.org/10.1038/nri798.

[3] Klein L, Kyewski B, Allen PM, Hogquist KA. Positive and negative selection of the T cell repertoire: what thymocytes see (and don't see). Nat Rev Immunol 2014;14:377–91. https://doi.org/10.1038/nri3667.

[4] Gardner JM, Fletcher AL, Anderson MS, Turley SJ. AIRE in the thymus and beyond. Curr Opin Immunol 2009;21:582–9. https://doi.org/10.1016/j.coi.2009.08.007.

[5] Melamed D, Nemazee D. Self-antigen does not accelerate immature B cell apoptosis, but stimulates receptor editing as a consequence of developmental arrest. Proc Natl Acad Sci U S A 1997;94:9267–72. https://doi.org/10.1073/pnas.94.17.9267.

[6] Mak TW, Saunders ME. The humoral response: B cell development and activation. In: The immune response. Elsevier; 2006. p. 209–45. https://doi.org/10.1016/b978-012088451-3.50011-9.

[7] Alpdogan O, Van Den Brink MRM. Immune tolerance and transplantation. Semin Oncol 2012;39:629–42. https://doi.org/10.1053/j.seminoncol.2012.10.001.

[8] Jain A, Pasare C. Innate control of adaptive immunity: beyond the three-signal paradigm. J Immunol 2017;198:3791–800. https://doi.org/10.4049/jimmunol.1602000.

[9] Rathmell JC, Thompson CB. Pathways of apoptosis in lymphocyte development, homeostasis, and disease. Cell 2002;109:97–107. https://doi.org/10.1016/S0092-8674(02)00704-3.

[10] Crow MK. Costimulatory molecules and T-cell-B-cell interactions. Rheum Dis Clin North Am 2004;30:175–91. https://doi.org/10.1016/S0889-857X(03)00111-X.

[11] Schnare M, Barton GM, Holt AC, Takeda K, Akira S, Medzhitov R. Toll-like receptors control activation of adaptive immune responses. Nat Immunol 2001;2:947–50. https://doi.org/10.1038/ni712.

[12] Pasare C, Medzhitov R. Toll-dependent control mechanisms of CD4 T cell activation. Immunity 2004;21:733–41. https://doi.org/10.1016/j.immuni.2004.10.006.

[13] Xing Y, Hogquist KA. T-cell tolerance: central and peripheral. Cold Spring Harb Perspect Biol 2012;4:1–15. https://doi.org/10.1101/cshperspect.a006957.

[14] Hartley SB, Crosbie J, Brink R, Kantor AB, Basten A, Goodnow CC. Elimination from peripheral lymphoid tissues of self-reactive B lymphocytes recognizing membrane-bound antigens. Nature 1991;353:765–9. https://doi.org/10.1038/353765a0.

[15] Lang J, Nemazee D. B cell clonal elimination induced by membrane-bound self-antigen may require repeated antigen encounter or cell competition. Eur J Immunol 2000;30:689–96. https://doi.org/10.1002/1521-4141(200002)30:2<689::AID-IMMU689>3.0.CO;2-I.

[16] Schwartz RH. T cell anergy. Annu Rev Immunol 2003;21:305–34. https://doi.org/10.1146/annurev.immunol.21.120601.141110.

[17] Wells AD, Liu Q-H, Hondowicz B, Zhang J, Turka LA, Freedman BD. Regulation of T cell activation and tolerance by phospholipase Cγ-1-dependent integrin avidity modulation. J Immunol 2003;170:4127–33. https://doi.org/10.4049/jimmunol.170.8.4127.

[18] Keir ME, Butte MJ, Freeman GJ, Sharpe AH. PD-1 and its ligands in tolerance and immunity. Annu Rev Immunol 2008;26:677–704. https://doi.org/10.1146/annurev.immunol.26.021607.090331.

[19] Walunas TL, Bakker CY, Bluestone JA. CTLA-4 ligation blocks CD28-dependent T cell activation. J Exp Med 1996;183:2541–50. https://doi.org/10.1084/jem.183.6.2541.

[20] Walunas TL, Lenschow DJ, Bakker CY, Linsley PS, Freeman GJ, Green JM, Thompson CB, Bluestone JA. CTLA-4 can function as a negative regulator of T cell activation. Immunity 1994;1:405–13. https://doi.org/10.1016/1074-7613(94)90071-X.

[21] Macián F, García-Cózar F, Im SH, Horton HF, Byrne MC, Rao A. Transcriptional mechanisms underlying lymphocyte tolerance. Cell 2002;109:719–31. https://doi.org/10.1016/S0092-8674(02)00767-5.
[22] Elizabeth Franks S, Cambier JC. Putting on the brakes: regulatory kinases and phosphatases maintaining B cell anergy. Front Immunol 2018;9:665. https://doi.org/10.3389/fimmu.2018.00665.
[23] Gauld SB, Benschop RJ, Merrell KT, Cambier JC. Maintenance of B cell anergy requires constant antigen receptor occupancy and signaling. Nat Immunol 2005;6:1160–7. https://doi.org/10.1038/ni1256.
[24] Schmetterer KG, Pickl WF. The IL-10/STAT3 axis: contributions to immune tolerance by thymus and peripherally derived regulatory T-cells. Eur J Immunol 2017;47:1256–65. https://doi.org/10.1002/eji.201646710.
[25] Mittal SK, Roche PA. Suppression of antigen presentation by IL-10. Curr Opin Immunol 2015;34:22–7. https://doi.org/10.1016/j.coi.2014.12.009.
[26] Moore KW, O'Garra A, Malefyt RW, Vieira P, Mosmann TR. Interleukin-10. Annu Rev Immunol 1993;11:165–90. https://doi.org/10.1146/annurev.iy.11.040193.001121.
[27] Avdic S, Cao JZ, McSharry BP, Clancy LE, Brown R, Steain M, Gottlieb DJ, Abendroth A, Slobedman B. Human cytomegalovirus interleukin-10 polarizes monocytes toward a deactivated M2c phenotype to repress host immune responses. J Virol 2013;87:10273–82. https://doi.org/10.1128/jvi.00912-13.
[28] Travis MA, Sheppard D. TGF-β activation and function in immunity. Annu Rev Immunol 2014;32:51–82. https://doi.org/10.1146/annurev-immunol-032713-120257.
[29] Liu M, Li S, Li MO. TGF-β control of adaptive immune tolerance: a break from Treg cells. Bioessays 2018;40: e1800063https://doi.org/10.1002/bies.201800063.
[30] Shevach EM, DiPaolo RA, Andersson J, Zhao D-M, Stephens GL, Thornton AM. The lifestyle of naturally occurring CD4+ CD25+ Foxp3+ regulatory T cells. Immunol Rev 2006;212:60–73. https://doi.org/10.1111/j.0105-2896.2006.00415.x.
[31] Fontenot JD, Gavin MA, Rudensky AY. Foxp3 programs the development and function of CD4+CD25+ regulatory T cells. Nat Immunol 2003;4:330–6. https://doi.org/10.1038/ni904.
[32] Schmidt A, Oberle N, Krammer PH. Molecular mechanisms oftreg-mediatedt cell suppression. Front Immunol 2012;3:51. https://doi.org/10.3389/fimmu.2012.00051.
[33] Shevach EM. Mechanisms of Foxp3+ T regulatory cell-mediated suppression. Immunity 2009;30:636–45. https://doi.org/10.1016/j.immuni.2009.04.010.
[34] Vignali DAA, Collison LW, Workman CJ. How regulatory T cells work. Nat Rev Immunol 2008;8:523–32. https://doi.org/10.1038/nri2343.
[35] Klein M, Bopp T. Cyclic AMP represents a crucial component of treg cell-mediated immune regulation. Front Immunol 2016;7:315. https://doi.org/10.3389/fimmu.2016.00315.
[36] Ohta A, Sitkovsky M. Extracellular adenosine-mediated modulation of regulatory T cells. Front Immunol 2014;5:304. https://doi.org/10.3389/fimmu.2014.00304.
[37] Hanahan D, Weinberg RA. Hallmarks of cancer: the next generation, Cell 2011;144:646–74. http://www.ncbi.nlm.nih.gov/pubmed/21376230.
[38] Pardoll D. Cancer and immune system: basic concepts and targets for intervention. Semin Oncol 2015;42:523–38. https://doi.org/10.1053/j.seminoncol.2015.05.003.
[39] Robbins PF, Lu YC, El-Gamil M, Li YF, Gross C, Gartner J, Lin JC, Teer JK, Cliften P, Tycksen E, Samuels Y, Rosenberg SA. Mining exomic sequencing data to identify mutated antigens recognized by adoptively transferred tumor-reactive T cells. Nat Med 2013;19:747–52. https://doi.org/10.1038/nm.3161.
[40] Burnet SM. Cancer – a biological approach. Br Med J 1957;779–86.
[41] Dunn GP, Old LJ, Schreiber RD, Louis S, Burnet FM, Thomas L. The immunobiology of cancer immunosurveillance and immunoediting. Immunity 2004;21:137–48.
[42] Smyth MJ, Dunn GP, Schreiber RD. Cancer immunosurveillance and immunoediting: the roles of immunity in suppressing tumor development and shaping tumor immunogenicity. Adv Immunol 2006;90:1–50. https://doi.org/10.1016/S0065-2776(06)90001-7.
[43] Vinay DS, Ryan EP, Pawelec G, Talib WH, Stagg J, Elkord E, Lichtor T, Decker WK, Whelan RL, Kumara HMCS, Signori E, Honoki K, Georgakilas AG, Amin A, Helferich WG, Boosani CS, Guha G, Ciriolo MR, Chen S, Mohammed SI, Azmi AS, Keith WN, Bilsland A, Bhakta D, Halicka D, Fujii H, Aquilano K, Ashraf SS, Nowsheen S, Yang X, Choi BK, Kwon BS. Immune evasion in cancer: mechanistic basis and therapeutic strategies. Semin Cancer Biol 2015;35:S185–98. https://doi.org/10.1016/j.semcancer.2015.03.004.

CHAPTER

2

Immunoconjugates as immune canoes to kill breast cancer cells

Mariana Segovia-Mendoza[a], *Cristina Lemini*[a], *Rocio García-Becerra*[b], *and Jorge Morales-Montor*[c]

[a]Departamento de Farmacología, Facultad de Medicina, Universidad Nacional Autónoma de México, Ciudad de México, México [b]Departamento de Biología Molecular y Biotecnología, Instituto de Investigaciones Biomédicas, Universidad Nacional Autónoma de México, Ciudad de México, México [c]Departamento de Inmunología, Instituto de Investigaciones Biomédicas, Universidad Nacional Autónoma de México, Ciudad de México, México

Abstract

Breast cancer remains a deadly disease, even with all the recent technological advancements for its detection and treatment. The resistance of breast cancer therapy is a critical problem that is not fully understood. In this sense, the molecular classification of this pathology determines the treatment that the patient needs. Different immune and non-immune therapies have been enforced for this pathology. In this regard, in the present chapter, we will compile and describe the use of clinical immune options such as immunoconjugates as an alternative for avoiding breast cancer resistance to conventional therapy. In addition, the present work is focused on addressing the in vitro approaches until clinical studies about the effects of these drugs in breast cancer.

Abbreviations

BDZs	benzodiazepines
CDR	complementary-determining region
DM1, DM4	maytansinoids drugs
DARs	drug-antibody ratios
D10	dolastatin 10
DS-8201a	trastuzumab deruxtecan
ER	estrogen receptor
FDA	Food and Drug Administration
HER2	epidermal growth factor receptor type 2
Ig	immunoglobulin
Ims	immunoconjugates

Immunotherapy in Resistant Cancer: From the Lab Bench Work to Its Clinical Perspectives
https://doi.org/10.1016/B978-0-12-822028-3.00006-6

KDa	kilodaltons
MMAE	monomethyl auristatin E
NK	natural killer cells
nM	nanomolar
PD-1	programmed death receptor
SYD985	trastuzumab-duocarmycin
T-DM1	trastuzumab-DM1
TNBC	triple-negative breast cancer

Conflict of interest

No potential conflicts of interest were disclosed by the authors.

Introduction

Breast cancer classification

Breast cancer is the most frequent malignancy in women worldwide [1]. It can be defined as a group of biologically and molecularly heterogeneous diseases originated from the breast. Within the large group of diverse breast carcinomas, there are several types of it based on their invasiveness, primary tumor sites, and the molecular classification. Traditionally, this last feature refers to three different groups: estrogen receptor (ER)-positive, epidermal growth factor receptor type 2 (HER2)-positive, and triple-negative breast cancer (TNBC) [2]. The therapies for the first two types have been targeted against established molecular markers: ER and aromatase inhibitors for ER-positive phenotype and tyrosine kinase blockers or monoclonal antibodies against HER2 protein [3–5]. In contrast, the triple-negative phenotype does not have targeted therapies available today [6]. Despite the inhibition of biological markers with targeted drugs, resistance to conventional treatments remains a clinical challenge. As discussed earlier, the employment of combined therapies has been an important clinical tool to strengthen cancer cell inhibition. More specifically, the combination of monoclonal antibodies with cytotoxic agents has reduced breast cancer resistance to conventional therapies [7].

Immunoconjugates

Immunoconjugates (Ims) are specific and highly effective anticancer therapies. They are composed by three segments: an antibody that recognizes the cancer cell antigen, an effector molecule or the cytotoxic payload which kills the cancer cell, and a linker that ensures that the effector molecule remains attached to the antibody until it reaches the tumor cell (Fig. 1).

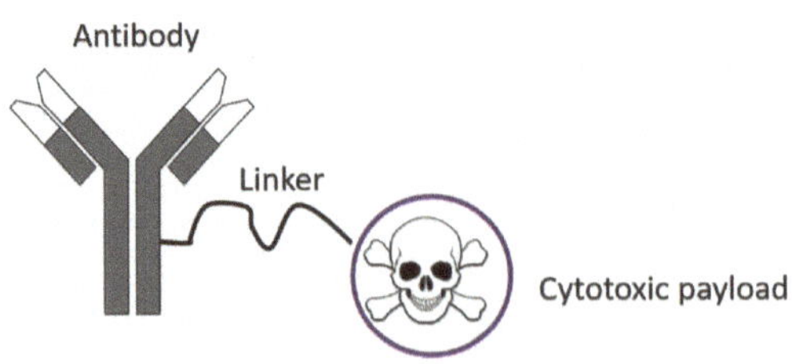

FIG. 1 Components of Ims.

The biological mechanism of this type of therapy is based on the specific detection from the antibody to a particular cellular epitope, with the subsequent drug delivery into the cancer cell [8–10]. The advantages of this immune carriers as compared to traditional chemotherapy are the improvement of the drug biodistribution and the reduction of the systemic toxicity [11]. However, several considerations for the design and use of this type of therapy must be appraised.

Mechanism of action of the Ims

After Ims are administered into the bloodstream, (1) the antibody recognizes and binds to the target cell surface antigen, (2) it is internalized by endocytosis, (3) processed within lysosomes, (4) release of the cytotoxic agent, (5) the payload disrupts DNA strands or microtubules, and (6) cell death is activated (Fig. 2) [12].

Immunoconjugate structure

Antibody portion

To better understand the function of the antibodies, the parts that make them up will first be described. The antibodies, also known as immunoglobulins (Igs), are globular structures which are synthesized by plasmatic cells. The Igs are present in the blood and other biological fluids. Specifically, they recognize other molecules (antigens) and form stable complexes with them. In mammals, five basic types of Igs are known: IgG, IgM, IgA, IgD, and IgE, and perform different functions. Of all of them, the IgG is the one that is found in the highest proportion in the bloodstream.

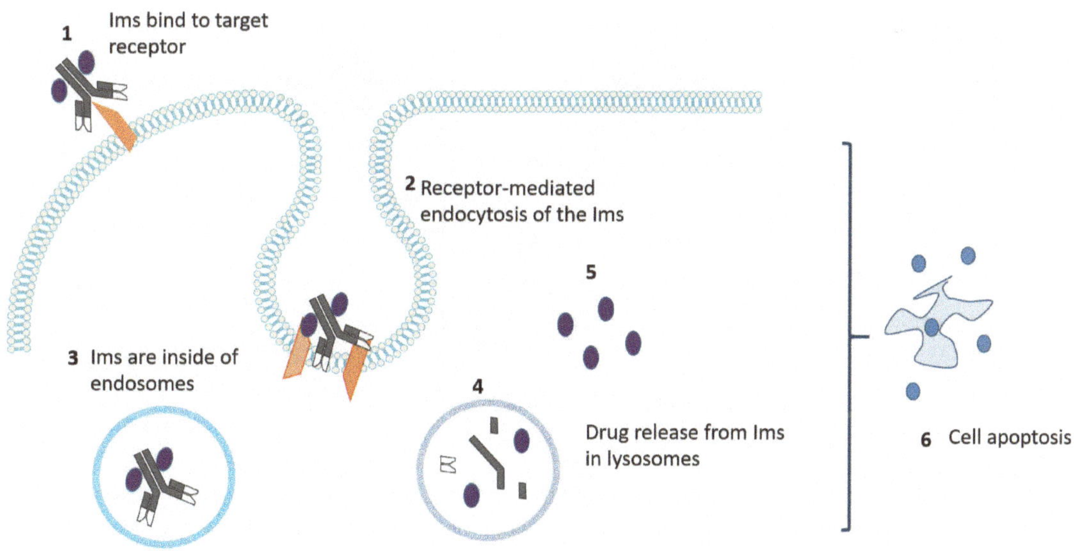

FIG. 2 Mechanism of action of Ims.

The antibodies are integrated by the Fab and the Fc regions, with a characteristic "Y" shape structure. They contain heterodimers consisting of two heavy chains and two light chains. Both have amino- and carboxyl-terminal ends and are linked by disulfide bonds. Both heavy and light chains are made up of two regions: the constant region (Fc) that interacts with cellular Fc receptors that regulate antibody-directed binding to immune cells and the variable region (Fab) that interacts with the target antigenic determinants. At the terminal end of both chains is located the specific epitope recognition site, also known as complementary-determining region (CDR) domain (Fig. 3) [10].

Routinely, the antibodies of the IgG class are most commonly employed for cancer therapy [13]. Their biodistribution properties have been extensively described. Nevertheless, it has been reported that low percentages of the injected dose of targeting IgG antibodies are retained by tumors in experimental murine models [14]. For this reason, different antibody variants, such as the mini- or nanoantibodies, have been used (Fig. 4), bispecific or bivalent antibodies, which cannot only be useful for therapy but can also be used for diagnostic purposes [15]. The greatest uses of them are the rapid blood circulation, good tissue penetration, low immunogenicity, low retention in kidneys, as well as, better infiltration in tumors. In addition, their molecular weight is located between 15 and 55 KDa, while the molecular weight of entire antibodies is around 150 KDa [16].

Despite the use of antibody fragments, Ims are a good strategy in the clinic, although some critical points should be considered in their use. The first consideration is that they need to bind a specific protein mostly found or overexpressed on cancer cells as compared to normal tissues, characteristic known as high antigen affinity. These target proteins are commonly surface antigens. Of note, the density of them and the Ims efficacy will depend on the type of cancer cell. The selection of the antibody needs other aspects to consider as the retention time in circulation, minimal immunogenic reaction, and low cross-reactivity. In addition, its

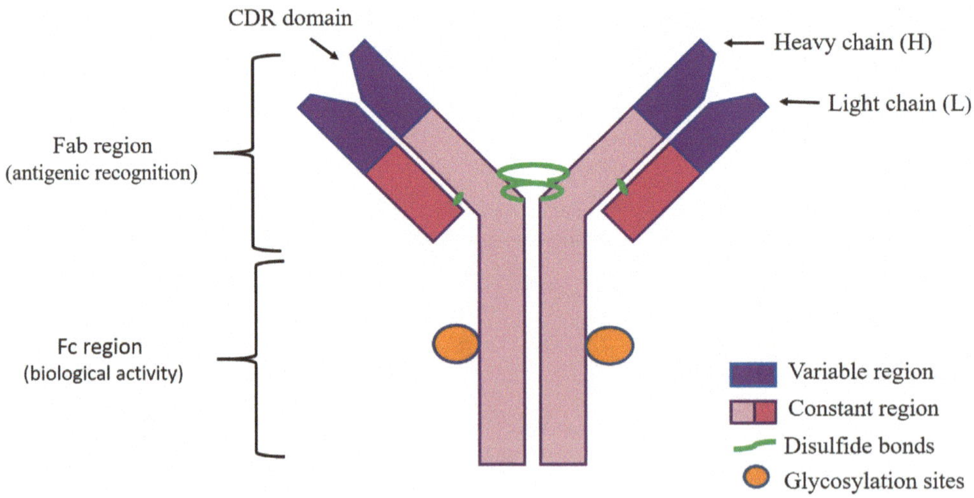

FIG. 3 Schematization of the structure of an antibody. Heavy chains are denoted in *pale pink* (H), while light chains are denoted in *dark pink* (L). The *purple region* indicates the antigen recognition site (CDR domain), which is located in the amino-terminal region of both chains. The disulfide bridges are outlined in *green* and in *orange* are denoted the glycosylation sites that contain the characteristic carbohydrates of different Igs.

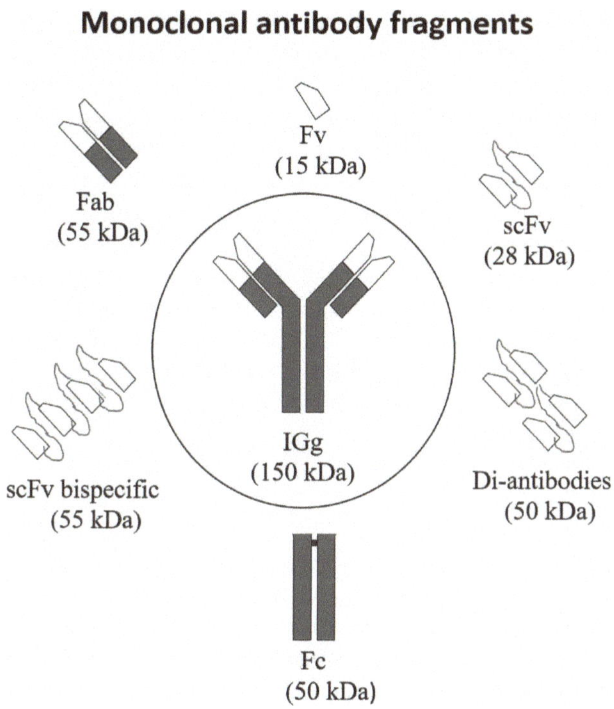

FIG. 4 Antibody variants. The different antibody variants derived from their conventional structure are outlined.

binding affinity needs to be between 0.1 and 1 nM [17]. It is important to look at that for the internalization of the Im, and the antibody fraction needs to enter quickly into the cancer cell and so release the cytotoxic agent.

Effector molecules

The therapeutic drugs employed in the Im construction must be highly toxic for cancer cells. The type of drugs used depends on the nature of the antibodies, considering that they need to be efficiently delivered from Ims into the target. In addition, the type and number of pharmaceutical agents are important factors that need consideration for their effectiveness. Usually, some reports suggest that the concentration of the cytotoxic drug must be between the picomolar range [18].

The following describes some main characteristics of the chemical and biological components mostly used for the creation of Im structure.

Chemical components

Auristatins

Auristatins are synthetic analogs of dolastatin 10 (D10), a highly cytotoxic antineoplastic agent employed for cancer chemotherapy, which is derived from a marine species known as *Dolabella auricularia* [19,20]. Initially, modifications in the C-terminal of the D10 chemical structure were made, generating compounds with less cytotoxic activity. Therefore, changes in the N-terminal end gave rise to the current auristatins used. Of note, they have lower toxic effects and similar antineoplastic action as compared to the original compound. Additionally, it was described that having a chemical change, specifically, a secondary amine at their N-terminus end improves the attachment to a linker and the subsequent conjugation to monoclonal antibodies [21]. An example of this type of drug is the monomethyl auristatin E (MMAE), which inhibits cell division by blocking the microtubule assembly and tubulin-dependent GTP hydrolysis, resulting in cell cycle arrest and apoptosis [22]. Brentuximab vedotin is the trade name of Im combined with MMAE through an enzyme-cleavable dipeptide linker. This antibody-drug conjugate is directed against the transmembrane cytokine receptor (CD30) from the lymphocytes and is used for the treatment of Hodgkin lymphoma [23]. As well as other Ims, after its union with the CD30, it is internalized and degraded via lysosomes where proteases cleave the peptide linker and release the MMAE into the cytosol carrying out its antineoplastic effects [24].

Benzodiazepines

Benzodiazepines (BDZs) are not commonly employed drugs in the design of Ims. These compounds are frequently occupied in the treatment of central nervous system disorders. Their biological mechanism is through inducing oxidative DNA damage in human cells [25]. In addition, BDZs inhibit the proliferation of human hypopharyngeal carcinoma cells and promote G_0/G_1 cell cycle arrest [26]. There are few reports about their effects on breast cancer cells. Of note, despite their cytotoxic activity, a recent article reported no significant association between benzodiazepine use and decreased survival in patients with breast cancer [27]. Nevertheless, modifications in the BDZ structure, giving a pyrrolo-BDZ synthetic compound, in combination with non-cleavable linkers in an Im conception have been shown to be effective for inhibiting the proliferation of breast cancer cells [28] even though its application in cancer patients has not yet been widely explored.

Calicheamicins

Calicheamicin is a highly potent anti-tumor antibiotic originally isolated from the actinomycete *Micromonospora echinospora*. It binds to the minor groove of DNA and cleaves double stranded [29]. Calicheamicin-Ims contain four carbohydrate residues, a hexasubstituted benzene ring, and an unusual N—O glycosidic linkage. After binding of calicheamicin to the minor groove of DNA, it is reduced by cellular thiols. The resultant product undergoes rearrangement to generate free radicals that bind hydrogen atoms from DNA. Calicheamicin $_{\gamma 1I}$ is the most toxic member of the family; however, the *N*-acetyl-calicheamicin $_{\gamma 1I}$ is considered more chemically stable for the Im payload designs [30,31]. Gemtuzumab ozogamicin/Mylotarg, inotuzumab ozogamicin/CMC544, PF-06647263, CMD-193, and CMB-401 are examples of this type of Ims tested in clinical trials and useful for the treatment of leukemia, non-Hodgkin's lymphoma, breast cancer, and among other neoplasms [30,32–35].

Camptothecins and deruxtecan

Both drugs are topoisomerase I inhibitors designed to interfere with the action of topoisomerase enzymes and regulate DNA topology. They interfere with transcription and replication causing DNA damage, failure to repair strand breaks, and cell death [36].

Camptothecin (CPT) is a plant alkaloid first identified from the Chinese tree, *Camptotheca acuminata*, that has been used in the treatment of several types of cancer [37]. Of note, CPT has been effective for inhibiting the proliferation of different subtypes of breast cancer [38]. This compound and its derivatives are currently used as a second- or third-line treatment for patients with endocrine-resistant breast cancer [38]. Regarding deruxtecan, it is a derivative of the CPT analog, which showed to be highly effective for inhibiting breast cancer cells, and improved the objective response rate in patients in clinical phases [39]. In fact, its combination with trastuzumab has been recently approved for metastatic breast cancer.

Cryptophycins

Cryptophycins, a large class of peptolides, were originally isolated from the cyanobacteria *Nostoc* sp. They are potent tubulin-binding antimitotic agents. The most abundant component, cryptophycin-1, has excellent antiproliferative activity against a broad spectrum of solid tumors; its effects are mediated by cellular arrest at the G2/M phase via apoptotic cascade activation [40]. An advantage of this type of drugs is that they are effective in different types of cancer cells, and they are still active in multiple drug-resistant cancer cell lines. Nevertheless, its stability and the susceptibility to hydrolysis have limited their use [41]. In this regard, several synthetic analogs have been developed [42,43]. However, these compounds did not show a significant effect in clinical studies [44,45]. Thus, their clinical employment is still under study.

Duocarmycin

Duocarmycins are members of a series of related natural products that were first isolated from *Streptomyces* bacteria in 1978. They are noted for their extreme cytotoxicity and thus represent an exceptionally potent class of anti-tumor drugs. Their mechanism of action is via DNA minor groove binding [46]. Several analogs of duocarmycin have been developed to enhance its cytotoxic potency further, favor its chemical stability, and decrease some of its side effects [47]. Currently, clinical studies are carried out in breast cancer patients.

Doxorubicin

Doxorubicin (DOX), or adriamycin, is an anthracycline antibiotic discovered as a chemotherapeutic drug several decades ago. DOX interacts with the DNA by intercalation, thus inhibiting the progression of the enzyme topoisomerase II, and relaxes supercoils in DNA. In addition, depending on its formulation, this drug can also form free radicals that induce DNA and cell membrane damage [48]. It is still one of the most effective medications used for treating various types of cancers, and different drug formulations have been employed, the liposomal DOX-based combination is an example of this. However, severe side effects in its clinical use have been reported regarding cardiovascular toxicity is the most frequent health consequence [49]. Moreover, generating resistance to this drug has led to the search for different cytotoxic agents as a therapeutic alternative [50]. It has been reported that tumor cells overexpressing HER2 molecule are more responsive to the cytotoxic effects of DOX, making it a good candidate to treat breast cancer patients with this tumor phenotype [51].

Hemiasterlins

Hemiasterlin is a natural cytotoxic product that was initially identified from marine sponges. This compound and its analogs have tubulin-based antimitotic mechanism inducing G2/M arrest and apoptosis [52,53]. In the particular case of this type of compounds, a novel design of Ims has been fabricated, which is known as antibody/drug-conjugated micelle system. This modification is shown to be effective on in vitro and in vivo experiments. In addition, the administration of E7974, a synthetic analog of hemiasterlin, in patients with solid tumors, offered good clinic results [54], although the use of this type of Ims in cancer patients still needs to be deepened.

Maytansinoids

Maytansinoid drugs, commonly known as DM1 and DM4, are microtubulin polymerization inhibitors, which are derived from maytansine, a natural benzoansamacrolide product isolated from the bark of the African shrub, *Maytenus ovatus* [55]. Both DM1 or DM4 act similar to vinca alkaloids, eliciting the mitotic arrest.

There are two antibody-maytansinoid conjugates that have shown good response in the clinic. Trastuzumab-DM1 (T-DM1) is a maytansinoid (emtansine) conjugated to the anti-HER2 therapeutic antibody trastuzumab, which is employed in the treatment for metastatic breast cancer [56]. On the contrary, lorvotuzumab mertansine has shown promising results in solid and liquid tumors that overexpress the neural cell adhesion molecule (CD56) [57]. T-DM1 contains a thioether-based linker, while lorvotuzumab mertansine utilizes a disulfide-based linker. Both types of linkers will be described in the later section.

Taxanes

Paclitaxel and docetaxel are terpenes produced by plants of the genus Taxus; for this reason, they belong to the family of taxol drugs. Both compounds suppress microtubule dynamics. Also, they have a remarkable efficacy against advanced solid tumors such as ovarian and breast cancer, alone or in combination fashion [58–60]. Concerning their therapeutic characteristics, they show severe side effects due to their lack of specificity. In this regard, many research groups have proposed chemical coupling with Ims to achieve them into the cellular target and deliver the cytotoxic agent. Combination with paclitaxel and the DOX-based Ims (BR96-DOX) has improved its cytotoxic activity in a broad spectrum of cancer cell lines and on in vivo studies [61]. Recently, several Ims with ester linkage containing taxol drugs have shown efficacy for inhibiting the proliferation and tumor growth of human cancer cells expressing the epidermal growth factor receptor [62,63]. There are no reports about the use of paclitaxel Ims in clinical phases. In the case of docetaxel Ims, similar to the BR96-doxorubicin study, a recent work reported that the administration of the SGN-15, a DOX Im, plus docetaxel improved the cytotoxic effects of the compounds and was well-tolerated in patients with non-small cell lung cancer [64]. However, there are no clinical studies about taxol Ims in the treatment of breast cancer patients.

Tubulysins

Tubulysins are antimitotic tetrapeptides originally isolated from myxobacteria *Archangium gephyra* and *Angiococcus disciformis*. These compounds exhibit cytostatic properties against human cancer cell lines. They cause depolymerization of microtubules, impacting on the cytoskeleton integrity [65]. Interestingly, these compounds have shown to be highly effective in picomolar concentrations against multidrug-resistant tumor cells, because they are not effluxed by drug transmembrane cellular pumps. Additionally, these compounds and their derivatives own antiangiogenic properties that might also be useful in cancer therapy [66,67]. Similar to the two drugs described previously, they can also bind to the vinca domain [68]. On in vitro experiments and phases of clinical trials, the combination of tubulysins with antibodies that recognize the folate receptor has shown good results, which sustain their therapeutic use [69,70]. Recently, glucuronide-linked antibodies with tubulysin conjugates have demonstrated that they have good effectiveness against tumor cells. In addition, this type of linker has been proposed as a promising tool to avoid resistance to Ims due to multidrug-resistant proteins [68,71].

Biological components

Beginning in the 21st century, the use of biological agents started to be incorporated into the formulations of Ims, even though their use is scarce. The characteristics of some tested biological agents are described later.

Cytokines

Cytokines are proteins normally secreted by immune cells with pleiotropic actions depend on the microenvironment context. Different Im constructions plus one or more cytokines have been developed and tested in cancer studies, mainly in melanoma diseases [72–75]. Typically, cytokines are fused with the C-terminus of the Ig light chain of the antibody, rather than fusing to the heavy chain. The aim of these Ims' setting-up offers longer circulating cytokine half-life and increased cytokine uptake by the cancer cells [76]. Up to now, the cytokine Ims have not been proved in breast cancer clinical trials.

Effector immune cells

As described, the immune system has cells, different cellular subtypes with different activities. There is a particular type responsible for killing tumor cells, such as natural killer cells (NK); they are called effector cells. In this regard, it is important to highlight that most of the cancer patients normally have an imbalance in the functioning of immune system cells [77]. Taking this into account, some studies have evaluated their concomitant cell administration with therapeutic antibodies. However, the structure of the Im is not conserved in the strict sense but may give rise to new formulations, for instance, the NCT00941928 trial, which was composed of the combination of epratuzumab, a monoclonal antibody against CD22, with haploidentical NK cells. This study evaluated the perfusion of this combination in leukemia cancer patients; the results were promising in comparison with the administration of the antibody alone. Another clinical research with encouraging clinical outcomes for the treatment of recurrent solid tumors was the NCT02843204 trial. This study was based on the administration of NK cells with nivolumab, a therapeutic antibody against the programmed death receptor (PD-1). Other studies have employed antigen-presenting cells such as dendritic cells for the combination schemes (NCT03190811). As well as combinations of Ims with cytokines, these types of therapeutic constructs have not yet been investigated in breast cancer patients [7].

RNases and toxins

A novel combination that can be applied to the manufacture of Ims is the substitution of cytotoxic payload moiety; in other words, the cytotoxic drugs are substituted by RNase molecules, which specifically bind and selectively kill the target cells. The mechanism by which the RNases act is settled by the catalytic cleavage of exposed RNA, the cellular action of the products of its hydrolysis, and the electrostatic interaction of the exogenous enzyme with different cellular components [78]. Anti-HER RNase-based structures have shown outstanding results to inhibit the proliferation of different cancer cells that overexpress this receptor, pointing out as attractive anticancer agents for breast cancer patients overexpressing this molecule [79,80]. Different studies have used other types of RNases and have been effective in CD22-positive malignancies [81]. In addition, RNases have been combined with toxins in order to synergize its effects; the ribotoxin α-sarcin is an example of this implementation [82]. Similar to the two types of Ims mentioned earlier, no reports of this type of therapeutic alternative have been documented in breast cancer clinical trials. In Table 1, the drugs used in the payloads described in this section are listed.

Linkers

As was mentioned before, the linkers are the molecules that ensure that the therapeutic agent can be released into the cancer cells with the subsequent disengagement of the antibody. Thioether, hydrazine linkers, disulfide bonds as well as several peptides are examples of used linkers. The choice for the individual use lies in the type of Im and the stability that it will confer for its internalization to the cell. Of note, the identity, number, and type of linker

TABLE 1 Mechanism of action of the drugs most used as payloads in Ims.

Chemical payloads		Biological payloads		
Microtubule inhibition agents	**DNA damaging agents**	**Cytokines**	**Effector immune cells**	**RNases and toxins**
Auristatins	Benzodiazepines	IL-12	NK cells	Anti-HER RNase-based structures
Cryptophycins	Calicheamicins	Fusion of IL-19/IL-2	Dendritic cells	Ribotoxin α-sarcin
Hemiasterlins	Camptothecin and deruxtecan	Fusion of IL-19/TNF-α		
Maytansinoids	Duocarmycin			
Taxanes (docetaxel and paclitaxel)	Doxorubicin			
Tubulysins				

attachment sites is a critical issue. According to this point, the linkers are divided into two types, cleaved and no-cleaved linkers.

Cleavable linkers

As the name implies, this class of linkers leads the release of the payload after bond cleavage. These types of molecules are designed to respond to environmental changes between extracellular and intracellular contexts (pH, redox potential, hypoxia, lysosomal enzymes, and among others) [83]. They can act in three different mechanisms:

Lysosomal protease-sensitive linkers

This strategy utilizes lysosomal proteases that recognize and cleave a dipeptide bond to release the free drug from the conjugate. Cathepsin B is the most used lysosomal protease linker; it recognizes specific sequences such as phenylalanine-lysine and valine-citrulline and cleaves a peptide bond on the C-terminal side of such sequences. Upon internalization through endocytosis and transportation to lysosomes, cathepsin B cleaves linkers with these motifs and releases the cytotoxic payloads [84,85].

Acid-sensitive linkers

This class of linkers takes the advantage of the low pH in the lysosomal compartment to trigger the hydrolysis of an acid-labile group within the linker and release the drug payload. Hydrazone linker is a classic example of this type of linker [85].

Glutathione-sensitive disulfide linker

This therapeutic strategy takes advantage of the higher concentration of thiol groups between intra- and extracellular cell environments. In this sense, disulfiders are the most outstanding class of this type of linker employed in Im constructs, which are embedded within the linker and resist the reductive cleavage in circulation. However, upon internalization, abundant glutathione enzyme is present; then, the linker suffers a reductive cleaves in the disulfide bond, and the payload molecule is released [85,86].

Non-cleavable linkers

Non-cleavable linkers consist of stable bonds that are resistant to proteolytic degradation. They are also known as chemical conjugation linkers. They offer greater stability in plasma than cleavable linkers. Non-cleavable linkers suffer complete degradation of the antibody by cytosolic and lysosomal proteases, and then, the payload is released. The cytotoxic agents used in this type of linkers are drugs capable of exerting their anti-tumor effects despite being chemically modified. Of note, the linker needs to supply the rapidly released of the toxic payload once it is internalized into the target tumor cell [83,85].

Lysine amide coupling is a useful conjugation method worn in the antibody lysine residues for linkers with activated carboxylic acid esters. In addition, it is known that around 80 lysine residues are on a typical antibody. However, only about 10% can be available for these purposes, due to their accessibility. On the contrary, it has been described that they also participate in antigen recognition, which can reduce the binding affinity to the target cell and its effectiveness [87].

Cysteine-based conjugation methods are another example of non-cleavable linkers, and they are made between cysteine residues of the antibody and the thiol functional group of the payload. Notably, cysteine residues, coming from the antibody, form interchain disulfide bonds that need to be reduced for its manipulation. It is of standing out that cysteine-based conjugation is superior to lysine conjugation yielding better drug-antibody ratios (DARs). Of note, high DARs increase not only potency but also the risk of aggregation, clearance rate, and premature release of the toxic agent during circulation; thus, the distribution needs to be carefully controlled [87].

Ims proved in breast cancer

In this section, we will summarize the Ims used in breast cancer therapeutics, which are listed in Table 2.

TABLE 2 Ims evaluated in different phenotypes of breast cancer in preclinical and clinical studies.

Im	Payload	Target antigen	Population	Response	Type of cleave	Reference
METASTATIC BREAST CANCER						
BR96-DOX	Doxorubicin	Lewis-Y antigen	Metastatic breast cancer patients	Limited anti-tumor activity	Cleavable linker: acid-labile hydrazine linker	[88]
Im with ricin A-260F9	Ricin A-260F9	Human transferrin receptor	Metastatic breast cancer patients	Not conclusive for cancer treatment, severe side effects	Cleavable linker: disulfide bond	[89]
HER2 BREAST CANCER						
Trastuzumab emtansine[a]	DM1	HER2	HER2-positive metastatic breast cancer patients	FDA approved Improving the progression-free survival and overall survival	Non-cleavable linker: lysine conjugated	[90]
DS-8201a Trastuzumab-deruxtecan[a]	Deruxtecan	HER2	HER2-positive metastatic breast cancer patients	FDA approved Increasing the progression-free survival	Cleavable linker: peptide-based linker	[91]

TABLE 2 Ims evaluated in different phenotypes of breast cancer in preclinical and clinical studies—cont'd

Im	Payload	Target antigen	Population	Response	Type of cleave	Reference
SYD985 Trastuzumab-duocarmycin	Duocarmycin	HER2	HER2-positive breast cancer patients	Still under study	Cleavable linker: valine-citrulline-seco-group	[92]
ZHER2:289	MMAE	HER2	Breast cancer cells	Inhibition of cell proliferation	Not specified	
HER2-affitoxin	Affitoxin	HER2	HER2-positive breast cancer cells	Inhibition of cell proliferation	Cleavable linker	[90]
H5B14 and PCM5B14	Duocarmycin and MMAE	RON tyrosine kinase receptor	Breast cancer cells	Inhibition of cell proliferation and spheroid formation Inhibition of tumor growth	Cleavable linker	[91]
TN BREAST CANCER						
Bevacizumab and Trastuzumab with selenium	Selenium	EGFR/HER2 VEGF	TNBC cells	Inhibition of cell proliferation and induction of apoptosis	Not specified	[93]
Glembatumumab vedotin MMAE	MMAE	GPNMB	TNBC patients	Prolonging progression-free survival	Cleavable linker: valine-citrulline enzyme	[92,94]
PF-06647263 calicheamicin	Calicheamicin	Antigen ephrin A4	TNBC patients	Limited anti-tumor activity	Cleavable linker: hydrazone	[95]

[a] *Approved therapy in the clinic.*

As we mention before, HER2 are overexpressed membrane surface proteins in breast cancer cells with an important role in proliferation and survival. For this reason, some Ims look them at as significant targets. In this regard, since the 1990s, a chimeric antibody conjugated with DOX (BR96-DOX) was tested in metastatic breast cancer patients. This Im was developed for recognition of the Lewis-Y antigen, which is expressed in several cancers, including breast cancer [96]. It exhibited limited clinical anti-tumor activity in metastatic breast cancer and provided important guidelines to develop more specific Ims [88]. More recently, trastuzumab emtansine Im was approved by the Food and Drug Administration (FDA) in

February 2013 for the treatment of patients with HER2-positive metastatic breast cancer. It is shaped by the anti-HER2 monoclonal antibody trastuzumab and the maytansinoid derivative 1 (DM1). Broadly, its effects in cancer patients have been the improvement both the progression-free survival and overall survival [97]. Recently, to improve the therapeutic options of T-DM1, different Ims based on trastuzumab have been tested, for instance, trastuzumab-deruxtecan (DS-8201a) or trastuzumab-duocarmycin (SYD985). They have promising anti-tumor activity in various preclinical models. In fact, on 2019, the FDA approved to DS-8201a for patients with unresectable or metastatic HER2-positive breast cancer who have received two or more prior anti-HER2-based regimens in the metastatic setting [98]. Meanwhile, SYD985 has shown notable clinical results in breast cancer patients with HER2-expressing breast metastatic cancer, including HER2-positive T-DM1-resistant individuals, as has been reported in phase I studies [39,99–102]. However, despite the clinical efficacy of Ims based on trastuzumab, offbeat modifications in the Ims recognition are currently being developed and tested in vitro to obtain better clinical outcomes.

The implementation of antibody variants is an example of it. This is the case of ZHER2:289, an anti-HER2 affibody combined with MMAE. This shift in the antibody structure showed to be highly effective for inhibiting breast cancer cell proliferation in a procedure that combines potency and selectivity with inexpensive and relatively straightforward manufacturing [103]. In addition, in the same case, the combination of HER2-affibodies with toxins, such as affitoxin, an exotoxin coming from *Pseudomonas aeruginosa* species, has also been resulted in a promising area for the treatment of breast cancer cells [90]. Nevertheless, its toxicological effects on in vivo models are still pending. In this regard, there are few studies about toxins combined with Im formulations. A phase I study with metastatic breast cancer patients was proved an Im combined with ricin A-260F9, a toxin extracted from castor beans. The results were not favorable since many patients experienced severe side effects. However, the study population was very small [89]. In addition, the study of H5B14 and PCM5B14, a human and a monkey antibody, respectively, is another precedent of the implementation of antibody variants. In this case, both molecules were modified and joined, resulting in a humanized biological particle against a specific domain against the tumorigenic factor RON, and conjugated with two payloads, MMAE and duocarmycin. This Im inhibited the proliferation and spheroid formation of HER2-positive breast cancer cells. Furthermore, it also inhibited the tumor growth of breast xenografts with the same phenotype in a remarkable manner [91]. However, the search for new combined treatment alternatives for this subtype continues to be the subject of study by various research groups.

On the contrary, TNBC is known to be the breast neoplasm that presents the least therapeutic options due to its aggressiveness and lack of molecular targets that can be blocked. Regarding this notion, some Ims have been investigated. Firstly, based on in vitro studies, approved humanized monoclonal antibodies such as bevacizumab and trastuzumab have been combined with selenium and evaluated in TNBC cells. They showed higher ranges of cellular inhibition and apoptosis induction of cancer cells in comparison with treatments alone. Their effects are mediated by the production of superoxide anions and for interfering with the HER2/EGFR and VEGF signaling. The previously mentioned gives rise to the search for new combinations of Ims not only with cytotoxic drugs but also with essential dietary supplements [93]. In relation to clinical studies, glembatumumab vedotin is an Im proved in this cancer subtype whose molecular target is the glycoprotein non-metastatic melanoma protein B (GPNMB), a molecule overexpressed in breast cancer cells that confer metastatic phenotype by activation of metalloproteinases and several protein tyrosine kinases. This Im contains MMAE as payload, and the results are related to prolonging progression-free survival [92,94]. Another Im developed for the application in this breast cancer phenotype is PF-06647263, which comprises a humanized antibody that is directed against the breast cancer antigen ephrin A4 and is conjugated to calicheamicin. It has shown promising efficacy in decreasing the proliferation and eradication of tumor-initiating cells in breast cancer [95]. Nevertheless, preliminary clinical evidence suggests limited anti-tumor activity in patients with TNBC [33].

Conclusion

Classically, the concept of Ims as magic bullets pointed out by Paul Ehrlich in 1913, which was defined as a selective drug delivery into the tumor cells, has been widely studied for breast cancer treatment as described in the previous section. Also, several authors have established that the selection of the antibody format, the appropriate antigen target, and the conjugation method for the payload are crucial factors for their effectiveness. However, it would be advisable to deepen the studies on this last concept, since it is still not very clear which type of linker is better in certain situations.

Globally, a large number of research works have been recognized that this biological therapy has overcome both the efficacy and the development of the drug resistance of conventional and immune therapies employed in cancer [104]. The previously mentioned is sustained because the copy number of the specific antigen is much higher in cancer cells compared to normal cells, conferring a specific cell selection [105].

Despite the fact that Ims have been shown to have unmatched clinical efficacy, these therapies have to face several challenges in multiple areas such as biological, chemical, pharmacological, and clinical aspects. We envisioned that to further increase the effectiveness of the Ims, they should be administered intratumorally, in addition, to systematically in breast cancer tumors, which could potentially increase their effectiveness and lead to attack the core of the tumor mass, instead of just the superficial cellular.

Acknowledgments

MSM and CL thank the Faculty of Medicine of the National Autonomous University of Mexico (UNAM). This study was supported by the grants from Programa de Apoyo a Proyectos de Investigación e Innovación Tecnológica (PAPIIT), Dirección General de Asuntos del Personal Académico (DGAPA), Universidad Nacional Autónoma de México (UNAM), grant/award number IN-209719, and from Fronteras en la Ciencia, Consejo Nacional de Ciencia y Tecnología (CONACYT), Grant No FC 2016 2125, both to Jorge Morales-Montor. This work was supported by the grants from Consejo Nacional de Ciencia y Tecnología (CONACYT), México. No. 256994 and from Programa de Apoyo a Proyectos de Investigación e Innovación Tecnológica (PAPIIT), Dirección General de Asuntos del Personal Académico (DGAPA), Universidad Nacional Autónoma de México (UNAM) No. IN208520 corresponding to Rocío García-Becerra.

References

[1] Harbeck N, Penault-Llorca F, Cortes J, Gnant M, Houssami N, Poortmans P, Ruddy K, Tsang J, Cardoso F. Breast cancer. Nat Rev Dis Primers 2019;5:66.

[2] Feng Y, Spezia M, Huang S, Yuan C, Zeng Z, Zhang L, Ji X, Liu W, Huang B, Luo W, Liu B, Lei Y, Du S, Vuppalapati A, Luu HH, Haydon RC, He TC, Ren G. Breast cancer development and progression: risk factors, cancer stem cells, signaling pathways, genomics, and molecular pathogenesis. Genes Dis 2018;5:77–106.

[3] Tremont A, Lu J, Cole JT. Endocrine therapy for early breast cancer: updated review. Ochsner J 2017;17:405–11.

[4] Segovia-Mendoza M, Gonzalez-Gonzalez ME, Barrera D, Diaz L, Garcia-Becerra R. Efficacy and mechanism of action of the tyrosine kinase inhibitors gefitinib, lapatinib and neratinib in the treatment of HER2-positive breast cancer: preclinical and clinical evidence. Am J Cancer Res 2015;5:2531–61.

[5] Wilson FR, Coombes ME, Brezden-Masley C, Yurchenko M, Wylie Q, Douma R, Varu A, Hutton B, Skidmore B, Cameron C. Herceptin(R) (trastuzumab) in HER2-positive early breast cancer: a systematic review and cumulative network meta-analysis. Syst Rev 2018;7:191.

[6] Mehanna J, Haddad FG, Eid R, Lambertini M, Kourie HR. Triple-negative breast cancer: current perspective on the evolving therapeutic landscape. Int J Womens Health 2019;11:431–7.

[7] Corraliza-Gorjon I, Somovilla-Crespo B, Santamaria S, Garcia-Sanz JA, Kremer L. New strategies using antibody combinations to increase cancer treatment effectiveness. Front Immunol 2017;8:1804.

[8] Smaglo BG, Aldeghaither D, Weiner LM. The development of immunoconjugates for targeted cancer therapy. Nat Rev Clin Oncol 2014;11:637–48.

[9] Sharkey RM, Goldenberg DM. Use of antibodies and immunoconjugates for the therapy of more accessible cancers. Adv Drug Deliv Rev 2008;60:1407–20.

[10] Strohl WR. Current progress in innovative engineered antibodies. Protein Cell 2018;9:86–120.

[11] Tsuchikama K, An Z. Antibody-drug conjugates: recent advances in conjugation and linker chemistries. Protein Cell 2018;9:33–46.

[12] Chari RV, Miller ML, Widdison WC. Antibody-drug conjugates: an emerging concept in cancer therapy. Angew Chem Int Ed Engl 2014;53:3796–827.

[13] Brezski RJ, Georgiou G. Immunoglobulin isotype knowledge and application to Fc engineering. Curr Opin Immunol 2016;40:62–9.

[14] Kuo WY, Lin JJ, Hsu HJ, Chen HS, Yang AS, Wu CY. Noninvasive assessment of characteristics of novel anti-HER2 antibodies by molecular imaging in a human gastric cancer xenograft-bearing mouse model. Sci Rep 2018;8:13735.

[15] Ahmad ZA, Yeap SK, Ali AM, Ho WY, Alitheen NB, Hamid M. scFv antibody: principles and clinical application. Clin Dev Immunol 2012;2012:980250.

[16] Xenaki KT, Oliveira S, van Bergen en Henegouwen PMP. Antibody or antibody fragments: implications for molecular imaging and targeted therapy of solid tumors. Front Immunol 2017;8:1287.

[17] Damelin M, Zhong W, Myers J, Sapra P. Evolving strategies for target selection for antibody-drug conjugates. Pharm Res 2015;32:3494–507.

[18] Chari RV. Targeted cancer therapy: conferring specificity to cytotoxic drugs. Acc Chem Res 2008;41:98–107.

[19] Maderna A, Leverett CA. Recent advances in the development of new auristatins: structural modifications and application in antibody drug conjugates. Mol Pharm 2015;12:1798–812.

[20] Newman DJ. The "utility" of highly toxic marine-sourced compounds. Mar Drugs 2019;17(6):324.

[21] Doronina SO, Toki BE, Torgov MY, Mendelsohn BA, Cerveny CG, Chace DF, DeBlanc RL, Gearing RP, Bovee TD, Siegall CB, Francisco JA, Wahl AF, Meyer DL, Senter PD. Development of potent monoclonal antibody auristatin conjugates for cancer therapy. Nat Biotechnol 2003;21:778–84.

[22] Francisco JA, Cerveny CG, Meyer DL, Mixan BJ, Klussman K, Chace DF, Rejniak SX, Gordon KA, DeBlanc R, Toki BE, Law CL, Doronina SO, Siegall CB, Senter PD, Wahl AF. cAC10-vcMMAE, an anti-CD30-monomethyl auristatin E conjugate with potent and selective antitumor activity. Blood 2003;102:1458–65.

[23] Donato EM, Fernandez-Zarzoso M, Hueso JA, de la Rubia J. Brentuximab vedotin in Hodgkin lymphoma and anaplastic large-cell lymphoma: an evidence-based review. Onco Targets Ther 2018;11:4583–90.

[24] Deng C, Pan B, O'Connor OA. Brentuximab vedotin. Clin Cancer Res 2013;19:22–7.

[25] Ekonomopoulou MT, Akritopoulou K, Mourelatos C, Iakovidou-Kritsi Z. A comparative study on the cytogenetic activity of three benzodiazepines in vitro. Genet Test Mol Biomarkers 2011;15:373–8.

[26] Dou Y, Li Y, Chen J, Wu S, Xiao X, Xie S, Tang L, Yan M, Wang Y, Lin J, Zhu W, Yan G. Inhibition of cancer cell proliferation by midazolam by targeting transient receptor potential melastatin 7. Oncol Lett 2013;5:1010–6.

[27] O'Donnell SB, Nicholson MK, Boland JW. The association between benzodiazepines and survival in patients with cancer: a systematic review. J Pain Symptom Manage 2019;57:999–1008. e1011.

[28] Gregson SJ, Masterson LA, Wei B, Pillow TH, Spencer SD, Kang GD, Yu SF, Raab H, Lau J, Li G, Lewis Phillips GD, Gunzner-Toste J, Safina BS, Ohri R, Darwish M, Kozak KR, Dela Cruz-Chuh J, Polson A, Flygare JA, Howard PW. Pyrrolobenzodiazepine dimer antibody-drug conjugates: synthesis and evaluation of noncleavable drug-linkers. J Med Chem 2017;60:9490–507.

[29] Gredicak M, Jeric I. Enediyne compounds – new promises in anticancer therapy. Acta Pharm 2007;57: 133–50.

[30] Damle NK. Tumour-targeted chemotherapy with immunoconjugates of calicheamicin. Expert Opin Biol Ther 2004;4:1445–52.

[31] Zein N, Poncin M, Nilakantan R, Ellestad GA. Calicheamicin gamma 1I and DNA: molecular recognition process responsible for site-specificity. Science 1989;244:697–9.

[32] Kantarjian HM, DeAngelo DJ, Stelljes M, Martinelli G, Liedtke M, Stock W, Gokbuget N, O'Brien S, Wang K, Wang T, Paccagnella ML, Sleight B, Vandendries E, Advani AS. Inotuzumab ozogamicin versus standard therapy for acute lymphoblastic leukemia. N Engl J Med 2016;375:740–53.

[33] Garrido-Laguna I, Krop I, Burris 3rd HA, Hamilton E, Braiteh F, Weise AM, Abu-Khalaf M, Werner TL, Pirie-Shepherd S, Zopf CJ, Lakshminarayanan M, Holland JS, Baffa R, Hong DS. First-in-human, phase I study of PF-06647263, an anti-EFNA4 calicheamicin antibody-drug conjugate, in patients with advanced solid tumors. Int J Cancer 2019;145:1798–808.

[34] Herbertson RA, Tebbutt NC, Lee FT, MacFarlane DJ, Chappell B, Micallef N, Lee ST, Saunder T, Hopkins W, Smyth FE, Wyld DK, Bellen J, Sonnichsen DS, Brechbiel MW, Murone C, Scott AM. Phase I biodistribution and pharmacokinetic study of Lewis Y-targeting immunoconjugate CMD-193 in patients with advanced epithelial cancers. Clin Cancer Res 2009;15:6709–15.

[35] Chan SY, Gordon AN, Coleman RE, Hall JB, Berger MS, Sherman ML, Eten CB, Finkler NJ. A phase 2 study of the cytotoxic immunoconjugate CMB-401 (hCTM01-calicheamicin) in patients with platinum-sensitive recurrent epithelial ovarian carcinoma. Cancer Immunol Immunother 2003;52:243–8.

[36] Binaschi M, Zunino F, Capranico G. Mechanism of action of DNA topoisomerase inhibitors. Stem Cells 1995;13:369–79.

[37] Pommier Y. Topoisomerase I inhibitors: camptothecins and beyond. Nat Rev Cancer 2006;6:789–802.

[38] Tesauro C, Simonsen AK, Andersen MB, Petersen KW, Kristoffersen EL, Algreen L, Hansen NY, Andersen AB, Jakobsen AK, Stougaard M, Gromov P, Knudsen BR, Gromova I. Topoisomerase I activity and sensitivity to camptothecin in breast cancer-derived cells: a comparative study. BMC Cancer 2019;19:1158.

[39] Tamura K, Tsurutani J, Takahashi S, Iwata H, Krop IE, Redfern C, Sagara Y, Doi T, Park H, Murthy RK, Redman RA, Jikoh T, Lee C, Sugihara M, Shahidi J, Yver A, Modi S. Trastuzumab deruxtecan (DS-8201a) in patients with advanced HER2-positive breast cancer previously treated with trastuzumab emtansine: a dose-expansion, phase 1 study. Lancet Oncol 2019;20:816–26.

[40] Shih C, Teicher BA. Cryptophycins: a novel class of potent antimitotic antitumor depsipeptides. Curr Pharm Des 2001;7:1259–76.

[41] Smith CD, Zhang X, Mooberry SL, Patterson GM, Moore RE. Cryptophycin: a new antimicrotubule agent active against drug-resistant cells. Cancer Res 1994;54:3779–84.

[42] Weiss C, Sammet B, Sewald N. Recent approaches for the synthesis of modified cryptophycins. Nat Prod Rep 2013;30:924–40.

[43] Varie DL, Shih C, Hay DA, Andis SL, Corbett TH, Gossett LS, Janisse SK, Martinelli MJ, Moher ED, Schultz RM, Toth JE. Synthesis and biological evaluation of cryptophycin analogs with substitution at C-6 (fragment C region). Bioorg Med Chem Lett 1999;9:369–74.

[44] Stevenson JP, Sun W, Gallagher M, Johnson R, Vaughn D, Schuchter L, Algazy K, Hahn S, Enas N, Ellis D, Thornton D, O'Dwyer PJ. Phase I trial of the cryptophycin analogue LY355703 administered as an intravenous infusion on a day 1 and 8 schedule every 21 days. Clin Cancer Res 2002;8:2524–9.

[45] Sessa C, Weigang-Kohler K, Pagani O, Greim G, Mora O, De Pas T, Burgess M, Weimer I, Johnson R. Phase I and pharmacological studies of the cryptophycin analogue LY355703 administered on a single intermittent or weekly schedule. Eur J Cancer 2002;38:2388–96.

[46] Elgersma RC, Coumans RG, Huijbregts T, Menge WM, Joosten JA, Spijker HJ, de Groot FM, van der Lee MM, Ubink R, van den Dobbelsteen DJ, Egging DF, Dokter WH, Verheijden GF, Lemmens JM, Timmers CM, Beusker PH. Design, synthesis, and evaluation of linker-duocarmycin payloads: toward selection of HER2-targeting antibody-drug conjugate SYD985. Mol Pharm 2015;12:1813–35.
[47] Patil PC, Satam V, Lee M. A short review on the synthetic strategies of duocarmycin analogs that are powerful DNA alkylating agents. Anticancer Agents Med Chem 2015;15:616–30.
[48] Rivankar S. An overview of doxorubicin formulations in cancer therapy. J Cancer Res Ther 2014;10:853–8.
[49] Kalyanaraman B. Teaching the basics of the mechanism of doxorubicin-induced cardiotoxicity: have we been barking up the wrong tree? Redox Biol 2020;29:101394.
[50] Christowitz C, Davis T, Isaacs A, van Niekerk G, Hattingh S, Engelbrecht AM. Mechanisms of doxorubicin-induced drug resistance and drug resistant tumour growth in a murine breast tumour model. BMC Cancer 2019;19:757.
[51] Campiglio M, Somenzi G, Olgiati C, Beretta G, Balsari A, Zaffaroni N, Valagussa P, Menard S. Role of proliferation in HER2 status predicted response to doxorubicin. Int J Cancer 2003;105:568–73.
[52] Crews P, Farias JJ, Emrich R, Keifer PA. Milnamide A, an unusual cytotoxic tripeptide from the Marine sponge auletta cf. constricta. J Org Chem 1994;59:2932–4.
[53] Kuznetsov G, TenDyke K, Towle MJ, Cheng H, Liu J, Marsh JP, Schiller SE, Spyvee MR, Yang H, Seletsky BM, Shaffer CJ, Marceau V, Yao Y, Suh EM, Campagna S, Fang FG, Kowalczyk JJ, Littlefield BA. Tubulin-based antimitotic mechanism of E7974, a novel analogue of the marine sponge natural product hemiasterlin. Mol Cancer Ther 2009;8:2852–60.
[54] Rocha-Lima CM, Bayraktar S, Macintyre J, Raez L, Flores AM, Ferrell A, Rubin EH, Poplin EA, Tan AR, Lucarelli A, Zojwalla N. A phase 1 trial of E7974 administered on day 1 of a 21-day cycle in patients with advanced solid tumors. Cancer 2012;118:4262–70.
[55] Kupchan SM, Komoda Y, Branfman AR, Sneden AT, Court WA, Thomas GJ, Hintz HP, Smith RM, Karim A, Howie GA, Verma AK, Nagao Y, Dailey Jr. RG, Zimmerly VA, Sumner Jr. WC. The maytansinoids. Isolation, structural elucidation, and chemical interrelation of novel ansa macrolides. J Org Chem 1977;42:2349–57.
[56] Montemurro F, Ellis P, Anton A, Wuerstlein R, Delaloge S, Bonneterre J, Quenel-Tueux N, Linn SC, Irahara N, Donica M, Lindegger N, Barrios CH. Safety of trastuzumab emtansine (T-DM1) in patients with HER2-positive advanced breast cancer: primary results from the KAMILLA study cohort 1. Eur J Cancer 2019;109:92–102.
[57] Van Acker HH, Capsomidis A, Smits EL, Van Tendeloo VF. CD56 in the immune system: more than a marker for cytotoxicity? Front Immunol 2017;8:892.
[58] Wei Y, Pu X, Zhao L. Preclinical studies for the combination of paclitaxel and curcumin in cancer therapy (review). Oncol Rep 2017;37:3159–66.
[59] Gamucci T, Pizzuti L, Natoli C, Mentuccia L, Sperduti I, Barba M, Sergi D, Iezzi L, Maugeri-Sacca M, Vaccaro A, Magnolfi E, Gelibter A, Barchiesi G, Magri V, D'Onofrio L, Cassano A, Rossi E, Botticelli A, Moscetti L, Omarini C, Fabbri MA, Scinto AF, Corsi D, Carbognin L, Mazzotta M, Bria E, Foglietta J, Samaritani R, Garufi C, Mariani L, Barni S, Mirabelli R, Sarmiento R, Graziano V, Santini D, Marchetti P, Tonini G, Di Lauro L, Sanguineti G, Paoletti G, Tomao S, De Maria R, Veltri E, Paris I, Giotta F, Latorre A, Giordano A, Ciliberto G, Vici P. A multicenter REtrospective observational study of first-line treatment with PERtuzumab, trastuzumab and taxanes for advanced HER2 positive breast cancer patients. RePer Study. Cancer Biol Ther 2019;20:192–200.
[60] Azarenko O, Smiyun G, Mah J, Wilson L, Jordan MA. Antiproliferative mechanism of action of the novel taxane cabazitaxel as compared with the parent compound docetaxel in MCF7 breast cancer cells. Mol Cancer Ther 2014;13:2092–103.
[61] Trail PA, Willner D, Bianchi AB, Henderson AJ, TrailSmith MD, Girit E, Lasch S, Hellstrom I, Hellstrom KE. Enhanced antitumor activity of paclitaxel in combination with the anticarcinoma immunoconjugate BR96-doxorubicin. Clin Cancer Res 1999;5:3632–8.
[62] Wu X, Ojima I. Tumor specific novel taxoid-monoclonal antibody conjugates. Curr Med Chem 2004;11:429–38.
[63] Ojima I, Geng X, Wu X, Qu C, Borella CP, Xie H, Wilhelm SD, Leece BA, Bartle LM, Goldmacher VS, Chari RV. Tumor-specific novel taxoid-monoclonal antibody conjugates. J Med Chem 2002;45:5620–3.
[64] Ross HJ, Hart LL, Swanson PM, Rarick MU, Figlin RA, Jacobs AD, McCune DE, Rosenberg AH, Baron AD, Grove LE, Thorn MD, Miller DM, Drachman JG, Rudin CM. A randomized, multicenter study to determine the safety and efficacy of the immunoconjugate SGN-15 plus docetaxel for the treatment of non-small cell lung carcinoma. Lung Cancer 2006;54:69–77.

[65] Nicolaou KC, Yin J, Mandal D, Erande RD, Klahn P, Jin M, Aujay M, Sandoval J, Gavrilyuk J, Vourloumis D. Total synthesis and biological evaluation of natural and designed tubulysins. J Am Chem Soc 2016;138:1698–708.

[66] Kaur G, Hollingshead M, Holbeck S, Schauer-Vukasinovic V, Camalier RF, Domling A, Agarwal S. Biological evaluation of tubulysin a: a potential anticancer and antiangiogenic natural product. Biochem J 2006;396: 235–42.

[67] Rath S, Liebl J, Furst R, Ullrich A, Burkhart JL, Kazmaier U, Herrmann J, Muller R, Gunther M, Schreiner L, Wagner E, Vollmar AM, Zahler S. Anti-angiogenic effects of the tubulysin precursor pretubulysin and of simplified pretubulysin derivatives. Br J Pharmacol 2012;167:1048–61.

[68] Staben LR, Yu SF, Chen J, Yan G, Xu Z, Del Rosario G, Lau JT, Liu L, Guo J, Zheng B, Cruz-Chuh JD, Lee BC, Ohri R, Cai W, Zhou H, Kozak KR, Xu K, Lewis Phillips GD, Lu J, Wai J, Polson AG, Pillow TH. Stabilizing a tubulysin antibody-drug conjugate to enable activity against multidrug-resistant tumors. ACS Med Chem Lett 2017;8:1037–41.

[69] Reddy JA, Dorton R, Bloomfield A, Nelson M, Dircksen C, Vetzel M, Kleindl P, Santhapuram H, Vlahov IR, Leamon CP. Pre-clinical evaluation of EC1456, a folate-tubulysin anti-cancer therapeutic. Sci Rep 2018;8:8943.

[70] Reddy JA, Westrick E, Santhapuram HK, Howard SJ, Miller ML, Vetzel M, Vlahov I, Chari RV, Goldmacher VS, Leamon CP. Folate receptor-specific antitumor activity of EC131, a folate-maytansinoid conjugate. Cancer Res 2007;67:6376–82.

[71] Burke PJ, Hamilton JZ, Pires TA, Lai HWH, Leiske CI, Emmerton KK, Waight AB, Senter PD, Lyon RP, Jeffrey SC. Glucuronide-linked antibody-tubulysin conjugates display activity in MDR(+) and heterogeneous tumor models. Mol Cancer Ther 2018;17:1752–60.

[72] Charych DH, Hoch U, Langowski JL, Lee SR, Addepalli MK, Kirk PB, Sheng D, Liu X, Sims PW, VanderVeen LA, Ali CF, Chang TK, Konakova M, Pena RL, Kanhere RS, Kirksey YM, Ji C, Wang Y, Huang J, Sweeney TD, Kantak SS, Doberstein SK. NKTR-214, an engineered cytokine with biased IL2 receptor binding, increased tumor exposure, and marked efficacy in mouse tumor models. Clin Cancer Res 2016;22:680–90.

[73] Danielli R, Patuzzo R, Di Giacomo AM, Gallino G, Maurichi A, Di Florio A, Cutaia O, Lazzeri A, Fazio C, Miracco C, Giovannoni L, Elia G, Neri D, Maio M, Santinami M. Intralesional administration of L19-IL2/L19-TNF in stage III or stage IVM1a melanoma patients: results of a phase II study. Cancer Immunol Immunother 2015;64:999–1009.

[74] Danielli R, Patuzzo R, Ruffini PA, Maurichi A, Giovannoni L, Elia G, Neri D, Santinami M. Armed antibodies for cancer treatment: a promising tool in a changing era. Cancer Immunol Immunother 2015;64:113–21.

[75] Spitaleri G, Berardi R, Pierantoni C, De Pas T, Noberasco C, Libbra C, Gonzalez-Iglesias R, Giovannoni L, Tasciotti A, Neri D, Menssen HD, de Braud F. Phase I/II study of the tumour-targeting human monoclonal antibody-cytokine fusion protein L19-TNF in patients with advanced solid tumours. J Cancer Res Clin Oncol 2013;139:447–55.

[76] Gillies SD. A new platform for constructing antibody-cytokine fusion proteins (immunocytokines) with improved biological properties and adaptable cytokine activity. Protein Eng Des Sel 2013;26:561–9.

[77] Gonzalez H, Hagerling C, Werb Z. Roles of the immune system in cancer: from tumor initiation to metastatic progression. Genes Dev 2018;32:1267–84.

[78] Il'inskaia ON, Makarov AA. Why ribonucleases cause death of cancer cells. Mol Biol (Mosk) 2005;39:3–13.

[79] De Lorenzo C, Nigro A, Piccoli R, D'Alessio G. A new RNase-based immunoconjugate selectively cytotoxic for ErbB2-overexpressing cells. FEBS Lett 2002;516:208–12.

[80] Borriello M, Laccetti P, Terrazzano G, D'Alessio G, De Lorenzo C. A novel fully human antitumour immunoRNase targeting ErbB2-positive tumours. Br J Cancer 2011;104:1716–23.

[81] Weber T, Mavratzas A, Kiesgen S, Haase S, Botticher B, Exner E, Mier W, Grosse-Hovest L, Jager D, Arndt MA, Krauss J. A humanized anti-CD22-onconase antibody-drug conjugate mediates highly potent destruction of targeted tumor cells. J Immunol Res 2015;2015:561814.

[82] Tome-Amat J, Ruiz-de-la-Herran J, Martinez-del-Pozo A, Gavilanes JG, Lacadena J. Alpha-sarcin and RNase T1 based immunoconjugates: the role of intracellular trafficking in cytotoxic efficiency. FEBS J 2015;282:673–84.

[83] Dan N, Setua S, Kashyap VK, Khan S, Jaggi M, Yallapu MM, Chauhan SC. Antibody-drug conjugates for cancer therapy: chemistry to clinical implications. Pharmaceuticals (Basel) 2018;11(2):32.

[84] Dubowchik GM, Firestone RA. Cathepsin B-sensitive dipeptide prodrugs. 1. A model study of structural requirements for efficient release of doxorubicin. Bioorg Med Chem Lett 1998;8:3341–6.

[85] Bargh JD, Isidro-Llobet A, Parker JS, Spring DR. Cleavable linkers in antibody-drug conjugates. Chem Soc Rev 2019;48:4361–74.

[86] Saito G, Swanson JA, Lee KD. Drug delivery strategy utilizing conjugation via reversible disulfide linkages: role and site of cellular reducing activities. Adv Drug Deliv Rev 2003;55:199–215.

[87] Sochaj AM, Swiderska KW, Otlewski J. Current methods for the synthesis of homogeneous antibody-drug conjugates. Biotechnol Adv 2015;33:775–84.

[88] Tolcher AW, Sugarman S, Gelmon KA, Cohen R, Saleh M, Isaacs C, Young L, Healey D, Onetto N, Slichenmyer W. Randomized phase II study of BR96-doxorubicin conjugate in patients with metastatic breast cancer. J Clin Oncol 1999;17:478–84.

[89] Weiner LM, O'Dwyer J, Kitson J, Comis RL, Frankel AE, Bauer RJ, Konrad MS, Groves ES. Phase I evaluation of an anti-breast carcinoma monoclonal antibody 260F9-recombinant ricin A chain immunoconjugate. Cancer Res 1989;49:4062–7.

[90] Zielinski R, Lyakhov I, Jacobs A, Chertov O, Kramer-Marek G, Francella N, Stephen A, Fisher R, Blumenthal R, Capala J. Affitoxin–a novel recombinant, HER2-specific, anticancer agent for targeted therapy of HER2-positive tumors. J Immunother 2009;32:817–25.

[91] Tong XM, Feng L, Suthe SR, Weng TH, Hu CY, Liu YZ, Wu ZG, Wang MH, Yao HP. Therapeutic efficacy of a novel humanized antibody-drug conjugate recognizing plexin-semaphorin-integrin domain in the RON receptor for targeted cancer therapy. J Immunother Cancer 2019;7:250.

[92] Vaklavas C, Forero A. Management of metastatic breast cancer with second-generation antibody-drug conjugates: focus on glembatumumab vedotin (CDX-011, CR011-vcMMAE). BioDrugs 2014;28:253–63.

[93] Khandelwal S, Boylan M, Spallholz JE, Gollahon L. Cytotoxicity of selenium immunoconjugates against triple negative breast cancer cells. Int J Mol Sci 2018;19:.

[94] Taya M, Hammes SR. Glycoprotein non-metastatic melanoma protein B (GPNMB) and cancer: a novel potential therapeutic target. Steroids 2018;133:102–7.

[95] Damelin M, Bankovich A, Park A, Aguilar J, Anderson W, Santaguida M, Aujay M, Fong S, Khandke K, Pulito V, Ernstoff E, Escarpe P, Bernstein J, Pysz M, Zhong W, Upeslacis E, Lucas J, Nichols T, Loving K, Foord O, Hampl J, Stull R, Barletta F, Falahatpisheh H, Sapra P, Gerber HP, Dylla SJ. Anti-EFNA4 calicheamicin conjugates effectively target triple-negative breast and ovarian tumor-initiating cells to result in sustained tumor regressions. Clin Cancer Res 2015;21:4165–73.

[96] Westwood JA, Murray WK, Trivett M, Haynes NM, Solomon B, Mileshkin L, Ball D, Michael M, Burman A, Mayura-Guru P, Trapani JA, Peinert S, Honemann D, Miles Prince H, Scott AM, Smyth MJ, Darcy PK, Kershaw MH. The Lewis-Y carbohydrate antigen is expressed by many human tumors and can serve as a target for genetically redirected T cells despite the presence of soluble antigen in serum. J Immunother 2009;32:292–301.

[97] Amiri-Kordestani L, Blumenthal GM, Xu QC, Zhang L, Tang SW, Ha L, Weinberg WC, Chi B, Candau-Chacon R, Hughes P, Russell AM, Miksinski SP, Chen XH, McGuinn WD, Palmby T, Schrieber SJ, Liu Q, Wang J, Song P, Mehrotra N, Skarupa L, Clouse K, Al-Hakim A, Sridhara R, Ibrahim A, Justice R, Pazdur R, Cortazar P. FDA approval: ado-trastuzumab emtansine for the treatment of patients with HER2-positive metastatic breast cancer. Clin Cancer Res 2014;20:4436–41.

[98] Batoo S, Bayraktar S, Al-Hattab E, Basu S, Okuno S, Gluck S. Recent advances and optimal management of human epidermal growth factor receptor-2-positive early-stage breast cancer. J Carcinog 2019;18:5.

[99] Beck A, Goetsch L, Dumontet C, Corvaia N. Strategies and challenges for the next generation of antibody-drug conjugates. Nat Rev Drug Discov 2017;16:315–37.

[100] Xu Z, Guo D, Jiang Z, Tong R, Jiang P, Bai L, Chen L, Zhu Y, Guo C, Shi J, Yu D. Novel HER2-targeting antibody-drug conjugates of trastuzumab beyond T-DM1 in breast cancer: trastuzumab deruxtecan(DS-8201a) and (Vic-) trastuzumab duocarmazine (SYD985). Eur J Med Chem 2019;183:111682.

[101] Banerji U, van Herpen CML, Saura C, Thistlethwaite F, Lord S, Moreno V, Macpherson IR, Boni V, Rolfo C, de Vries EGE, Rottey S, Geenen J, Eskens F, Gil-Martin M, Mommers EC, Koper NP, Aftimos P. Trastuzumab duocarmazine in locally advanced and metastatic solid tumours and HER2-expressing breast cancer: a phase 1 dose-escalation and dose-expansion study. Lancet Oncol 2019;20:1124–35.

[102] Modi S, Saura C, Yamashita T, Park YH, Kim SB, Tamura K, Andre F, Iwata H, Ito Y, Tsurutani J, Sohn J, Denduluri N, Perrin C, Aogi K, Tokunaga E, Im SA, Lee KS, Hurvitz SA, Cortes J, Lee C, Chen S, Zhang L, Shahidi J, Yver A, Krop I. Trastuzumab Deruxtecan in previously treated HER2-positive breast cancer. N Engl J Med 2020;382:610–21.

[103] Sochaj-Gregorczyk AM, Serwotka-Suszczak AM, Otlewski J. A novel affibody-auristatin E conjugate with a potent and selective activity against HER2+ cell lines. J Immunother 2016;39:223–32.
[104] Perez HL, Cardarelli PM, Deshpande S, Gangwar S, Schroeder GM, Vite GD, Borzilleri RM. Antibody-drug conjugates: current status and future directions. Drug Discov Today 2014;19:869–81.
[105] Clark T, Han X, King L, Barletta F. Insights into antibody-drug conjugates: bioanalysis and biomeasures in discovery. Bioanalysis 2013;5:985–7.

CHAPTER

3

New intratumoral immunotherapeutic approaches to inhibit the tumor growth and metastasis in breast cancer

Carmen T. Gomez de Leon and Jorge Morales-Montor

Departamento de Inmunología, Instituto de Investigaciones Biomédicas, Universidad Nacional Autónoma de México, Ciudad de México, México

Abstract

Breast cancer presents several unique challenges, in part, because the few tumor antigens that have been identified are expressed in only a relatively small fraction of the tumors. Additionally, the majority of immunotherapeutic approaches have focused on the post-operative setting; nevertheless, strategies that utilize autologous tumor are the most promising of the immunotherapeutic approaches, generating a response against relevant tumor antigens. Even more, benefits of systemic intravenous therapies have been observed only in a small population of patients and have been associated with toxicity profiles. A local intratumoral delivery reduces toxic effects and ensures their arrival.

Several intratumoral therapies with differing mechanisms of action have successfully entered clinical trials and shown promising results. These mechanisms involve the triggering of immune response using pathogens; enhance immune response using recombinant cytokines; inhibit immune checkpoints using monoclonal antibodies against CTLA4, PD1, or PDL1; or the combination of two or more of these strategies. Nevertheless, there is still only limited clinical use of nonsystemic intratumoral chemotherapy for breast cancer.

In this chapter, we are going to resume the intratumoral immunotherapeutic approaches that have been successful in the inhibition of tumoral growth and/or metastasis in breast cancer, paying attention to the delivery systems that have been developed.

Abbreviations

Ad	adenovirus
APCs	antigen presenting cells
CAR T cells	chimeric antigen receptor-modified T cells
CTLs	cytolytic T lymphocytes
CY	cyclophosphamide
DCs	dendritic cells

Immunotherapy in Resistant Cancer:
From the Lab Bench Work to Its Clinical Perspectives
https://doi.org/10.1016/B978-0-12-822028-3.00010-8

DHT	delayed-type hypersensitivity
GM-CSF	granulocyte-macrophage colony stimulating factor
IL-2	interleukyne-2
IL-4	interleukyne-4
IL-12	interleukyne-12
IL-18	interleukyne-18
iRAES	immune-related adverse events
i.t.	intratumoral
LAK	lymphokine-activated killer
NDES	nanofluidic-based drug eluting seed
PLAM	polylactic acid microspheres
scFv	single chain variable fragment
SFV	semliki forest virus
TME	tumor microenvironment
TNBC	triple negative breast cancer
TNF-α	tumor necrosis factor alpha
VLPs	virus-like particles

Conflict of interest

No potential conflicts of interest were disclosed by the authors.

Introduction

Cancer immunotherapy strategies may activate the immune system, but they can also push it into supraphysiological levels with a subsequent risk of increasing immune-related adverse events (irAEs); that is why targeted or localized drug delivery should be a major goal of chemotherapy [1]. In general, intratumoral immunotherapies aim to initiate local recruitment of immune cells into the tumor microenvironment and subsequently prime T cells for a systemic polyclonal antitumor response and enhance this response. Intratumoral strategy offers enhanced locoregional efficacy and reduced systemic toxicity by enabling high bioavailability of the agent at the injected tumor sites while limiting systemic exposure [2,3].

Since the first clinical success with cancer immunotherapy that was reported over a century ago in patients with malignant tumors treated with intratumoral inoculation of live bacteria [4], a lot of advances in this area have been made in both immune strategies and delivery methods. Despite the lot of information available on several types of cancer, the use of these new strategies in breast cancer has been poorly achieved and the translation of proved strategies, as well as the development of new specific ones, remains urgent.

Intratumoral injections

Bacteria

Animal testing

In early studies, Fisher et al. analyzed the effect of combining the cyclophosphamide (CY) chemotherapy with intratumoral (i.t.) injections of formalin-killed freeze-dried preparation of *Corynebacterium parvum* in a murine mammary tumor model. The systemic administration of *C. parvum* by itself failed in the inhibition of tumor growth, while its intratumoral administration caused tumor regression. The combined treatment of CY and *C. parvum*, administered i.p. or i.t., inhibited tumor growth; furthermore, when *C. parvum* was administered i.t., an inhibition of the growth of a distant tumor was also observed. Besides, an increment of macrophage colony production was reported when *C. parvum* was administered in both normal and tumor-bearing mice, exhibiting the more prolonged effect in the latter. In this work, the effect of *C. parvum* was compared with *Brucella abortus* extract and glucan, but these treatments were not as effective as *C. parvum* [5]. In a subsequent study, the same investigation group aimed to determine whether the growth of a distant tumor focus could be more effectively controlled by delaying the removal of a primary tumor in order to begin chemoimmunotherapy in its presence, rather than by immediately removing the tumor. The tumor used was a spontaneous mammary carcinoma arising in a C3H/HeJ female; they found that the combination of CY with *C. parvum* administered i.t. prior to primary tumor removal was superior in inhibiting distant tumor growth compared to the preoperative administration i.p. of *C. parvum* and the immediate operation. Even more, they found that prolonged periods of contact between tumor cells and *C. parvum* may be beneficial in the last 4 weeks of preoperative i.t. *C. parvum* with CY were more effective in controlling distant disease than 2 weeks of such therapy [6]. These approaches were determinant in the intratumoral immunotherapy, because they indicated the importance of contact between immunomodulators and tumor cells (putative tumor antigen) in order to augment the tumor resistance by the host.

Cytokines secreted by cells

Animal testing

The strategies that involve immune system stimulation not only include the use of pathogens as stimuli, but also attempt approaches that use cytokines in order to promote the immune response. One of the first studied cytokines was IL-2, because it stimulates the proliferation of cytotoxic T lymphocytes and enhances normal killer cell and lymphokine-activated killer (LAK) cell activity. Systemic biological therapy involving IL-2 results in significant antitumor effects in patients with advanced metastatic cancer; nevertheless, it also produces adverse effects. In the searching for more direct delivery, Deshmukh et al. used a cellular vaccine that consisted of allogeneic fibroblasts (H-2K) modified to secrete IL-2 in an intracerebrally metastasizing breast tumor model. They used EO771 breast cancer cells, and the animals were treated with a single intratumoral injection of 10^6H-2K fibroblasts secreting IL-2. Treatment resulted in prolonged survival of the animals; even more, when modified fibroblast was co-injected with EO771 cells, 40% of the animals didn't develop tumors; and when these animals were rechallenged with the cancer cells, they had a markedly prolonged survival and smaller tumors than the nontreated controls, which were associated with lymphocytic infiltrations [7].

Human research

Dendritic cells (DCs) are distributed widely and are highly efficient antigen-presenting cells. In contrast to other antigen-presenting cells, antigen-pulsed DCs can be administered in situ to prime naïve T-helper and cytolytic T lymphocytes (CTLs) without additional adjuvants [8]. There is a lot of evidence suggesting that i.t. DCs play an important role in antitumor immune responses and that the increase of the number of i.t. DCs in cancer patients with immunomodulators is beneficial [9]. The tumor microenvironment may lack the appropriate proinflammatory signals to differentiate DC precursors. Through productions of soluble factors, tumors may actively suppress DCs, which may be a central mechanism by which tumors escape immunosurveillance [10]. Triozzi et al. performed a study with three patients that exhibited breast carcinoma. Patients received granulocyte-macrophage-colony stimulating factor to increase the numbers of peripheral blood monocyte precursors. DCs were generated from monocytes with granulocyte-macrophage-colony stimulating factor and interleukin-4 in autologous plasma by phlebotomy. Tumors were injected at multiple sites with 30 million autologous DCs per tumor. Regression of the injected tumors, at 4 days after injection, was observed in two of three treated patients. Biopsies of regressing lesions showed lymphocyte infiltration associated with DCs and necrosis, without evidence of neutrophils and macrophages [11].

Chimeric cells

Animal testing

Chimeric antigen receptor-modified T cells (CAR T cells) are redirected effector immune cells genetically modified to deliver tumoricidal functions upon recognition of antigen. CAR T cells are effective in the treatment of several hematologic malignancies [12]. However, the effectiveness of CAR T cells in the treatment of solid tumors remains modest. Mainly because most tumor antigens are expressed, albeit at lower levels, in normal tissues, CAR T cells may lead to on-target/off-tumor effects. In addition, the microenvironment of solid tumors is immunosuppressive, which may limit the potency of CAR T cells [13]. Priceman et al. aimed to evaluate CAR design as well as the route of CAR T-cell administration for the treatment of HER2+ breast cancer that has metastasized to the brain. In this study, the author demonstrated robust antitumor responses in human xenograft models of HER2+ breast cancer metastasis to the brain after local intratumoral or regional intraventricular delivery of HER2-BBζ CAR T cells. In contrast, intravenous delivery of HER2-CAR T cells achieved only partial antitumor responses in mice even at 10-fold higher doses compared with local or regional delivery to the brain. These findings may also affect the development of CAR T cells for other HER2+ solid tumors, and their brain metastases and further studies should be achieved [14].

Human research

c-Met is a cell-surface protein tyrosine kinase expressed in a variety of solid tumors including 50% of breast cancer tumors [15,16]. A monovalent anti-c-Met antibody, onartuzumab, has been tested in a variety of patients with advanced-stage solid cancers in clinical trials [17]. Tchou et al. determined whether c-Met might serve as a target for CAR T cells by replacing the single-chain variable fragment (scFv) portion of the CD19 binding domain in a previous established CD19-CAR construct [18], with that of onartuzumab in order to evaluate the activity of CAR T cells directed against c-Met (c-Met-CAR T cells) in patients with metastatic breast cancer. The authors first evaluated the safety and feasibility of treating metastatic breast cancer with intratumoral administration of mRNA-transfected c-Met-CAR T cells in a phase 0 clinical trial (NCT01837602) [19]. Introducing the CAR construct via mRNA ensured safety by limiting the nontumor cell effects (on-target/off-tumor) of targeting c-Met. Patients with metastatic breast cancer with accessible cutaneous or lymph node metastases received a single intratumoral injection of 3 × 107 or 3 × 108 cells. CAR T mRNA was detectable in peripheral blood and in the injected tumor tissues after intratumoral injection in 2 and 4 patients, respectively. mRNA c-Met-CAR T cell injections were well tolerated, as none of the patients had study drug-related adverse effects greater than grade 1. Tumors treated with intratumoral injected mRNA c-Met-CAR T cells were excised and analyzed by immunohistochemistry, revealing extensive tumor necrosis at the injection site, cellular debris, loss of c-Met immunoreactivity, all surrounded by macrophages at the leading edges and within necrotic zones [20].

Cytokines and adjuvants

Animal testing

Researchers have put a lot of attention in the interleukin (IL)-12 immunotherapies as it has shown to control metastasis. IL-12 is a potent proinflammatory cytokine with the ability to induce rapid activation of innate and antiangiogenic mechanisms and to promote the development of a Th1-type cellular response [21–23]. In 2015, Ln Vo et al. aimed to develop a localized delivery of IL-12 co-formulated with chitosan (chitosan/IL-12). Chitosan is a biocompatible, unbranched copolymer of glucosamine and *N*-acetylglucosamine, derived primarily from the exoskeletons of crustaceans.

Authors first demonstrated that simple mixtures of IL-12 with solutions of chitosan (chitosan/IL-12) administered i.t. can significantly enhance local IL-12 retention and safely induce complete tumor regression and protective immunity in colorectal and pancreatic cancer models [24]. Even more, it has been shown that recombinant IL-12 co-formulated with chitosan is retained in the tumor microenvironment for at least 5–6 d following i.t. injection; in contrast, IL-12 injected alone is undetectable between 24 and 48 h after administration [25]. These strategies were tested in mice bearing spontaneously metastatic 4T1 mammary adenocarcinomas. Mice received intratumoral injections of chitosan/IL-12 prior to breast tumor resection. Chitosan/IL-12 i.t. immunotherapy resulted in long-term tumor-free survival in 67% of mice compared to only 24% or 0% of mice treated with IL-12 alone or chitosan alone, respectively. Antitumor responses following chitosan/IL-12 treatment were durable and provided complete protection against rechallenge with 4T1 cells, but not with RENCA renal adenocarcinoma cells. Lymphocytes from chitosan/IL-12-treated mice demonstrated robust tumor-specific lytic activity and interferon-γ production. Cell-mediated immune memory was confirmed in vivo via clinically relevant delayed-type hypersensitivity (DTH) assays. Comprehensive hematology and toxicology analyses revealed that chitosan/IL-12 induced transient, reversible leukopenia with no changes in critical organ function [26].

Cytokines contained in biodegradable microspheres

Animal testing

Sabel et al. hypothesized that, for a better antitumor immune response, the delivery of the cytokines must be not only local, but sustained. In different experimental sets, the authors evaluated the potential of combinations of IL-12, Tumor Necrosis Factor-alpha (TNF-α), granulocyte-macrophage colony stimulating factor (GM-CSF), and Interleukin-18 (IL-18) loaded in biodegradable polylactic acid microspheres (PLAM). Balb/c mice with established MT-901 or 4T1 mammary carcinomas were treated with a single intratumoral injection of BSA, IL-12, GM-CSF, TNF-α, and IL-18 loaded PLAM alone or in combination. Combined treatment with IL-12 and TNF-α PLAM was superior to all other treatments for ablating of the primary tumor, eradicating distant disease, and enhancing survival. In vivo lymphocyte depletion studies established that the effects of IL-12 alone were mediated primarily by NK cells, while the combination of IL-12 and TNF-α was dependent upon CD8+ T-cells. Only the combination of IL-12 and TNF-α resulted in an increase in both CD4+ and CD8+ T-cells and a reduction in CD4 + CD25+ cells. The combination also resulted in a dramatic increase in DCs maturation and antigen presentation. These findings suggested that i.t. IL-12/TNF-α- PLAM delivery resulted in tumor regression and the establishment of T-cell based antitumor immune response capable of eradicating disseminated disease [27,28].

Cytokines expressed in viral vectors

Animal testing

Another intratumoral delivery method that has been extensively studied is the gene transfer by adenoviral vectors. Adenoviral vectors are ideal delivery systems because they are highly infectious, yield high levels of transgene product, and the expression is transient [29]. In 2001 Gyorffy et al. used a human type 5 adenovirus (Ad) that expressed the cDNA for murine angiostatin, with or without IL-12, in a murine model of breast carcinoma. Angiostatin has cytotoxic effects that result in the inhibition of endothelial cell proliferation and migration, leading in cell apoptosis and, therefore, in the inhibition of primary and metastatic tumor growth [30,31].

When a vector that expressed angiostatin by itself was i.t. administered, they observed a delay, but not the eradication of the tumor growth. The direct intratumor administration of IL-12 by itself delayed the tumor growth and induced regression in 13% of treated animals. Interestingly, the combination of IL-12 with the angiostatin vector provoked a total tumor regression of the 54% of the animals and the development of strong CTL protection against tumor rechallenge. In theory, this therapy limits the tumor size by attacking the vasculature with angiostatin, and thereby allows IL-12 to mount a T cell-specific response against the tumor [32].

The other viral vector that has been effectively used in cancer gene therapy research is the alphavirus vector. Alphavirus vectors exhibit high-level expression of heterologous proteins in a broad host range of cells [33,34] and have been shown to induce apoptosis in infected cells [35]. Semliki Forest virus (SFV) is an alphavirus, enveloped positive-sense RNA of the family Togaviridae; it has been developed into a transient RNA-based expression vector system that consists of a vector in which foreign genes can be expressed and two helper vectors that encode the structural protein genes. Co-transfection of all the three vectors into cells results in the release of recombinant virus-like particles (VLPs) coding for the foreign gene [36,37]. SFV vectors have been successfully used to induce in vivo expression of a number of genes such as cytokines. It has been shown that recombinant SFV (rSFV) particles can efficiently infect tumor cells, and, that, intratumoral treatment with rSFV particles results in the induction of p53-independent apoptosis leading to significant tumor inhibition. However, in order to achieve complete regression in at least a proportion of treated mice, multiple treatments at relatively high doses are needed. The enhanced vector pSFV10-E has been developed in order to express foreign genes at levels up to 10 times higher than the original vector [35,38]. In 2005, Chikkanna-Gowda et al. cloned the two IL-12 gene subunits from mouse splenocytes and inserted them into the pSFV10-E and pSFV10 (nonenhanced) vectors. Both constructs that expressed and secreted biologically active murine IL-12 were i.t. administered in a spontaneously metastasizing 4T1 mouse mammary carcinoma model. The pSFV10-E-IL12 particles treated mice showed complete and permanent tumor regression, as well as the inhibition of metastasis formation [39].

Combined treatments

Cytokines expressed in viral vectors and bacteria

Animal testing

The SFV-IL-12 particles were used in combination with an aroC (−) *Salmonella Typhimurium* strain (LVR01). Treatment was injected in 4T1 tumor nodules orthotopically implanted in mice, and the tumors were surgically resected after treatment. Combined therapy inhibited lethal lung metastasis, while it prolonged survival in 90% of the mice. SFV-IL-12 without the LVR01 showed a potent antiangiogenic effect, being able to inhibit tumor growth and extend survival, but could not prevent the establishment of distant metastasis and the death of tumor-excised animals. On the other hand, when LVR01 was administered alone, it showed a significant, but limited, antitumor potential, despite its ability to invade breast cancer cells and induce granulocyte recruitment. The efficacy of the combined therapy depended on the order in which both factors were administered; the therapeutic effect was only observed when SFV-IL-12 was administered previous to LVR01. Moreover, pretreatment with LVR01 seemed to suppress the antiangiogenic effects of SFV-IL-12. The rechallenged mice were unable to reject a second 4T1 tumor; however, 100% of them could be totally cured by applying the same combined regimen [40].

Cytokines expressed in viral vectors and blockade immune checkpoints

Animal testing

The enhancement of the immune system has not been the only strategy studied in order to abolish cancer. The immune system recognizes and is poised to eliminate cancer, but is held in check by inhibitory receptors and ligands. These immune checkpoint pathways, which normally maintain self-tolerance and limit collateral tissue damage during antimicrobial immune responses, can be co-opted by cancer to evade immune destruction. Drugs that inhibit the immune checkpoints, such as anti-CTLA-4, anti-PD-1, anti-PD-L1, and others, in early development can unleash anti-tumor immunity and mediate durable cancer regressions [41]. Anti-CTLA-4 therapy caused drug-related irAEs in 15%–30% of treated patients, sometimes resulting in fatalities. These irAEs were associated with inflammation in normal tissues [42]. Despite these promising results, the majority of patients treated with anti-PD-1/PD-L1 monotherapies do not achieve objective responses, and most tumor regressions are partial rather than complete. Animal models suggest that treatment combinations based on the PD-1 pathway blockade may be synergistic [43]. ICI strategies have been combined with oncolytic viruses, because releasing the brakes of the immune system is critical to maximize the immunotherapeutic efficacy of oncolytic viruses [44]. JX-594 (pexastimogene devacirepvec, Pexa-vec) is an oncolytic vaccinia virus that is engineered to express an immune-activating transgene, GM-CSF, and that has the viral thymidine kinase gene disrupted [45]. Chon et al. used the mJX-594 (JX) oncolytic vaccinia virus as a strategy to remodel the tumor microenvironment (TME) and subsequently increase sensitivity to αPD-1 and/or αCTLA-4 immunotherapy. JX was intratumorally injected into implanted Renca kidney tumors or MMTV-PyMT transgenic mouse breast cancers with or without αPD-1 and/or αCTLA-4. Intratumoral injection of JX remodeled the TME, from noninflamed to inflamed, as was shown by the increased tumor-infiltrating T cells and upregulation of immune-related genes; it also increased the intratumoral infiltration of CD8+ T cells in distant tumors. Dual-combination therapy with intratumoral JX and systemic αPD-1 or αCTLA-4 further enhanced the anticancer immune response, regardless of various treatment schedules. Of note, triple combination immunotherapy with JX, αPD-1, and αCTLA-4 elicited the most potent anticancer immunity and induced complete tumor regression and long-term overall survival [46].

New theoretical strategies

Multi-needle injection

The success of the single injections strategies is related, in part, to the size of the tumor and the number of different single shots given. Histological observations of treated tumors have shown that cellular damage is limited to the tumor injected zone. Subbotin and Fiksel hypothesized that the core of the problem stems from the fact that the single-needle intratumoral injection forms a very localized, jet-like distribution of the drug that constitutes only a small fraction of the total volume of the tumor. Authors aimed to solve this limitation with a multi-needle injection with the creation of a model of injectant distribution in a solid tissue based on the traditional technique of single-needle injection and then extended that model to a case of simultaneous multi-needle injection. They also developed the model of drug delivery and transport in biological tissues by following a frequently used approach of modeling the diffusive transport of liquid through a porous media, considering the relation between the flow velocity, the pressure gradient, and the tissue permeability (Fig. 1). Their analysis demonstrated that a multi-needle injection setup provides a significantly more widespread and homogeneous injectant distribution within a solid tumor than that for a single needle injection for the same tumor size [47].

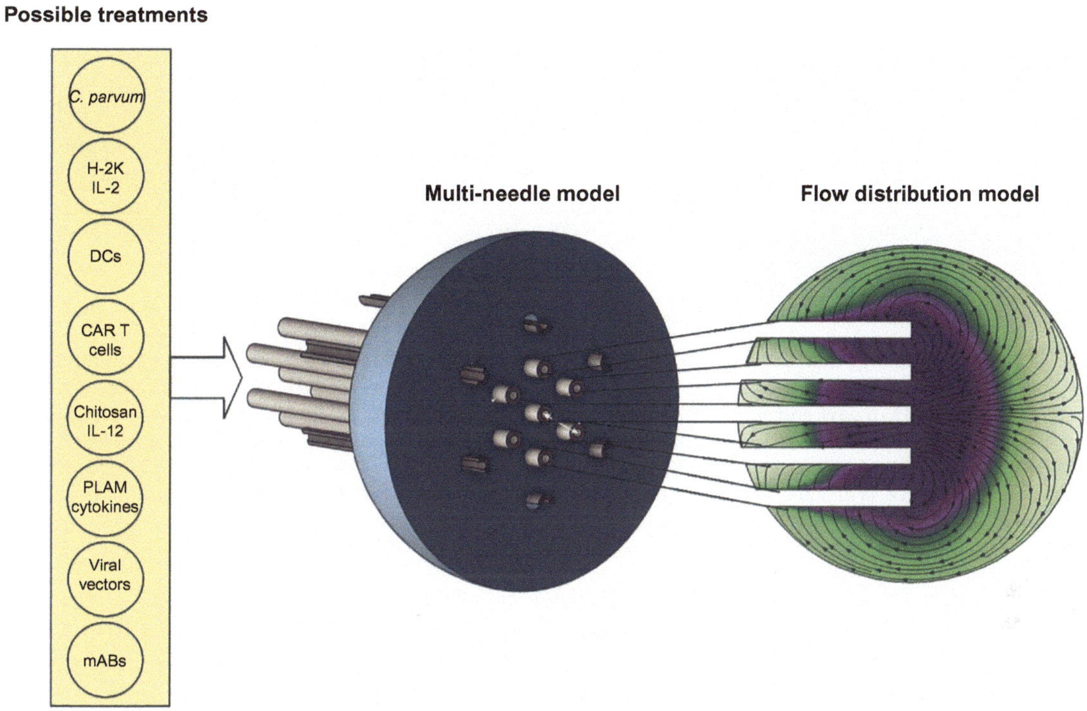

FIG. 1 Theoretical model of multiple needle setup with drainage needles. The theoretical model aims to solve the limitation of jet-like distribution of the drug that constitutes only a small fraction of the total volume of the tumor. The model considers seven needles and includes the velocity map and intratumoral flow streamlines intratumor. We put the possible treatments that have currently succeeded in single injection therapy. These treatments have not been proved with the multi-needle injection. *Modified from Subbotin V, et al. Modeling multi needle injection into solid tumor. Am J Cancer Res 2019;9(10):2209–2215.*

Nanotechnology

Immunostimulatory monoclonal antibodies delivered by implant

Animal testing

Although TNBC accounts for approximately 15% to 20% of all invasive breast cancers, it has a poor prognosis with high mortality and early relapse [48]. Effective and specific therapies for TNBC strategies are urgently needed. A new strategy that has been proved in TNBC therapy is the Nanofluidic-based Drug Eluting Seed (NDES) for sustained intratumoral delivery of immunotherapeutics. Nanochannels in the NDES control the passive release of immunotherapeutics via steric, electrostatic, and hydrodynamic hindrance on diffusing molecule, thereby achieving sustained delivery for months without the need for actuation or clinician intervention [49] (Fig. 2A). CD40 is a member of the tumor necrosis factor (TNF) receptor superfamily that co-stimulates protein expressed on APCs such as dendritic cells, macrophages, and B cells [48]. CD40 pathway activation is important for the maturation

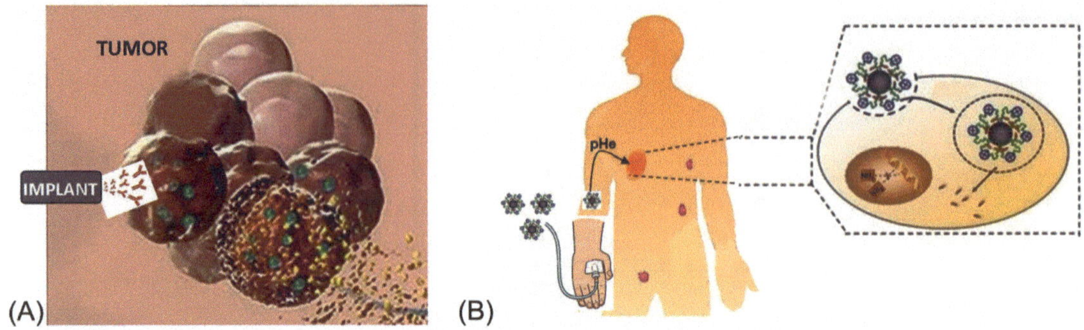

FIG. 2 Nanotechnology mechanisms. (A) The implant solves the limited time of contact between the immunotherapy and the tumor cells by sustained intratumoral delivery. Therapeutical effect lasts one month, while biodegradable spheres effect lasts 5 to 6 days and single injection effect lasts only for a few hours. (B) Nanoparticles are administered systemically, after their accumulation in tumors, and the pH of the tumor favor the release of their content. *(A) Modified from Hamid O, Ismail R, Puzanov I. Intratumoral immunotherapy—Update 2019. Oncologist 2019;25(3):e423–e438. (B) Modified from Yang XZ, et al. Rational design of polyion complex nanoparticles to overcome cisplatin resistance in cancer therapy. Adv Mater 2014;26(6):931–6.*

and activation of APCs that leads to T helper cell 1 (Th1) polarization and CTL priming [50,51]. Besides, CD40 is expressed by some tumors; therefore, its activation could result in direct cytotoxicity. OX40 (CD134) is a co-stimulatory molecule expressed by activated CD4 and CD8 T cells [52].

The NDES was inserted intratumorally using a minimally invasive trocar method similar to brachytherapy seed insertion in a 4T1 orthotopic murine mammary carcinoma model, which recapitulates triple-negative breast cancer. NDES-mediated intratumoral release of agonist monoclonal antibodies, OX40 and CD40, resulted in potentiation of local and systemic antitumor immune response and inhibition of tumor growth compared to control mice. Further, mice treated with NDES-CD40 demonstrated minimal liver damage compared to systemically treated mice. This strategy has an enormous potential clinical impact since NDES could be applicable to a broad spectrum of drugs and solid tumors [52].

Chemokines contained in tumor targeted nanoparticles

Animal testing

Another new strategy for local delivery that has been developed is the PIC nanoparticles. Nanoparticles exhibit prolonged circulation and enhanced drug accumulation in tumors. Subsequently, tumor pH leads to the release of nanoparticle content, which facilitates cellular uptake followed by rapid intracellular release [53] (Fig. 2B). This delivery strategy was also studied as an alternative for TNBC. In this proposal, authors take advantage that neither human nor mouse TNBCs express CCL25. The chemokine CCL25 is the only ligand for CCR9 [50] and is selectively expressed by medullary DCs and cortical epithelial cells in the thymus and the epithelium of the small intestine under normal conditions [51]. It has been reported that CCR9+ T cells have an increased potential to be activated [54] and to produce proinflammatory cytokines [55], and that, CCR9+ T helper cells can promote expansion and survival of CD8+ T cells [56]. Furthermore, CCL25/CCR9 signaling inhibits CD4+ T cell differentiation into regulatory T (Treg) cells, and treatment with anti-CCL25-neutralizing antibodies accelerated tumor growth in a CCL25-expressing mouse tumor model [57]. In this study, Chen et al. raised the hypothesis that intratumoral delivery of CCL25 may enhance antitumor immunotherapy in TNBCs. They determined whether this approach could enhance CD47-targeted immunotherapy using a tumor acidity-responsive nanoparticle delivery system (NP-siCD47/CCL25) to sequentially release CCL25 protein and CD47 small interfering RNA in tumor. BALB/c mice were inoculated with 4T1 tumor cells and treated with intravenous injection of PBS, NP-siNC, NP-siCD47, NP-siNC/CCL25, or NP-siCD47/CCL25. NP-siCD47/CCL25 significantly increased infiltration of CCR9 + CD8+ T cells and down-regulated CD47 expression in tumor, resulting in inhibition of tumor growth and metastasis through a T cell-dependent immunity. Furthermore, the antitumor effect of NP-siCD47/CCL25 was synergistically enhanced when used in combination with programmed cell death protein-1/programmed death ligand-1 blockades [58].

Conclusion

Intratumoral immunotherapy treatments definitely reduce toxic effects and enhance the aimed responses. One of the main concerns about intratumoral injections was the possibility to "open the door" to metastasis; nevertheless, the evidence points out the fact that presurgical intratumoral treatments avoid metastasis. The contact time between the immunotherapy and the tumor cells seems to be critical in order to trigger specific immunity against tumors. Effectiveness of single i.t. injections are related to the tumor size, the number, and the sites of given shots. That is why a homogenized distribution and long last delivery must be achieved in further researches in order to increase the success of i.t. treatments. In order to achieve this goal, the combination of the administration technologies with the delivery strategies will be critical. It also will be desirable to personalize the treatments as much as possible, as not all patients respond to the same strategies, even if they bear the same type of breast cancer.

Acknowledgments

Carmen T. Gómez de León is recipient of a Post-Doctoral fellowship from Grant FC2016-2125 from Fronteras en la Ciencia, Consejo Nacional de Ciencia y Tecnología (CONACYT). This study was supported by grants from Programa de Apoyo a Proyectos de Investigación e Innovación Tecnológica (PAPIIT), Dirección General de Asuntos del Personal Académico (DGAPA), Universidad Nacional Autónoma de México (UNAM), grant/award number IN-209719, and from Fronteras en la Ciencia, Consejo Nacional de Ciencia y Tecnología (CONACYT), Grant No FC 2016 2125, both to Jorge Morales-Montor.

References

[1] Sanmamed MF, Chen L. A paradigm shift in cancer immunotherapy: from enhancement to normalization. Cell 2018;175:313–26.

[2] Sloot S, Rashid OM, Zager JS. Intralesional therapy for metastatic melanoma. Expert Opin Pharmacother 2014;15:2629–39.

[3] Marabelle A, Tselikas L, de Baere T, Houot R. Intratumoral immunotherapy: using the tumor as the remedy. Ann Oncol 2017;28:xii33–43.
[4] Coley WB. The treatment of malignant tumors by repeated inoculations of erysipelas. With a report of ten original cases. 1893. Clin Orthop Relat Res 1991;3–11.
[5] Fisher B, Gebhardt M. Comparative effects of *Corynebacterium parvum*, *Brucella abortus* extract, Bacillus Calmette-Guerin, glucan, levamisole, and tilorone with or without cyclophosphamide on tumor growth, macrophage production, and macrophage cytotoxicity in a murine mammary tumor model. Cancer Treat Rep 1978;62:1919–30.
[6] Fisher B, Gunduz N. Further observations on the inhibition of tumor growth by *Corynebacterium parvum* with cyclophosphamide. X. Effect of treatment on tumor cell kinetics in mice. J Natl Cancer Inst 1979;62:1545–51.
[7] Deshmukh P, Glick RP, Lichtor T, Moser R, Cohen EP. Immunogene therapy with interleukin-2-secreting fibroblasts for intracerebrally metastasizing breast cancer in mice. J Neurosurg 2001;94:287–92.
[8] Hart DN. Dendritic cells: unique leukocyte populations which control the primary immune response. Blood 1997;90:3245–87.
[9] Tsujitani S, Okamura T, Baba H, Korenaga D, Haraguchi M, Sugimachi K. Endoscopic intratumoral injection of OK-432 and Langerhans' cells in patients with gastric carcinoma. Cancer 1988;61:1749–53.
[10] Gabrilovich DI, Chen HL, Girgis KR, Cunningham HT, Meny GM, Nadaf S, Kavanaugh D, Carbone DP. Production of vascular endothelial growth factor by human tumors inhibits the functional maturation of dendritic cells. Nat Med 1996;2:1096–103.
[11] Triozzi PL, Khurram R, Aldrich WA, Walker MJ, Kim JA, Jaynes S. Intratumoral injection of dendritic cells derived in vitro in patients with metastatic cancer. Cancer 2000;89:2646–54.
[12] Porter DL, Hwang WT, Frey NV, Lacey SF, Shaw PA, Loren AW, Bagg A, Marcucci KT, Shen A, Gonzalez V, Ambrose D, Grupp SA, Chew A, Zheng Z, Milone MC, Levine BL, Melenhorst JJ, June CH. Chimeric antigen receptor T cells persist and induce sustained remissions in relapsed refractory chronic lymphocytic leukemia. Sci Transl Med 2015;7: 303ra139.
[13] Beatty GL, O'Hara M. Chimeric antigen receptor-modified T cells for the treatment of solid tumors: defining the challenges and next steps. Pharmacol Ther 2016;166:30–9.
[14] Priceman SJ, Tilakawardane D, Jeang B, Aguilar B, Murad JP, Park AK, Chang WC, Ostberg JR, Neman J, Jandial R, Portnow J, Forman SJ, Brown CE. Regional delivery of chimeric antigen receptor-engineered T cells effectively targets HER2(+) breast cancer metastasis to the brain. Clin Cancer Res 2018;24:95–105.
[15] Bottaro DP, Rubin JS, Faletto DL, Chan AM, Kmiecik TE, Vande Woude GF, Aaronson SA. Identification of the hepatocyte growth factor receptor as the c-met proto-oncogene product. Science 1991;251:802–4.
[16] Ghoussoub RA, Dillon DA, D'Aquila T, Rimm EB, Fearon ER, Rimm DL. Expression of c-met is a strong independent prognostic factor in breast carcinoma. Cancer 1998;82:1513–20.
[17] Merchant M, Ma X, Maun HR, Zheng Z, Peng J, Romero M, Huang A, Yang NY, Nishimura M, Greve J, Santell L, Zhang YW, Su Y, Kaufman DW, Billeci KL, Mai E, Moffat B, Lim A, Duenas ET, Phillips HS, Xiang H, Young JC, Vande Woude GF, Dennis MS, Reilly DE, Schwall RH, Starovasnik MA, Lazarus RA, Yansura DG. Monovalent antibody design and mechanism of action of onartuzumab, a MET antagonist with anti-tumor activity as a therapeutic agent. Proc Natl Acad Sci U S A 2013;110:E2987–96.
[18] Porter DL, Levine BL, Kalos M, Bagg A, June CH. Chimeric antigen receptor-modified T cells in chronic lymphoid leukemia. N Engl J Med 2011;365:725–33.
[19] Kummar S, Rubinstein L, Kinders R, Parchment RE, Gutierrez ME, Murgo AJ, Ji J, Mroczkowski B, Pickeral OK, Simpson M, Hollingshead M, Yang SX, Helman L, Wiltrout R, Collins J, Tomaszewski JE, Doroshow JH. Phase 0 clinical trials: conceptions and misconceptions. Cancer J 2008;14:133–7.
[20] Tchou J, Zhao Y, Levine BL, Zhang PJ, Davis MM, Melenhorst JJ, Kulikovskaya I, Brennan AL, Liu X, Lacey SF, Posey Jr. AD, Williams AD, So A, Conejo-Garcia JR, Plesa G, Young RM, McGettigan S, Campbell J, Pierce RH, Matro JM, DeMichele AM, Clark AS, Cooper LJ, Schuchter LM, Vonderheide RH, June CH. Safety and efficacy of intratumoral injections of chimeric antigen receptor (CAR) T cells in metastatic breast cancer. Cancer Immunol Res 2017;5:1152–61.
[21] Smyth MJ, Taniguchi M, Street SE. The anti-tumor activity of IL-12: mechanisms of innate immunity that are model and dose dependent. J Immunol 2000;165:2665–70.
[22] Gee MS, Koch CJ, Evans SM, Jenkins WT, Pletcher Jr. CH, Moore JS, Koblish HK, Lee J, Lord EM, Trinchieri G, Lee WM. Hypoxia-mediated apoptosis from angiogenesis inhibition underlies tumor control by recombinant interleukin 12. Cancer Res 1999;59:4882–9.

[23] Trinchieri G. Interleukin-12: a cytokine at the interface of inflammation and immunity. Adv Immunol 1998;70:83–243.

[24] Zaharoff DA, Hance KW, Rogers CJ, Schlom J, Greiner JW. Intratumoral immunotherapy of established solid tumors with chitosan/IL-12. J Immunother 2010;33:697–705.

[25] Salem ML, Gillanders WE, Kadima AN, El-Naggar S, Rubinstein MP, Demcheva M, Vournakis JN, Cole DJ. Review: novel nonviral delivery approaches for interleukin-12 protein and gene systems: curbing toxicity and enhancing adjuvant activity. J Interferon Cytokine Res 2006;26:593–608.

[26] Vo JL, Yang L, Kurtz SL, Smith SG, Koppolu BP, Ravindranathan S, Zaharoff DA. Neoadjuvant immunotherapy with chitosan and interleukin-12 to control breast cancer metastasis. Onco Targets Ther 2014;3:e968001.

[27] Sabel MS, Skitzki J, Stoolman L, Egilmez NK, Mathiowitz E, Bailey N, Chang WJ, Chang AE. Intratumoral IL-12 and TNF-alpha-loaded microspheres lead to regression of breast cancer and systemic antitumor immunity. Ann Surg Oncol 2004;11:147–56.

[28] Sabel MS, Su G, Griffith KA, Chang AE. Intratumoral delivery of encapsulated IL-12, IL-18 and TNF-alpha in a model of metastatic breast cancer. Breast Cancer Res Treat 2010;122:325–36.

[29] Bramson JL, Graham FL, Gauldie J. The use of adenoviral vectors for gene therapy and gene transfer in vivo. Curr Opin Biotechnol 1995;6:590–5.

[30] Sim BK, O'Reilly MS, Liang H, Fortier AH, He W, Madsen JW, Lapcevich R, Nacy CA. A recombinant human angiostatin protein inhibits experimental primary and metastatic cancer. Cancer Res 1997;57:1329–34.

[31] Lucas R, Holmgren L, Garcia I, Jimenez B, Mandriota SJ, Borlat F, Sim BK, Wu Z, Grau GE, Shing Y, Soff GA, Bouck N, Pepper MS. Multiple forms of angiostatin induce apoptosis in endothelial cells. Blood 1998;92:4730–41.

[32] Gyorffy S, Palmer K, Podor TJ, Hitt M, Gauldie J. Combined treatment of a murine breast cancer model with type 5 adenovirus vectors expressing murine angiostatin and IL-12: a role for combined anti-angiogenesis and immunotherapy. J Immunol 2001;166:6212–7.

[33] Lundstrom K. Alphavirus vectors for gene therapy applications. Curr Gene Ther 2001;1:19–29.

[34] Lundstrom K, Schweitzer C, Rotmann D, Hermann D, Schneider EM, Ehrengruber MU. Semliki Forest virus vectors: efficient vehicles for in vitro and in vivo gene delivery. FEBS Lett 2001;504:99–103.

[35] Murphy AM, Morris-Downes MM, Sheahan BJ, Atkins GJ. Inhibition of human lung carcinoma cell growth by apoptosis induction using Semliki Forest virus recombinant particles. Gene Ther 2000;7:1477–82.

[36] Berglund P, Sjoberg M, Garoff H, Atkins GJ, Sheahan BJ, Liljestrom P. Semliki Forest virus expression system: production of conditionally infectious recombinant particles. Biotechnology (N Y) 1993;11:916–20.

[37] Smerdou C, Liljestrom P. Two-helper RNA system for production of recombinant Semliki forest virus particles. J Virol 1999;73:1092–8.

[38] Colmenero P, Chen M, Castanos-Velez E, Liljestrom P, Jondal M. Immunotherapy with recombinant SFV-replicons expressing the P815A tumor antigen or IL-12 induces tumor regression. Int J Cancer 2002;98:554–60.

[39] Chikkanna-Gowda CP, Sheahan BJ, Fleeton MN, Atkins GJ. Regression of mouse tumours and inhibition of metastases following administration of a Semliki Forest virus vector with enhanced expression of IL-12. Gene Ther 2005;12:1253–63.

[40] Kramer MG, Masner M, Casales E, Moreno M, Smerdou C, Chabalgoity JA. Neoadjuvant administration of Semliki Forest virus expressing interleukin-12 combined with attenuated Salmonella eradicates breast cancer metastasis and achieves long-term survival in immunocompetent mice. BMC Cancer 2015;15:620.

[41] Topalian SL, Drake CG, Pardoll DM. Immune checkpoint blockade: a common denominator approach to cancer therapy. Cancer Cell 2015;27:450–61.

[42] Topalian SL, Sharpe AH. Balance and imbalance in the immune system: life on the edge. Immunity 2014;41:682–4.

[43] Wolchok JD, Kluger H, Callahan MK, Postow MA, Rizvi NA, Lesokhin AM, Segal NH, Ariyan CE, Gordon RA, Reed K, Burke MM, Caldwell A, Kronenberg SA, Agunwamba BU, Zhang X, Lowy I, Inzunza HD, Feely W, Horak CE, Hong Q, Korman AJ, Wigginton JM, Gupta A, Sznol M. Nivolumab plus ipilimumab in advanced melanoma. N Engl J Med 2013;369:122–33.

[44] Liu Z, Ravindranathan R, Kalinski P, Guo ZS, Bartlett DL. Rational combination of oncolytic vaccinia virus and PD-L1 blockade works synergistically to enhance therapeutic efficacy. Nat Commun 2017;8:14754.

[45] Kirn DH, Thorne SH. Targeted and armed oncolytic poxviruses: a novel multi-mechanistic therapeutic class for cancer. Nat Rev Cancer 2009;9:64–71.

[46] Chon HJ, Lee WS, Yang H, Kong SJ, Lee NK, Moon ES, Choi J, Han EC, Kim JH, Ahn JB, Kim C. Tumor microenvironment remodeling by intratumoral oncolytic vaccinia virus enhances the efficacy of immune-checkpoint blockade. Clin Cancer Res 2019;25:1612–23.
[47] Subbotin V, Fiksel G. Modeling multi-needle injection into solid tumor. Am J Cancer Res 2019;9:2209–15.
[48] Bourgeois C, Rocha B, Tanchot C. A role for CD40 expression on CD8+ T cells in the generation of CD8+ T cell memory. Science 2002;297:2060–3.
[49] Ferrati S, Nicolov E, Zabre E, Geninatti T, Shirkey BA, Hudson L, Hosali S, Crawley M, Khera M, Palapattu G, Grattoni A. The nanochannel delivery system for constant testosterone replacement therapy. J Sex Med 2015;12:1375–80.
[50] Zaballos A, Gutierrez J, Varona R, Ardavin C, Marquez G. Cutting edge: identification of the orphan chemokine receptor GPR-9-6 as CCR9, the receptor for the chemokine TECK. J Immunol 1999;162:5671–5.
[51] Papadakis KA, Prehn J, Nelson V, Cheng L, Binder SW, Ponath PD, Andrew DP, Targan SR. The role of thymus-expressed chemokine and its receptor CCR9 on lymphocytes in the regional specialization of the mucosal immune system. J Immunol 2000;165:5069–76.
[52] Croft M. Control of immunity by the TNFR-related molecule OX40 (CD134). Annu Rev Immunol 2010;28:57–78.
[53] Yang XZ, Du XJ, Liu Y, Zhu YH, Liu YZ, Li YP, Wang J. Rational design of polyion complex nanoparticles to overcome cisplatin resistance in cancer therapy. Adv Mater 2014;26:931–6.
[54] Papadakis KA, Landers C, Prehn J, Kouroumalis EA, Moreno ST, Gutierrez-Ramos JC, Hodge MR, Targan SR. CC chemokine receptor 9 expression defines a subset of peripheral blood lymphocytes with mucosal T cell phenotype and Th1 or T-regulatory 1 cytokine profile. J Immunol 2003;171:159–65.
[55] Kathuria N, Kraynyak KA, Carnathan D, Betts M, Weiner DB, Kutzler MA. Generation of antigen-specific immunity following systemic immunization with DNA vaccine encoding CCL25 chemokine immunoadjuvant. Hum Vaccin Immunother 2012;8:1607–19.
[56] McGuire HM, Vogelzang A, Ma CS, Hughes WE, Silveira PA, Tangye SG, Christ D, Fulcher D, Falcone M, King C. A subset of interleukin-21+ chemokine receptor CCR9+ T helper cells target accessory organs of the digestive system in autoimmunity. Immunity 2011;34:602–15.
[57] Jacquelot N, Enot DP, Flament C, Vimond N, Blattner C, Pitt JM, Yamazaki T, Roberti MP, Daillere R, Vetizou M, Poirier-Colame V, Semeraro M, Caignard A, Slingluff Jr. CL, Sallusto F, Rusakiewicz S, Weide B, Marabelle A, Kohrt H, Dalle S, Cavalcanti A, Kroemer G, Di Giacomo AM, Maio M, Wong P, Yuan J, Wolchok J, Umansky V, Eggermont A, Zitvogel L. Chemokine receptor patterns in lymphocytes mirror metastatic spreading in melanoma. J Clin Invest 2016;126:921–37.
[58] Chen H, Cong X, Wu C, Wu X, Wang J, Mao K, Li J, Zhu G, Liu F, Meng X, Song J, Sun X, Wang X, Liu S, Zhang S, Yang X, Song Y, Yang YG, Sun T. Intratumoral delivery of CCL25 enhances immunotherapy against triple-negative breast cancer by recruiting CCR9(+) T cells. Sci Adv 2020;6:eaax4690.

CHAPTER

4

The Fas/FasL pathway as a target for enhancing anticancer adoptive cell therapy

Greta Volpedo[a,b,*], *Thalia Pacheco-Fernández*[c,*], *Marianna de Carvalho Clímaco*[d,*], *and Abhay R. Satoskar*[a,b]

[a]Department of Pathology, The Ohio State University Medical Center, Columbus, OH, United States [b]Department of Microbiology, The Ohio State University, Columbus, OH, United States [c]Unidad de Biomedicina, Facultad de Estudios Superiores Iztacala, Universidad Nacional Autónoma de México, Tlalnepantla, México [d]Instituto de Ciências Biológicas, Departamento de Parasitologia, Universidade Federal de Minas Gerais, Belo Horizonte, Minas Gerais, Brazil

Abstract

Adoptive cell therapy (ACT) is a promising cancer therapy in terms of safety and efficacy. In ACT, the expression of FasL by therapeutic lymphocytes can induce apoptosis in Fas-expressing tumor cells. However, malignant cells can adapt by decreasing Fas to escape from therapeutic lymphocytes and increasing FasL expression to induce apoptosis of immune cells, thereby mounting resistance to ACT. To overcome this problem, the expression of Fas and FasL can be modulated in the tumor microenvironment to immunosensitize tumor cells for ACT killing. This combined strategy represents better cost-effectiveness for cancer patients.

Abbreviations

5-FU fluorouracil, adrucil
ACT adoptive cell therapy
AICD activation-induced cell death
AIDS acquired immune deficiency syndrome
AP-1 activator protein 1
Bcl B-cell lymphoma

*Co-first authors.

Immunotherapy in Resistant Cancer: From the Lab Bench Work to Its Clinical Perspectives
https://doi.org/10.1016/B978-0-12-822028-3.00013-3

CAR chimeric antigen receptor
CDDP cisplatin
CIK cytokine-induced killer
CINK cytokine-induced natural killer
CLL chronic lymphocytic leukemia
CML chronic myeloid leukemia
CTL cytotoxic T lymphocytes
CTLA4 cytotoxic T-lymphocyte antigen 4
DC dendritic cells
FADD Fas-associated death domain
FcμR immunoglobulin M Fc receptor
FLICE FADD-like interleukin-1ß-converting enzyme
FLIP FLICE inhibitory protein
GVHD graft-versus-host disease
HCC hepatocarcinoma cells
IFN interferon
IL interleukin
IV intravenous
LAK lymphokine-activated killer
MDSC myeloid-derived suppressor cell
mFasL transmembrane FasL
MHC major histocompatibility complex
NF-κB nuclear factor-kappa B
NK natural killer cell
PBMC peripheral blood mononuclear cells
ROI radical oxygen intermediates
sFasL soluble FasL
TCR T cell receptor
TGF-β transforming growth factor-beta
TIL tumor-infiltrating lymphocytes
TNBC triple-negative breast cancer
TNF tumor necrosis factor
TRAIL TNF-related apoptosis-inducing ligand
Treg T regulatory cell

Conflict of interest

No potential conflicts of interest were disclosed by the authors.

Introduction

Apoptosis is a form of programmed cell death, a physiological process that plays a fundamental role in the development and homeostasis of an organism, whether during embryogenesis, organogenesis, or tissue cell renewal [1]. Additionally, this mechanism is used by cytotoxic T lymphocytes (CTLs) and natural killer (NK) cells as a response against intracellular pathogens and malignant cells [2]. Two different apoptotic pathways have been described: the intrinsic and extrinsic pathways. The intrinsic pathway is regulated by members of the B-cell lymphoma (Bcl)-2 family. This process is triggered by genotoxic agents, which cause mitochondria to release proteins responsible for caspase activation. Conversely, the extrinsic pathway is controlled by receptors (-R) of the tumor necrosis factor (TNF) family, such as Fas, TNF-related apoptosis-inducing ligand (TRAIL)-R, and TNF-α-R. The interaction between these receptors and their ligands, FasL, TRAIL, and TNF-α, respectively, also activates the caspase cascade and results in cell death [3,4].

Fas (CD95/APO-1)/FasL (CD95L) signaling is one of the best described pathways involved in apoptosis. The Fas gene is located on chromosome 10 in humans and 19 in mice [5]. In both organisms, the gene has nine exons, which encode for a transmembrane protein with a cysteine-rich region, essential for communication with its ligand. In general, the extracellular portion of the receptors that belong to the TNF family is quite conserved. However, the intracellular portion that constitutes the death domain and is responsible for the transduction of the apoptosis signal varies between its members [6,7].

The FasL gene is located on chromosome 1 in humans and mice. It has five exons that encode for two different forms of the protein [8]. The transmembrane form (mFasL) is composed of an intracellular region with a proline-rich domain and an extracellular portion that displays the TNF homology domain, which in turn is responsible for the interaction with Fas [6]. When cleaved by metalloproteinases, the soluble form of FasL (sFasL) is released from the membrane [9,10]. Despite being able to bind to the Fas receptor in the same way as its membrane-bound counterpart, sFasL plays a limited role in the apoptotic process and it has been implicated in different cellular interactions. FasL can also be found encapsulated into exosomes and stored in secretory vesicles, allowing for this ligand to participate in additional types of cellular communications [11–13].

Fas and FasL are involved in several signaling mechanisms, either by directly eliminating infected cells or by regulating the population levels of effector cells after the resolution of the immune response. Fas is found in numerous tissues, while FasL is primarily expressed by activated T cells, NK, and tumor cells [14]. Furthermore, FasL is expressed at high levels in immune-privileged sites, such as the eyes, testis, ovaries, and placenta [15,16]. The interaction of Fas and FasL results in conformational changes to the receptor, which consequently leads to the binding of the Fas-associating protein with a death domain (FADD) to the intracellular domain of Fas. Following this, FADD recruits procaspase-8 to form the death-inducing signaling complex (DISC) [17]. Once activated, caspase-8 propagates the apoptotic signal by cleaving its subunits into the cytosol [18] (Fig. 1). Although apoptosis is the most commonly described event following Fas-FasL coupling, many studies indicate different roles for these molecules. Fas has been implicated in T cell [19], hepatocyte [20], and neuron [21] proliferation, in addition to stimulating the differentiation of CD4+ and CD8+ T cells [22] and the recruitment of neutrophils and T cells [23].

The apoptotic and non-apoptotic outcomes of the Fas-FasL system play important roles in diseases, including Alzheimer's disease, Parkinson's disease, AIDS, and certain types of cancers [24]. While protective lymphocytes use this mechanism to induce tumor cell death, it is well established that cancer cells can acquire resistance to Fas-FasL-mediated apoptosis, creating an immunoprivileged environment where malignant cells can multiply freely [25]. Tumor evasion from the immune system is orchestrated by mechanisms such as a deletion mutation in the Fas gene or receptor downregulation to resist apoptosis. Additionally, tumor cells can increase their expression of membrane-bound FasL to induce apoptosis of protective infiltrating lymphocytes (Fig. 2) [26,27]. For instance, human breast carcinomas and breast

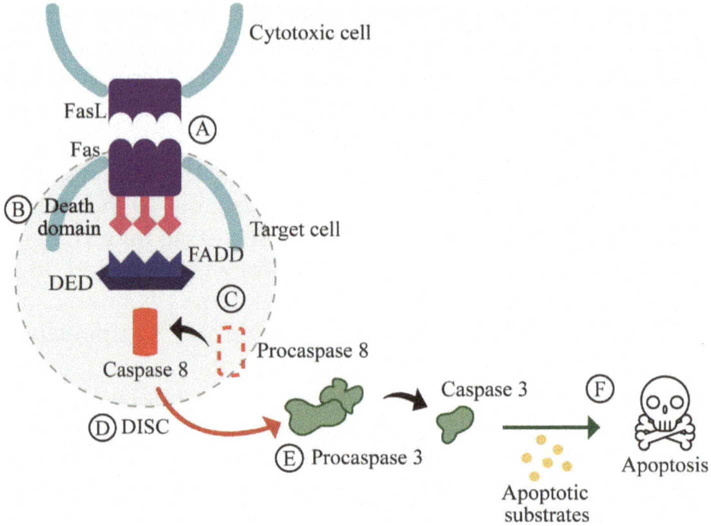

FIG. 1 Fas/FasL apoptosis pathway. (A) Cytotoxic lymphocytes express the Fas ligand (FasL) on their membrane, which binds to its receptor Fas expressed on the target cell. (B) Fas/FasL interaction results in the binding of the Fas-associating protein with a death domain (FADD) to the intracellular death domain of Fas. (C) Coupled to FADD, the death effector domain (DED) recruits procaspase-8, which converts into its active form caspase-8. The active Fas receptor, along with these associated proteins, forms the death-inducing signaling complex (DISC) (D). (E) Once released from the DISC, caspase-8 cleaves procaspase-3 into its active form, caspase-3. (F) Caspase-3 activates different apoptotic substrates which ultimately causes cell death.

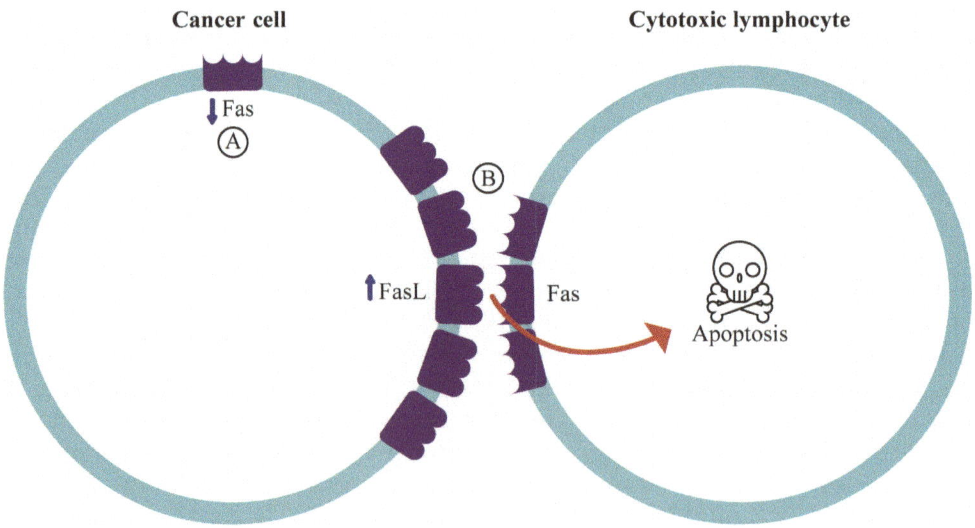

FIG. 2 Resistance to FasL-mediated apoptosis by cancer cells. (A) Cancer cells can develop resistance to ACT by downregulating the expression of Fas in their membrane, reducing its interaction with FasL from cytotoxic lymphocytes, and avoiding apoptosis. (B) Cancer cells can display a "counterattack mechanism" which consists in increased FasL expression. FasL interacts with its receptor on the cytotoxic lymphocyte leading to apoptosis and failure of the cell therapy.

cancer cell lines present an increased expression of FasL compared to the surrounding normal tissue. The Fas-expressing lymphocytic population surrounding the tumor is susceptible to FasL-mediated apoptosis induced by malignant cells [28]. Cancer cells are also capable of escaping death through the expression of deficient Fas receptors that are unable to transmit the apoptotic signal downstream [29], or through the production of proteins that inhibit the proteolytic cascade required to induce apoptosis (Fig. 2) [30,31]. These Fas-FasL molecular interference mechanisms have already been described for several types of cancers, such as melanomas and carcinomas, as well as liver, testicular, colon, hematopoietic, and breast cancer [32–38]. Furthermore, Fas and FasL have been implicated in tumor angiogenesis [39] and in the recruitment of myeloid cells to the site of inflammation, promoting their migration through the endothelial cell layer [40]. Fas also seems to be involved in cancer stem cell proliferation, which favors tumor growth and metastasis [41,42].

Due to the importance of Fas-FasL signaling for homeostasis and its fundamental role in the survival of different types of tumors, these molecules have become the target of many cancer therapies. Traditional cancer treatment including chemotherapy, radiation, and surgery exhibits only partial efficacy, accompanied by a wide array of side effects [43]. Adoptive cell therapy (ACT) has shown tremendous potential in a variety of cancers and it is based on the infusion of activated autologous or heterologous immune cells displaying antitumor properties.

This immunotherapy aims to eliminate malignant cells employing two different strategies: the non-specific and the specific therapies. The non-specific approach uses generic activation of effector cells, while the specific approach has the goal of activating lymphocytes in order to recognize particular antigens derived from tumor cells [44]. Non-specific ACTs include lymphokine-activated killer (LAK), cytokine-induced killer (CIK), and natural killer (NK) cell therapies [45–49]. Specific therapies consist in the use of tumor-infiltrating lymphocytes (TIL), specifically the cytotoxic T lymphocytes (CTL), engineered T cell receptor (TCR), and chimeric antigen receptor (CAR) T cell therapy [46,50,51]. These cells employ different mechanisms to induce antitumor cytotoxicity, including the Fas/FasL apoptotic pathway. As mentioned previously, malignant cells can also utilize this signaling mechanism to kill tumor-infiltrating lymphocytes (Fig. 2). This process of "tumor counterattack" also extends to adoptively transferred lymphocytes. For instance, in many tumor microenvironments the gene encoding FasL (*FASLG*) has been found to be overexpressed [52], and most cell types used for anticancer ACT have been shown to express Fas [53]. Once tumor cells acquire modifications in the Fas/FasL apoptotic pathway, they can become resistant to one or more cell therapies [54]. Due to the role of Fas/FasL signaling in tumor immune evasion, the combined use of ACTs with sensitizing agents that enhance the activation of the Fas/FasL pathway against tumor cells or block apoptosis in therapeutic cells can be a promising treatment against cancer.

In this chapter, we review different ACT-based therapeutic options and discuss the potential modulation of Fas-FasL signaling to enhance their efficacy and antitumor cytotoxicity. We will focus primarily on the use of soluble factors, cytokines, monoclonal antibodies, chemotherapeutic drugs, and small molecules inhibitors used to increase the Fas-dependent killing potential of ACT or decrease the counterattack of FasL-expressing tumor cells.

Non-specific cell therapies

Lymphokine-activated killer (LAK) cell therapy consists in isolating autologous or allogenic peripheral blood mononuclear cells (PBMCs) and stimulating them with IL-2 to obtain a heterogeneous population of NK (~40%), NKT, and T cells. All of these cells are cytotoxic and are not major histocompatibility complex (MHC) restricted, but LAK therapy relies mainly on NK antitumor activity. Unfortunately, the clinical trial results for LAK therapies were inconsistent and some seemed to suggest that IL-2 therapy alone was just as effective as LAK+IL-2 therapy [45,55,56]. In order to improve the antitumor activity of LAK cells, PBMCs can be further stimulated with anti-CD3 antibodies and interferon (IFN)-γ to expand and polarize cytokine-induced killer (CIK) cells. CIKs are comprised of a mixed population displaying the functional properties of both T lymphocytes (~80%) and NK cells (~20%). Expansion with anti-CD3 antibodies and IL-2, but without IFN-γ, in serum-free CellGro SCGM medium results in increased NK cell purity (~50%) in the mixture, generating cytokine-induced natural killer (CINK) cells. For an even higher NK purity, CD3 depletion and CD56 selection can be performed to obtain from 50% to >90% NK cells. Although donor NK cells in the mixture have been shown to suppress the proliferation and activity of alloreactive T cells, a lower proportion of T cells in the population is important to prevent graft-versus-host disease (GVHD) in the case of an allogenic transfer [45,48,49,57,58].

Role of Fas/FasL pathway in LAK cell therapies

The antitumor killing potential of non-specific ACTs is based on perforin and granzyme production as well the TRAIL and the Fas/FasL pathways [45,49]. It has been shown that LAK cell-derived FasL as well as granzyme B can induce activation of caspase-3 in squamous cell carcinoma cells primed with chemotherapeutic drugs such as 5-FU (Fluorouracil, Adrucil), cisplatin (CDDP, Platinol), and γ-rays. Additionally, treatment with LAK cells induced the production of radical oxygen intermediates (ROI), leading ultimately to apoptosis of the malignant cells. Interestingly, this effect was diminished by the addition of anti-Fas antibodies, suggesting a role for the Fas/FasL pathway in ROI generation, as well as apoptosis [59]. 5-FU is a widely used chemotherapeutic drug that has been proved to enhance tumor Fas expression up to 50% in the 46 h after its administration, suggesting its potential as a combination therapy alongside ACTs [60]. This is important because for Fas-mediated apoptosis to be effective, malignant cells need to abundantly express Fas. Increasing the expression of this receptor on cancer cells could augment the cytotoxicity of ACTs. Along with 5-FU, other drugs can induce this effect. Cisplatin was shown to upregulate Fas expression in esophageal cancer cells, rendering them more susceptible to LAK cytotoxicity [61]. Low expression of Fas in tumor cells is not a result of the loss of the components during mutagenesis, but only its downregulation [62]. Chemotherapeutic agents such as cisplatin, 5-FU, doxorubicin (adriamycin, Rubex), as well as IFN-γ, can increase Fas expression in tumor cells by damaging the tumor's DNA. As a response to DNA damage, cells initiate apoptosis by activating transcription factors (AP-1 and NF-κB) that lead to the transcription of Fas mRNA [62,63].

Role of Fas/FasL pathway in CIK cell therapies

CIK therapies also become more effective when combined with Fas/FasL modulators such as 5-FU and cisplatin. Intravenous (IV) administration of 5-FU is currently being evaluated in phase II clinical trial in combination with autologous tumor lysate-pulsed dendritic and CIK therapies against post-surgery stage I-III colorectal cancer [64]. Additionally, a clinical study for non-small cell lung cancer patients has been conducted with cisplatin administered IV in combination with paclitaxel and CIK cell therapy [65]. Cisplatin in combination with CIK therapy and gemcitabine (Gemzar), which sensitizes malignant cells to FasL-mediated apoptosis by upregulating caspase-3 and caspase-8, has been assessed in a phase II clinical trial for nasopharyngeal carcinoma [66]. Furthermore, etoposide (VP-16, VePesid, etopophos, toposar, and etoposide phosphate), which has been shown to immunosensitize tumor cells and increase their susceptibility to FasL-mediated killing, is under investigation in phase I clinical trials in combination with either CIK or allogenic lymphocyte therapy in patients with leukemia and multiple myeloma [67,68]. Another modulator is trans-cinnamaldehyde, which induces Fas expression on human leukemia cells, enhancing the antitumor properties of CIK therapy [69].

As mentioned previously, CIK cells are comprised of a mixture of NK and T cells, of which approximately half are CD4+ and the other half CD8+. Interestingly, a Th1 polarization was observed in the CD4+/CIK population compared with CD4+/PBMCs. When incubated with B-cell leukemia (Raji) cells, which are sensitive to Fas-mediated cell death, this CD4+/CIK sub-population was able to induce apoptosis, while CD4+/PBMCs did not. Yu et al. demonstrated that this action was dependent on cell-to-cell contact via the Fas/FasL pathway, as CD4+/CIK supernatant alone did not induce apoptosis. After co-incubation with CD4+/CIK cells, malignant Raji cells upregulated Fas expression by almost 25%, suggesting a role for the Fas/FasL pathway in CD4+/CIK-mediated apoptosis. Interestingly, the incubation of the breast cancer MDA-MB-231 cell line with CD4+/CIK supernatant abrogated MDA-MB-231 resistance to Fas-mediated apoptosis due to the interaction with soluble CD40L, a member of the TNF superfamily, and partially to the presence of IFN-γ. These results show that Th1-polarized CD4+ cells can enhance the antitumor activities of CIK in different cancer models [70].

Role of Fas/FasL pathway in NK cell therapies

Fas and FasL are highly expressed on NK cells, and it has been shown that primary human NK cells mainly employ this pathway to target and destroy malignant cells, although granzyme and perforin are also used, especially for Fas-negative tumors [71,72]. The Fas/FasL mechanism relies on cell-to-cell contact and it occurs slowly, often requiring more than one interaction, while the use of perforin and granzyme is faster and can occur after a single contact [72,73]. Understanding when each of these pathways is used and modulating them could augment the anticancer properties of NK cells. Several compounds have been shown to modulate Fas-mediated cytotoxicity and therefore make promising immunosensitizing candidates for combinational therapies alongside NK cell ACT. For instance, the tyrosine kinase inhibitor radotinib has been used to target BCR-ABL1 in chronic myeloid leukemia (CML) cells and kill them directly [74]. A recent study showed that radotinib enhances

NK cell cytotoxicity in Fas+, but not Fas- solid tumor cell lines (lung, breast, and melanoma cells), suggesting a role for the Fas/FasL pathway in radotinib-mediated cytotoxicity [75]. Radotinib and rebastinib, both tyrosine kinase inhibitors, are currently undergoing clinical trials in combination with other therapies in a myeloid leukemia model [76] as well as locally advanced metastatic solid tumors [77] and a breast cancer [78] model, respectively.

IL-2 administration has been shown to aid NK-mediated cytotoxicity by facilitating target recognition and enhancing the frequency of NK-malignant cell interactions [72,73]. The modulation of the tumor microenvironment by the administration of IL-2 along with anti-TGF-β neutralizing antibodies resulted in higher numbers of NK and CD8 cells in a murine lung carcinoma model. Interestingly, the authors showed a direct regulation of CD8 cells by NK cells mediated by the Fas/FasL pathway after immunotherapy, which in turn affected their antitumor effects [79]. The delivery of a fusion protein containing the pro-inflammatory cytokine IL-12 and the transmembrane and intracellular domains of Fas to human cervical carcinoma HELA cells via lentiviral transduction in vitro resulted in enhanced activation and antitumor action of NK cells, as well as in increased caspase-3 activation in the tumor cells, which ultimately lead to reduced tumor growth [80]. Along with IL-2 and IL-12, IFN-α is another important cytokine in cancer therapy, as it activates NK cells and augments their antitumor activity. IFN-α has been used as an immunotherapy for hematological as well as solid tumors because it promotes antitumor activity of immune cells including T cells and dendritic cells. It also downregulates oncogene expression and it induces tumor suppressor gene. Jiang et al. genetically modified a human NK cell line (NKL) to constitutively secrete IFN-α (NKL-IFNα). NKL-IFNα treatment inhibited human hepatocarcinoma cells (HCC) growth more efficiently than regular NKL. The stable secretion of IFN-α increased the transcription of genes related to cytotoxicity like perforin, granzyme B, and Fas ligand in NK cells as well as the expression of Fas on HCCs [81]. These results show that modulation of the tumor microenvironment can both potentiate the antitumor activity of NK cells and immunosensitize malignant cells by enhancing their susceptibility to Fas-mediated killing.

Other compounds have shown to immunosensitize cancer cells to ACTs by upregulating Fas expression on malignant cells and FasL production in NK cells. Selenium-containing ruthenium complex stimulation upregulated Fas expression in two different prostate cancer cell lines, making them more susceptible to NK-mediated killing [82], while 1,25-dihydroxyvitamin D_3 also enhanced Fas expression in human melanoma cells via the heat shock protein Hsp60, increasing the sensitivity of this tumor to NK apoptosis [83]. Furthermore, rice hull polysaccharides upregulated FasL production in NK cells, resulting in enhanced cytotoxic activity against colon cancer in a murine model [84]. Lastly, low-dose ionizing radiation induced the activation of NK cells by upregulating the expression of FasL, in a P38-MAP kinase (MAPK)-dependent manner [85]. These results suggest that targeting the Fas/FasL pathway with immune or other therapies could act as a combination treatment to enhance the antitumor effect of ACTs. It is important to note that manipulating Fas/FasL signaling in both the tumor microenvironment and the therapeutic lymphocytes might be necessary to achieve optimal results. For instance, delivery of FasL into the tumor did not improve the survival in dogs with osteosarcoma with low lymphocyte infiltrate, but it did on those with higher inflammatory infiltrate [86], thus demonstrating that microenvironment modulation by itself may not be enough to improve survival outcome.

Specific cell therapies

Role of Fas/FasL pathway in tumor-infiltrating lymphocyte (TIL) therapy

Specific ACTs have the goal of activating lymphocytes to recognize particular antigens present in tumor cells. These therapies include cytotoxic T lymphocytes (CTL), tumor-infiltrating lymphocytes (TILs), and T cell therapy with modified T cell receptor (TCR) and with chimeric antigen receptors (CAR). Cancer cell therapy is based on the infiltration and activation of specific antitumor NK and T cells, and the effective trafficking of tumor-infiltrating lymphocytes (TILS) has been associated with a positive outcome in patients. TILs in the tumor microenvironment, particularly CD8 T cells, recognize cancer antigenic peptides and, once activated, lead to the elimination of tumor cells [43]. The tumor microenvironment is rich in immunosuppressive molecules which can stop or slow down the antitumor response of TILs produced by the patient. TIL therapy is based on obtaining the T cells from tumor biopsies, which are already specific against tumor antigens, and boost them with cytokines (such as IL-2) before re-introducing them into the patient's peripheral vasculature. These activated TILs will then return to the tumor and promote cancerous cell lysis and tumor regression [87]. This adoptive immunotherapy can only be carried out if the TILs display cancer-specific reactivity and can be resected from the tumor itself or isolated from the peripheral blood [88]. TIL therapy was the first T cell-based therapy to be described in 1988, and it has had a lot of success with a clinical response rate of more than 50%. Since then, other therapies have emerged such as engineered TCR therapy and CAR therapy [87].

The death receptor Fas and its ligand FasL are expressed by activated T cells to maintain homeostasis via apoptotic pathways. IFN-γ can also induce Fas and FasL expression to prevent immunopathology mediated by disproportionate T cell activity [71]. Interestingly, the Fas/FasL mechanism also plays a role in cancer immune surveillance and is used by T cells to mediate FasL-induced tumor cell death [89]. Increasing expression of FasL on TILs can enhance therapy efficacy. Symes et al. showed that murine T cells retrovirally transduced to overexpress FasL-exhibited augmented cytotoxic abilities in a prostate cancer model compared to the control group [90]. While this is an interesting application, it is important to understand that excessive upregulation of FasL can also cause damage to non-malignant bystander cells. As T cells infiltrate the tumor microenvironment, they can eliminate one another by FasL-mediated apoptosis before they are able to engage in antitumor activity, in a process called activation-induced cell death (AICD). This normally important mechanism for maintaining homeostasis can be very detrimental in the context of cancer. Histone deacetylase inhibitors suppress FasL expression on activated CD4 T cells and therefore enhanced anti-tumor immune responses. The use of histone deacetylase inhibitors in combination with anticytotoxic T-lymphocyte antigen 4 (CTLA4) therapy further augmented antitumor immunity in a melanoma model [91].

Activated CD8 T cells express FasL, which can bind to the tumor-expressed Fas receptor, triggering apoptosis of the malignant cell [89]. Because of this interaction, tumor expression of Fas has been identified as a prognostic marker for recurrence and as a potential marker for the overall survival of triple-negative breast cancer (TNBC) patients [92]. In order to escape T cell-mediated apoptosis, some tumors can use Fas transcriptional silencing as a mechanism of immune evasion. Epigenetic inhibitors such as decitabine (Dacogen) and vorinostat (Zolinza), were shown to reverse this silencing by promoting the upregulation of Fas in tumor cells and restore sensitization to Fas-mediated apoptosis by ACT therapy in a metastatic colon carcinoma model [89,93]. Decitabine was found to have a similar effect on human gliomas [94], and it has been evaluated in phase I and II clinical trials against acute myeloid leukemia due to its potential to enhance the ACTs by upregulating the expression of Fas in tumor cells as well as co-stimulatory molecules, adhesion molecules, and MHC, while reducing GVHD by promoting T regulatory cell (Treg) differentiation [95]. In addition to this, decitabine induces the expression of cancer antigens, making malignant cells more susceptible to ACTs [96]. Decitabine administered IV is under investigation in phase I trial for recurrent fallopian tube, ovarian, and primary peritoneal carcinoma, in combination with genetically engineered NY-ESO-1-specific T cells. It is thought that decitabine could increase the expression of NY-ESO-1, therefore enhancing the tumor-killing ability of these T cells [97].

Because of these observations, interfering with the Fas/FasL pathway could enhance the efficacy of TILs and other cancer immunotherapies [71]. Interestingly, malignant cells and TILs are not the only cells in the tumor microenvironment to use Fas/FasL interactions to promote apoptosis. IL-2/αCD40 therapy was able to promote Fas-mediated Treg and myeloid-derived suppressor cell (MDSC) death in two different murine renal carcinoma models. The removal of these cells is imperative for efficient and successful antitumor immunotherapy [98]. Furthermore, adoptive transfer with CD8 T cells, engineered to produce IL-12, induced Fas upregulation in macrophages, MDSCs, and dendritic cells residing in the tumor microenvironment, leading to the collapse of the stroma in a murine melanoma model. The stroma forms a physical barrier protecting the tumor from protective infiltrating lymphocytes. The collapse of this structure leaves the tumor exposed and allows for natural and adoptively transferred lymphocytes to access the tumor microenvironment and engage in cytotoxic activities [99].

As previously mentioned, malignant cells are not always effectively destroyed by the immune system and can also express FasL to mediate apoptosis of lymphocytes infiltrating into the tumor microenvironment. FasL-expressing tumors exhibit high malignancy and metastatic potential [71,98]. For instance, tumor-derived exosomes can express FasL and selectively mediate apoptosis in T cells [100]. The modulation of tumor counterattack can also be a potential therapeutic target. In the TirP model of resistant melanoma, FasL-tumor expression was shown to mediate the evasion of TIL therapy. Susceptibility to TIL, due to the abrogation of tumor counterattack mechanisms, was restored upon the blockage of Fas/FasL interactions [101]. Similarly, endothelial cells can trigger apoptosis of antitumor CD8+ T cells [71]. Interestingly, downregulation of FasL in a malignant glioma model resulted in enhanced T cell infiltration and antitumor activity [102]. Along with targeting FasL in the tumor, it is important to protect T cells from Fas-mediated cell death. Different cancers overexpress the gene coding for FasL [52], which can result in the recruitment of Fas-associated death domain (FADD) and activation of caspase-8, caspase-3, caspase-6, and caspase-7 [103], ultimately compromising the efficacy and survival of therapeutic cells. FLIP (FLICE [FADD-like interleukin-1ß-converting-enzyme]-inhibitory protein) is a viral antiapoptotic molecule which can interfere with DISC (death-inducing signaling complex) and block the signaling cascade leading to activate caspase-8 and therefore inhibit apoptosis mediated via virtually all death receptors, including Fas. Steiert et al. demonstrated that Jurkat T cells transfected with FLIP were less susceptible to Fas-mediated apoptosis when incubated with colorectal adenocarcinoma in vitro [104]. Furthermore, engineered Fas variants with impaired binding to FADD impeded the FasL-dependent apoptosis when adoptive lymphocytes were exposed to melanoma and leukemia cells [52].

Malignant cells resistant to FasL-mediated apoptosis can be targeted by chemotherapeutic drugs. Cisplatin, doxorubicin, and etoposide have shown to immunosensitize prostate tumor cells that are resistant to TIL therapy to become sensitive to Fas- and TNF-α-mediated killing [54,105]. The potential of cisplatin as an immunosensitizing drug has also been demonstrated on resistant melanoma. MART-1 is an antigen found in >90% of human melanomas, so when CTLs were designed against MART-1, some melanoma cells could still escape CTL killing. The sensitization of melanoma cells with cisplatin potentiated FasL-mediated killing by CTLs [106]. Similar results were shown in non-small cell lung carcinoma cells, where the use of gemcitabine (commonly used for this type of cancer) sensitized malignant cells to Fas-mediated apoptosis by CTLs via the upregulation of caspase-3 and caspase-8 activation [107]. Precision T cells specific to a personalized neoantigen have been used in combination with gemcitabine in phase II clinical trial for advanced biliary tract malignant tumor [108]. Gemcitabine has also been evaluated in combination with autologous EBV-specific CTLs in nasopharyngeal carcinoma patients in phase III clinical trial [109]. Lastly, allogenic stem cell infusion and adoptive immunotherapy following IV administration of gemcitabine are under investigation in phase II clinical trial for Hodgkin's disease [110].

Fas/FasL signaling can also mediate non-apoptotic pathways [39], which have been implicated in the impairment of adoptive cell transfer therapies. In particular, when mixed with memory CD8 T cells, naïve CD8 T cells exhibit reduced antitumor activity due to precocious differentiation. This early differentiation is mediated by non-apoptotic Fas/Akt signaling and can be abrogated by FasL neutralization. Disruption of precocious synchronized differentiation by interfering with the Fas/FasL pathway could enhance T cell-based therapies [111,112].

Role of Fas/FasL pathway in engineered T cell receptor (TCR) therapy

Despite their success, TIL therapies have some limitations including restricted survival and migration to the tumor site, as well as cancer immune evasion. In order to overcome these challenges, scientists have developed genetically engineered T cell therapies. Just like TIL therapies, engineering therapies start with the isolation of T cells from the patient's blood. After activation and amplification in vitro, the T cells are genetically altered to encode TCRs specific for the recognition of cancer antigens and to modify genes that promote survival and infiltration into the tumor site. After this modification, the cells are transfused back into the patient. TCR-T and CAR-T cell therapies are generated with genetic engineering technology. TCR-T cell technology is based on the recognition of intracellular cancer antigens presented by MHC molecules and its goal is to enhance TCR specificity recognition, affinity, and binding to these tumor peptides. In order to do that, a library of cancer antigens is selected and tested to screen out any possible cross-reaction with normal tissue polypeptides [113].

Infiltrating T cells can be induced into apoptosis by FasL-expressing tumor cells. A recent study showed that TCR-T cells and CAR-T cells co-engineered to express a Fas variant incapable of binding to FADD, were able to resist FasL-triggered apoptosis. This modified Fas receptor conferred protection after adoptive cell transfer, resulting in augmented T cell persistence and antitumor activity against both solid and hematologic cancers. Importantly, the transfer of these co-engineered T cells did not result in autoimmunity or aberrant expansion [52].

Role of Fas/FasL pathway in chimeric antigen receptors (CAR)-T cell therapy

CAR-T cell therapy is a type of genetically engineered T cell therapy using a recombinant receptor that has tumor antigen binding and T cell-activating functions. These modified T cells are not limited by MHC class and therefore can recognize a wider variety of cell surface targets than those the natural TCR would. There are 5 generations of CAR-T technologies, all using a single-chain variable antibody extracellular domain to directly recognize tumor-specific antigens. All of the generations also have the intracellular signal component CD3ζ, but in the second generation, a co-stimulatory molecule was added, and the third has an additional co-stimulatory molecule to enhance proliferation and survival. The fourth generation is linked to a downstream transcription factor capable of inducing cytokine production upon antigen recognition, while the fifth is designed to use gene editing to remove both α and β chains of the TCR. CAR-T cell therapy has shown potential in a variety of cancers including B-cell lymphoma and leukemia [113].

T cells perform their anticancer activity through perforin and granzyme actions, cytokine release, and the Fas/FasL pathway. CAR-T cells have originally been shown to mainly use the perforin and granzyme degranulation, although more recent evidence suggests an important contribution of the Fas/FasL pathway as well [114,115]. Engineered murine CTLs with a defective granular exocytosis pathway, but a functional Fas/FasL pathway, can be directed to malignant cells with antibodies and were able to induce effective tumor cell cytotoxicity through the Fas/FasL pathway in a lymphoma in vitro model [115]. The antitumor activity of CTLs adoptive transfer therapy can be enhanced via stimulation with IL-6-expressing engineered dendritic cells. These myeloid cells play an important role in bridging the innate and adaptive antitumor immunity by antigen presentation and activation of T lymphocytes. CTLs incubated with IL-6-expressing dendritic cells showed an increase in FasL expression, as well as augmented survival and cytotoxicity in a thymoma model. This technique can be applied to other T cell therapies [116]. Malignant cells such as chronic lymphocytic leukemia (CLL) cells upregulate the expression of Fas apoptotic inhibitory molecule-3, also called immunoglobulin M Fc receptor (FcμR), which protects them from FasL-mediated apoptosis. The enhanced expression of FcμR distinguishes leukemic from regular B cells and can be exploited as specific targets for CAR-T cell therapy. FcμR-specific CAR-T cells were able to effectively eliminate malignant cells through FasL and other mechanisms [117].

Recently, the potential use of Fas/FasL-modulator drugs together with CAR-T therapies has been translated into clinical trials. CAR-T cell therapy in combination with drugs such as acalabrutinib [118], fludarabine, and cyclophosphamide [119–126]. Rituximab and lenalidomide [127] have been evaluated in numerous clinical trials. Most of them are on phase I and II and are designed to treat different types of B-cell lymphomas. Additionally, CAR-T cell safety is being evaluated in association with cyclophosphamide to treat pancreatic cancer [128] and in combination with pembrolizumab [129,130] or ibrutinib [131,132] to treat B-cell lymphomas. Several phase I and II clinical trials have also been evaluating CAR-T cell safety and efficacy in association with cyclophosphamide and fludarabine to treat B-cell lymphoma, acute or chronic lymphocytic leukemia [133–138]. Lastly, a phase III study has been comparing the benefits and risks of tisagenlecleucel, blinatumomab, or inotuzumab in adult patients with relapsed or refractory acute lymphoblastic leukemia [139].

Conclusion

Cancer is the second cause of death worldwide, and the leading one in high-income countries [140]. Chemotherapeutic drugs have been proven to reduce tumor size in several types of cancers, yet many patients develop toxic side effects as well as resistance against these treatments [43]. A promising approach for bypassing these issues is to target the immune system with sensitizing agents and ACTs to potentiate the body's natural response to the tumor. Additionally, modulating the tumor microenviroment, even after the development of resistance against chemotherapy, could enhance the killing potential of ACTs and improve disease outcome. This dual approach might be necessary to obtain optimal results.

In this chapter, we have described the pivotal role of Fas/FasL signaling in cancer elimination by naturally and adoptively transferred protective lymphocytes. The relationship between the modulation of Fas/FasL pathway and its impact on the T cell populations is becoming more relevant. For example, a clinical trial with epithelial ovarian carcinoma patients will attempt to identify the different subpopulations of T cells differentiated after chemotherapy together with Fas/FasL modulation [141]. The Fas/FasL signaling mechanism can also be exploited by malignant cells to evade and destroy protective lymphocytes. Modulation of this pathway in both the therapeutic cells and the tumor cells is a promising strategy to achieve better disease prognosis and increase the patient's overall survival. In order to increase the efficacy of ACTs, it is crucial to target tumor cells to make them sensitive to Fas-mediated apoptosis, particularly by downregulating tumor FasL or increasing Fas expression (Fig. 3). On the other hand, the use of specific T cells can be potentiated by directly preventing tumor counterattack, allowing these therapeutic T cells to survive long enough to display an effective antitumor immune response. This delicate equilibrium would determine the success of ACTs.

The enhancement of specific T cell therapies can also be achieved through the transfer of dendritic cells (DCs) that specifically activate T cells by presenting a particular tumor antigen. Administration of doxorubicin, which increases Fas expression in tumor cells, is currently being evaluated in phase I and II clinical trials in combination with cyclophosphamide and adoptive DC transfer in breast cancer patients [142]. Cytotoxic lymphocytes and dendritic cells specific for iAPA are also being evaluated in pancreatic ductal carcinoma patients in phase II clinical trial [143]. Interestingly, dendritic cell exosomes have shown the potential to enhance the normal anticancer immune response as well as ACTs. In phase I and II clinical trials, it was shown that dendritic cell exosome therapy stimulated NK cells and improved the clinical outcome in patients with advanced non-small cell lung cancer [144] and metastatic melanoma [145]. Tumor-derived exosomes have also shown potential in a variety of clinical trials by serving as an antigen delivery system to T cells [146,147].

The use of different permutations of combination therapies, including dendritic cell therapy, ACT, and Fas/FasL modulators, has been proposed as one of the most promising

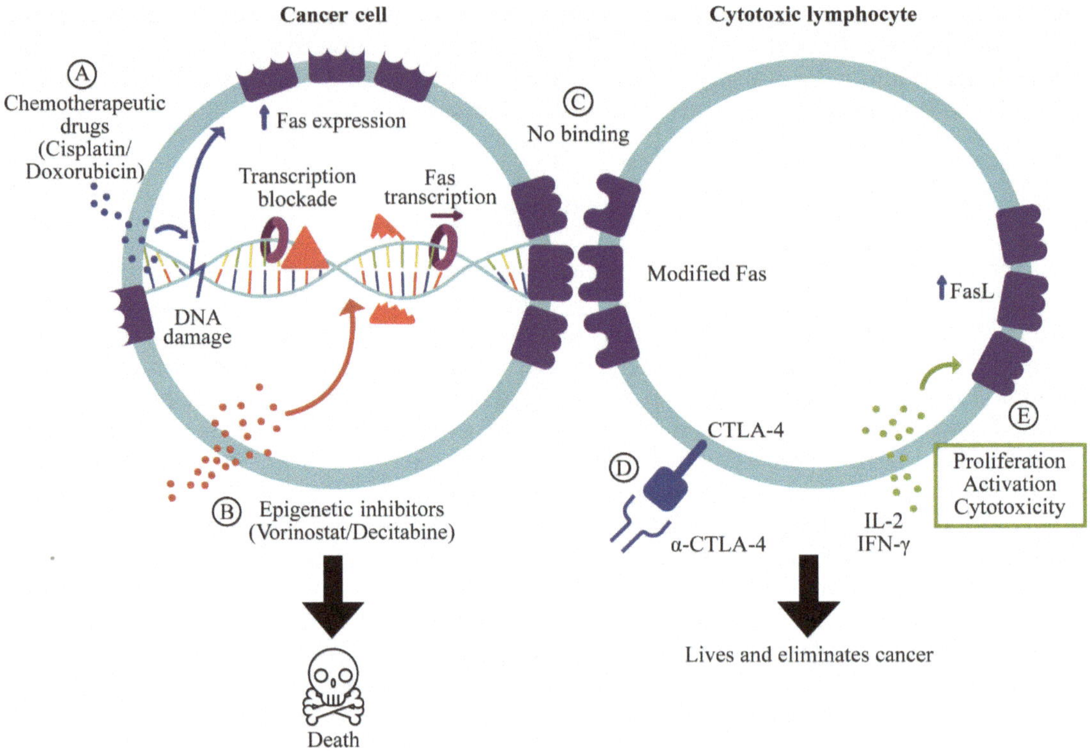

FIG. 3 Immunosensitization mechanisms to Fas-mediated apoptosis. (A) The low expression of Fas in cancer cells can be overcome with the use of chemotherapeutic drugs (such as cisplatin or doxorubicin). Chemotherapies cause DNA damage which leads to the increase of Fas transcription and expression as a mechanism of stress-induced apoptosis. (B) The expression of the Fas gene can be blocked by epigenetic mechanisms in the cancer cell. The use of epigenetic inhibitors like vorinostat or gemcitabine restores the transcription and expression of Fas, making the cancer cell more susceptible to cytotoxic lymphocytes. (C) Therapeutic T cells with genetically modified Fas can avoid the binding of Fas to the highly expressed FasL in cancer cells. (D) Even when FasL is expressed in cytotoxic cells, the expression of CTLA-4 stops the effector activity of these lymphocytes. Then, the use of anti-CTLA-4 antibodies can stop the action of this inhibitor molecule and allow the therapeutic cell to display its cytotoxic activity. (E) Finally, the administration of IL-2 and IFN-γ to therapeutic cells promotes activation and proliferation and enhances their cytotoxic activity via the increased expression of FasL. All those mechanisms result in the survival of the therapeutic cells as well as increased efficiency of its effector activity and result in cancer cell death.

strategies to combat tumor resistance [148]. The use of combination therapies limits the chances for a specific cancer to develop resistance by simultaneously targeting the immune system and the tumor itself. Additionally, the availability of different treatment strategies allows for physicians to design an appropriate and personalized therapeutic plan, tailored to each individual patient. Lastly, limiting the use of chemotherapic drugs in favor of other treatments reduces the exposure to toxic agents and the development of adverse side effects.

An issue to consider is the cost-effectiveness of these combination therapies. There is a clear improvement in the quality of life and overall survival of patients treated with ACTs compared to those treated with chemotherapeutic drugs alone or even biological therapies

[149]. On the other hand, the cost of ACTs is higher and may not be affordable for most patients. For example, the use of CAR-T cell therapy for B-cell lymphomas would double the life year expectancy of patients compared to the chemotherapeutic treatment, but it also costs at least three times more [150]. Through this chapter, we discussed many chemotherapeutic drugs that can enhance the potential of ACT. One of the most addressed is cisplatin, which is one of the cheapest chemotherapeutic drugs in the market [151]. Efficacy as well as cost should both be considered when designing a treatment plan to maximize therapeutic efficacy and affordability for a broader patient population.

This chapter highlights the importance of the regulation of the Fas/FasL pathway to immunosensitize tumor cells and ultimately increases the efficacy of ACT. Combination therapies utilizing ACT and Fas/FasL modulators are able to overcome cancer resistance while improving cost-effectiveness compared to ACT themselves.

Acknowledgments

Thalia Pacheco-Fernández is a doctoral student from the Programa de Doctorado en Ciencias Biomédicas, Universidad Nacional Autónoma de México (UNAM), and received a fellowship and a visiting research student scholarship from the National Council of Science and Technology of Mexico (CONACYT) (CVU 694650).

Marianna de Carvalho Clímaco is a doctoral student from the Programa de Pós Graduação em Parasitologia, Universidade Federal de Minas Gerais (UFMG), and received a fellowship as visiting scholar from Programa de Internacionalização, Coordenação de Aperfeiçoamento de Pessoal de Nível Superior (CAPES-PrInt) (88887.364670/2019-00).

References

[1] Jacobson MD, Weil M, Raff MC. Programmed cell death in animal development. Cell 1997;88:347–54.
[2] Krammer PH. CD95's deadly mission in the immune system. Nature 2000;407:789–95.
[3] Nagata S. Apoptosis and clearance of apoptotic cells. Annu Rev Immunol 2018;36:489–517.
[4] Hajra KM, Liu JR. Apoptosome dysfunction in human cancer. Apoptosis 2004;9:691–704.
[5] Nagata S, Golstein P. The Fas death factor. Science (New York, NY) 1995;267:1449–56.
[6] Villa-Morales M, Fernández-Piqueras J. Targeting the Fas/FasL signaling pathway in cancer therapy. Expert Opin Ther Targets 2012;16:85–101.
[7] Oehm A, Behrmann I, Falk W, et al. Purification and molecular cloning of the APO-1 cell surface antigen, a member of the tumor necrosis factor/nerve growth factor receptor superfamily. Sequence identity with the Fas antigen. J Biol Chem 1992;267:10709–15.
[8] Takahashi T, Tanaka M, Inazawa J, Abe T, Suda T, Nagata S. Human Fas ligand: gene structure, chromosomal location and species specificity. Int Immunol 1994;6:1567–74.
[9] Tanaka M, Suda T, Takahashi T, Nagata S. Expression of the functional soluble form of human Fas ligand in activated lymphocytes. EMBO J 1995;14:1129–35.
[10] Kayagaki N, Kawasaki A, Ebata T, et al. Metalloproteinase-mediated release of human Fas ligand. J Exp Med 1995;182:1777–83.
[11] O'Reilly LA, Tai L, Lee L, et al. Membrane-bound Fas ligand only is essential for Fas-induced apoptosis. Nature 2009;461:659–63.
[12] Schneider P, Holler N, Bodmer JL, et al. Conversion of membrane-bound Fas(CD95) ligand to its soluble form is associated with downregulation of its proapoptotic activity and loss of liver toxicity. J Exp Med 1998;187:1205–13.
[13] Abusamra AJ, Zhong Z, Zheng X, et al. Tumor exosomes expressing Fas ligand mediate CD8+ T-cell apoptosis. Blood Cells Mol Dis 2005;35:169–73.
[14] Nagata S. Apoptosis by death factor. Cell 1997;88:355–65.

[15] Griffith TS, Brunner T, Fletcher SM, Green DR, Ferguson TA. Fas ligand-induced apoptosis as a mechanism of immune privilege. Science (New York, NY) 1995;270:1189–92.
[16] Lee H, Ferguson TA. Biology of FasL. Cytokine Growth Factor Rev 2003;14:325–35.
[17] Hughes MA, Harper N, Butterworth M, Cain K, Cohen GM, MacFarlane M. Reconstitution of the death-inducing signaling complex reveals a substrate switch that determines CD95-mediated death or survival. Mol Cell 2009;35:265–79.
[18] Kruidering M, Evan GI. Caspase-8 in apoptosis: the beginning of "the end"? IUBMB Life 2000;50:85–90.
[19] Alderson MR, Armitage RJ, Maraskovsky E, et al. Fas transduces activation signals in normal human T lymphocytes. J Exp Med 1993;178:2231–5.
[20] Desbarats J, Newell MK. Fas engagement accelerates liver regeneration after partial hepatectomy. Nat Med 2000;6:920–3.
[21] Desbarats J, Birge RB, Mimouni-Rongy M, Weinstein DE, Palerme J-S, Newell MK. Fas engagement induces neurite growth through ERK activation and p35 upregulation. Nat Cell Biol 2003;5:118–25.
[22] Rensing-Ehl A, Völkl S, Speckmann C, et al. Abnormally differentiated CD4+ or CD8+ T cells with phenotypic and genetic features of double negative T cells in human Fas deficiency. Blood 2014;124:851–60.
[23] Guo Z, Zhang M, Tang H, Cao X. Fas signal links innate and adaptive immunity by promoting dendritic-cell secretion of CC and CXC chemokines. Blood 2005;106:2033–41.
[24] Thompson CB. Apoptosis in the pathogenesis and treatment of disease. Science (New York, NY) 1995;267:1456–62.
[25] Mellor AL, Munn DH. Creating immune privilege: active local suppression that benefits friends, but protects foes. Nat Rev Immunol 2008;8:74–80.
[26] Ivanov VN, Lopez Bergami P, Maulit G, Sato T-A, Sassoon D, Ronai Z. FAP-1 association with Fas (Apo-1) inhibits Fas expression on the cell surface. Mol Cell Biol 2003;23:3623–35.
[27] Jones CL, Wain EM, Chu C-C, et al. Downregulation of Fas gene expression in Sézary syndrome is associated with promoter hypermethylation. J Invest Dermatol 2010;130:1116–25.
[28] Gutierrez LS, Eliza M, Niven-Fairchild T, Naftolin F, Mor G. The Fas/Fas-ligand system: a mechanism for immune evasion in human breast carcinomas. Breast Cancer Res Treat 1999;54:245–53.
[29] Cascino I, Papoff G, De Maria R, Testi R, Ruberti G. Fas/Apo-1 (CD95) receptor lacking the intracytoplasmic signaling domain protects tumor cells from Fas-mediated apoptosis. J Immunol (Baltimore, Md: 1950) 1996;156:13–7.
[30] Irmler M, Thome M, Hahne M, et al. Inhibition of death receptor signals by cellular FLIP. Nature 1997;388:190–5.
[31] Pitti RM, Marsters SA, Lawrence DA, et al. Genomic amplification of a decoy receptor for Fas ligand in lung and colon cancer. Nature 1998;396:699–703.
[32] Natoli G, Ianni A, Costanzo A, et al. Resistance to Fas-mediated apoptosis in human hepatoma cells. Oncogene 1995;11:1157–64.
[33] Robertson MJ, Manley TJ, Pichert G, et al. Functional consequences of APO-1/Fas (CD95) antigen expression by normal and neoplastic hematopoietic cells. Leuk Lymphoma 1995;17:51–61.
[34] O'Connell J, O'Sullivan GC, Collins JK, Shanahan F. The Fas counterattack: Fas-mediated T cell killing by colon cancer cells expressing Fas ligand. J Exp Med 1996;184:1075–82.
[35] Lebel M, Bertrand R, Mes-Masson AM. Decreased Fas antigen receptor expression in testicular tumor cell lines derived from polyomavirus large T-antigen transgenic mice. Oncogene 1996;12:1127–35.
[36] Hahne M, Rimoldi D, Schröter M, et al. Melanoma cell expression of Fas(Apo-1/CD95) ligand: implications for tumor immune escape. Science (New York, NY) 1996;274:1363–6.
[37] Keane MM, Ettenberg SA, Lowrey GA, Russell EK, Lipkowitz S. Fas expression and function in normal and malignant breast cell lines. Cancer Res 1996;56:4791–8.
[38] Barnhart BC, Legembre P, Pietras E, Bubici C, Franzoso G, Peter ME. CD95 ligand induces motility and invasiveness of apoptosis-resistant tumor cells. EMBO J 2004;23:3175–85.
[39] Le Gallo M, Poissonnier A, Blanco P, Legembre P. CD95/Fas, non-apoptotic signaling pathways, and kinases. Front Immunol 2017;8:1216.
[40] Letellier E, Kumar S, Sancho-Martinez I, et al. CD95-ligand on peripheral myeloid cells activates Syk kinase to trigger their recruitment to the inflammatory site. Immunity 2010;32:240–52.
[41] Ceppi P, Hadji A, Kohlhapp FJ, et al. CD95 and CD95L promote and protect cancer stem cells. Nat Commun 2014;5:5238.

[42] Drachsler M, Kleber S, Mateos A, et al. CD95 maintains stem cell-like and non-classical EMT programs in primary human glioblastoma cells. Cell Death Dis 2016;7:e2209.
[43] Mayor P, Starbuck K, Zsiros E. Adoptive cell transfer using autologous tumor infiltrating lymphocytes in gynecologic malignancies. Gynecol Oncol 2018;150:361–9.
[44] Dudley ME, Rosenberg SA. Adoptive-cell-transfer therapy for the treatment of patients with cancer. Nat Rev Cancer 2003;3:666–75.
[45] Torabi-Rahvar M, Aghayan H-R, Ahmadbeigi N. Antigen-independent killer cells prepared for adoptive immunotherapy: one source, divergent protocols, diverse nomenclature. J Immunol Methods 2020;477:112690.
[46] Rosenberg SA, Restifo NP. Adoptive cell transfer as personalized immunotherapy for human cancer. Science (New York, NY) 2015;348:62–8.
[47] Auber ML, DeHaven JI, Raich PC, et al. IL-2/LAK cell treatment for advanced cancers with emphasis on a novel administration. W V Med J 1991;87:344–6.
[48] Hontscha C, Borck Y, Zhou H, Messmer D, Schmidt-Wolf IGH. Clinical trials on CIK cells: first report of the international registry on CIK cells (IRCC). J Cancer Res Clin Oncol 2011;137:305–10.
[49] Farhan S, Lee DA, Champlin RE, Ciurea SO. NK cell therapy: targeting disease relapse after hematopoietic stem cell transplantation. Immunotherapy 2012;4:305–13.
[50] Yang JC, Rosenberg SA. Adoptive T-cell therapy for cancer. Adv Immunol 2016;130:279–94.
[51] Yee C. Adoptive T cell therapy: points to consider. Curr Opin Immunol 2018;51:197–203.
[52] Yamamoto TN, Lee P-H, Vodnala SK, et al. T cells genetically engineered to overcome death signaling enhance adoptive cancer immunotherapy. J Clin Investig 2019;129:1551–65.
[53] Dhodapkar MV. Navigating the Fas lane to improved cellular therapy for cancer. J Clin Invest 2019;129:1522–3.
[54] Frost P, Ng CP, Belldegrun A, Bonavida B. Immunosensitization of prostate carcinoma cell lines for lymphocytes (CTL, TIL, LAK)-mediated apoptosis via the Fas-Fas-ligand pathway of cytotoxicity. Cell Immunol 1997;180:70–83.
[55] Law TM, Motzer RJ, Mazumdar M, et al. Phase III randomized trial of interleukin-2 with or without lymphokine-activated killer cells in the treatment of patients with advanced renal cell carcinoma. Cancer 1995;76:824–32.
[56] Rosenberg SA, Lotze MT, Muul LM, et al. Observations on the systemic administration of autologous lymphokine-activated killer cells and recombinant interleukin-2 to patients with metastatic cancer. N Engl J Med 1985;313:1485–92.
[57] Carlens S, Gilljam M, Chambers BJ, et al. A new method for in vitro expansion of cytotoxic human CD3-CD56+ natural killer cells. Hum Immunol 2001;62:1092–8.
[58] Olson JA, Leveson-Gower DB, Gill S, Baker J, Beilhack A, Negrin RS. NK cells mediate reduction of GVHD by inhibiting activated, alloreactive T cells while retaining GVT effects. Blood 2010;115:4293–301.
[59] Yamamoto T, Yoneda K, Ueta E, Doi S, Osaki T. Enhanced apoptosis of squamous cell carcinoma cells by interleukin-2-activated cytotoxic lymphocytes combined with radiation and anticancer drugs. Eur J Cancer (Oxford, England: 1990) 2000;36:2007–17.
[60] Backus HH, Dukers DF, van Groeningen CJ, et al. 5-Fluorouracil induced Fas upregulation associated with apoptosis in liver metastases of colorectal cancer patients. Ann Oncol 2001;12:209–16.
[61] Matsuzaki I, Suzuki H, Kitamura M, Minamiya Y, Kawai H, Ogawa J. Cisplatin induces fas expression in esophageal cancer cell lines and enhanced cytotoxicity in combination with LAK cells. Oncology 2000;59:336–43.
[62] Micheau O, Solary E, Hammann A, Martin F, Dimanche-Boitrel MT. Sensitization of cancer cells treated with cytotoxic drugs to fas-mediated cytotoxicity. J Natl Cancer Inst 1997;89:783–9.
[63] Debatin K-M. The role of CD95 system in chemotherapy. Drug Resist Updat 1999;2:85–90.
[64] Shenzhen Hornetcorn Bio-technology Company, LTD, Jingzhou Central Hospital. Adoptive cell therapy plus chemotherapy and radiation after surgery in treating patients with colorectal cancer, ClinicalTrials.gov; 2014. https://clinicaltrials.gov/ct2/show/NCT02202928. (Accessed 28 February 2020).
[65] Jinling Hospital. Preconditioning chemotherapy combination with cytokine induced killer cell (CIK) immunotherapy, ClinicalTrials.gov; 2013. https://clinicaltrials.gov/ct2/show/NCT01902875. (Accessed 28 February 2020).
[66] Sun Yat-sen University, Li J. GC regimen chemotherapy plus CIK cells for metastatic nasopharyngeal carcinoma, ClinicalTrials.gov; 2012. https://clinicaltrials.gov/ct2/show/NCT01655628. (Accessed 28 February 2020).
[67] Arai S, Standford Universit. Post-transplant autologous cytokine-induced killer (CIK) cells for treatment of high risk hematologic malignancies, ClinicalTrials.gov; 2007. https://clinicaltrials.gov/ct2/show/NCT00477035. (Accessed 28 February 2020).

[68] Fred Hutchinson Cancer Research Center. Cellular adoptive immunotherapy in treating patients with acute myeloid leukemia, acute lymphoblastic leukemia, or myelodysplastic syndromes that relapsed after donor stem cell transplant, ClinicalTrials.gov; 2005. https://clinicaltrials.gov/ct2/show/NCT00107354. (Accessed 28 February 2020).

[69] Zhang J, Liu L, He Y, Kong W, Huang S. Cytotoxic effect of trans-cinnamaldehyde on human leukemia K562 cells. Acta Pharmacol Sin 2010;31:861–6.

[70] Yu J, Ren X, Cao S, Zhang W, Hao X. Th1 polarization and apoptosis-inducing activity of CD4+ T-cells in cytokine-induced killers might favor the antitumor cytotoxicity of cytokine-induced killers in vivo. Cancer Biother Radiopharm 2006;21:276–84.

[71] Zhu J, Petit P-F, Van den Eynde BJ. Apoptosis of tumor-infiltrating T lymphocytes: a new immune checkpoint mechanism. Cancer Immunol Immunother 2019;68:835–47.

[72] Zhu Y, Huang B, Shi J. Fas ligand and lytic granule differentially control cytotoxic dynamics of natural killer cell against cancer target. Oncotarget 2016;7:47163–72.

[73] Akiyama K, Chen C, Wang D, et al. Mesenchymal stem cell-induced Immunoregulation involves Fas ligand/Fas-mediated T cell apoptosis. Cell Stem Cell 2012;10:544–55.

[74] Eskazan AE, Keskin D. Radotinib and its clinical potential in chronic-phase chronic myeloid leukemia patients: an update. Ther Adv Hematol 2017;8:237–43.

[75] Kim KE, Park S, Cheon S, et al. Novel application of radotinib for the treatment of solid tumors via natural killer cell activation. J Immunol Res 2018;2018:9580561.

[76] Il-Yang Pharm. Co., Ltd. A phase 3 study for the efficacy and safety of radotinib in CP-CML patients with failure or intolerance to previous TKIs, ClinicalTrials.gov; 2018. https://clinicaltrials.gov/ct2/show/NCT03459534. (Accessed 25 February 2020).

[77] Deciphera Pharmaceuticals LLC. A study of rebastinib (DCC-2036) in combination with carboplatin in patients with advanced or metastatic solid tumors, ClinicalTrials.gov; 2020. https://clinicaltrials.gov/ct2/show/NCT03717415. (Accessed 25 February 2020).

[78] Montefiore Medical Center, Deciphera Pharmaceuticals LLC, Albert Einstein College of Medicine, Jesus Anampa Mesias, Montefiore Medical Center. Rebastinib plus antitubulin therapy with paclitaxel or eribulin in metastatic breast cancer, ClinicalTrials.gov; 2017. https://clinicaltrials.gov/ct2/show/NCT02824575. (Accessed 25 February 2020).

[79] Alvarez M, Bouchlaka MN, Sckisel GD, Sungur CM, Chen M, Murphy WJ. Increased antitumor effects using IL-2 with anti-TGF-β reveals competition between mouse NK and CD8 T cells. J Immunol (Baltimore, Md: 1950) 2014;193:1709–16.

[80] Yang X, Yu X, Wei Y. Lentiviral delivery of novel fusion protein IL12/FasTI for cancer immune/gene therapy. PLoS ONE 2018;13:e0201100.

[81] Jiang W, Zhang C, Tian Z, Zhang J. hIFN-α gene modification augments human natural killer cell line anti-human hepatocellular carcinoma function. Gene Ther 2013;20:1062–9.

[82] Lai H, Zeng D, Liu C, Zhang Q, Wang X, Chen T. Selenium-containing ruthenium complex synergizes with natural killer cells to enhance immunotherapy against prostate cancer via activating TRAIL/FasL signaling. Biomaterials 2019;219:119377.

[83] Lee JH, Park S, Cheon S, et al. 1,25-Dihydroxyvitamin D_3 enhances NK susceptibility of human melanoma cells via Hsp60-mediated FAS expression. Eur J Immunol 2011;41:2937–46.

[84] Yang L-C, Lai C-Y, Hsieh C-C, Lin W-C. Natural killer cell-mediated anticancer effects of an arabinogalactan derived from rice hull in CT26 colon cancer-bearing mice. Int J Biol Macromol 2019;124:368–76.

[85] Yang G, Kong Q, Wang G, et al. Low-dose ionizing radiation induces direct activation of natural killer cells and provides a novel approach for adoptive cellular immunotherapy. Cancer Biother Radiopharm 2014;29:428–34.

[86] Modiano JF, Bellgrau D, Cutter GR, et al. Inflammation, apoptosis, and necrosis induced by neoadjuvant fas ligand gene therapy improves survival of dogs with spontaneous bone cancer. Mol Ther 2012;20:2234–43.

[87] Wu R, Forget M-A, Chacon J, et al. Adoptive T-cell therapy using autologous tumor-infiltrating lymphocytes for metastatic melanoma: current status and future outlook. Cancer J (Sudbury, Mass) 2012;18:160–75.

[88] Goff SL, Smith FO, Klapper JA, et al. Tumor infiltrating lymphocyte therapy for metastatic melanoma: analysis of tumors resected for TIL. J Immunother (Hagerstown, Md: 1997) 2010;33:840–7.

[89] Paschall AV, Yang D, Lu C, et al. H3K9 trimethylation silences Fas expression to confer colon carcinoma immune escape and 5-fluorouracil chemoresistance. J Immunol (Baltimore, Md: 1950) 2015;195:1868–82.

[90] Symes JC, Siatskas C, Fowler DH, Medin JA. Retrovirally transduced murine T lymphocytes expressing FasL mediate effective killing of prostate cancer cells. Cancer Gene Ther 2009;16:439–52.
[91] Cao K, Wang G, Li W, et al. Histone deacetylase inhibitors prevent activation-induced cell death and promote anti-tumor immunity. Oncogene 2015;34:5960–70.
[92] Blok EJ, van den Bulk J, Dekker-Ensink NG, et al. Combined evaluation of the FAS cell surface death receptor and CD8+ tumor infiltrating lymphocytes as a prognostic biomarker in breast cancer. Oncotarget 2017;8:15610–20.
[93] Yang D, Torres CM, Bardhan K, Zimmerman M, McGaha TL, Liu K. Decitabine and Vorinostat cooperate to sensitize colon carcinoma cells to Fas ligand-induced apoptosis in vitro and tumor suppression in vivo. J Immunol 2012;188:4441–9.
[94] Konkankit VV, Kim W, Koya RC, et al. Decitabine immunosensitizes human gliomas to NY-ESO-1 specific T lymphocyte targeting through the Fas/Fas ligand pathway. J Transl Med 2011;9:192.
[95] Chinese PLA General Hospital, Navy General Hospital, Beijing, Yu L. Decitabine followed by donor lymphocyte infusion for patients with relapsed acute myeloblastic leukemia (AML) after allogeneic stem cell transplantation, ClinicalTrials.gov; 2013. https://clinicaltrials.gov/ct2/show/NCT01758367. (Accessed 28 February 2020).
[96] Chinese PLA General Hospital, Navy General Hospital, Beijing, Yu L. Decitabine combining modified CAG followed by HLA haploidentical peripheral blood mononuclear cells infusion for elderly patients with acute myeloid leukemia (AML), ClinicalTrials.gov; 2012. https://clinicaltrials.gov/ct2/show/NCT01690507. (Accessed 28 February 2020).
[97] Roswell Park Cancer Institute, National Cancer Institute (NCI). Genetically modified T cells and decitabine in treating patients with recurrent or refractory ovarian, primary peritoneal, or fallopian tube cancer, ClinicalTrials.gov; 2019. https://clinicaltrials.gov/ct2/show/NCT03017131. (Accessed 28 February 2020).
[98] Weiss JM, Subleski JJ, Back T, et al. Regulatory T cells and myeloid-derived suppressor cells in the tumor microenvironment undergo Fas-dependent cell death during IL-2/αCD40 therapy. J Immunol (Baltimore, Md: 1950) 2014;192:5821–9.
[99] Kerkar SP, Leonardi AJ, van Panhuys N, et al. Collapse of the tumor stroma is triggered by IL-12 induction of Fas. Mol Ther 2013;21:1369–77.
[100] Chen W, Jiang J, Xia W, Huang J. Tumor-related exosomes contribute to tumor-promoting microenvironment: an immunological perspective. J Immunol Res 2017;2017:1073947.
[101] Zhu J, Powis de Tenbossche CG, Cané S, et al. Resistance to cancer immunotherapy mediated by apoptosis of tumor-infiltrating lymphocytes. Nat Commun 2017;8.
[102] Jansen T, Tyler B, Mankowski JL, et al. FasL gene knock-down therapy enhances the antiglioma immune response. Neuro Oncol 2010;12:482–9.
[103] Krammer PH, Arnold R, Lavrik IN. Life and death in peripheral T cells. Nat Rev Immunol 2007;7:532–42.
[104] Steiert AE, Sendler D, Burke WF, Choi CY, Reimers K, Vogt PM. Attack the tumor counterattack-c-FLIP expression in Jurkat-T-cells protects against apoptosis induced by coculture with SW620 colorectal adenocarcinoma cells. J Surg Res 2012;176:133–40.
[105] Frost PJ, Belldegrun A, Bonavida B. Sensitization of immunoresistant prostate carcinoma cell lines to Fas/Fas ligand-mediated killing by cytotoxic lymphocytes: independence of de novo protein synthesis. Prostate 1999;41:20–30.
[106] Frost PJ, Butterfield LH, Dissette VB, Economou JS, Bonavida B. Immunosensitization of melanoma tumor cells to non-MHC Fas-mediated killing by MART-1-specific CTL cultures. J Immunol (Baltimore, Md: 1950) 2001;166:3564–73.
[107] Siena L, Pace E, Ferraro M, et al. Gemcitabine sensitizes lung cancer cells to Fas/FasL system-mediated killing. Immunology 2014;141:242–55.
[108] Second Military Medical University. Immunotherapy using precision T cells specific to personalized neoantigen for the treatment of advanced malignant tumor of biliary tract, ClinicalTrials.gov; 2015. https://clinicaltrials.gov/ct2/show/NCT02632019. (Accessed 28 February 2020).
[109] Tessa Therapeutics. A phase III trial evaluating chemotherapy and immunotherapy for advanced nasopharyngeal carcinoma (NPC) patients, ClinicalTrials.gov; 2015. https://clinicaltrials.gov/ct2/show/NCT02578641. (Accessed 28 February 2020).
[110] M.D. Anderson Cancer Center. Allogeneic blood stem cell transplantation and adoptive immunotherapy for Hodgkin's disease, ClinicalTrials.gov; 2018. https://clinicaltrials.gov/ct2/show/NCT00385788. (Accessed 28 February 2020).

[111] Hinrichs CS, Borman ZA, Cassard L, et al. Adoptively transferred effector cells derived from naive rather than central memory CD8+ T cells mediate superior antitumor immunity. Proc Natl Acad Sci U S A 2009;106:17469–74.
[112] Klebanoff CA, Scott CD, Leonardi AJ, et al. Memory T cell-driven differentiation of naive cells impairs adoptive immunotherapy. J Clin Invest 2016;126:318–34.
[113] Zhao L, Cao YJ. Engineered T cell therapy for cancer in the clinic. Front Immunol 2019;10:2250.
[114] Benmebarek M-R, Karches CH, Cadilha BL, Lesch S, Endres S, Kobold S. Killing mechanisms of chimeric antigen receptor (CAR) T cells. Int J Mol Sci 2019;20:.
[115] D'Aloia MM, Caratelli S, Palumbo C, et al. T lymphocytes engineered to express a CD16-chimeric antigen receptor redirect T-cell immune responses against immunoglobulin G-opsonized target cells. Cytotherapy 2016;18:278–90.
[116] Kalyanasundaram Bhanumathy K, Zhang B, Xie Y, Xu A, Tan X, Xiang J. Potent immunotherapy against well-established thymoma using adoptively transferred transgene IL-6-engineered dendritic cell-stimulated CD8+ T-cells with prolonged survival and enhanced cytotoxicity. J Gene Med 2015;17:153–60.
[117] Faitschuk E, Hombach AA, Frenzel LP, Wendtner C-M, Abken H. Chimeric antigen receptor T cells targeting Fc μ receptor selectively eliminate CLL cells while sparing healthy B cells. Blood 2016;128:1711–22.
[118] University of Washington, National Cancer Institute (NCI), AstraZeneca. Acalabrutinib and anti-CD19 CAR T-cell therapy for the treatment of B-cell lymphoma, ClinicalTrials.gov; 2020. https://clinicaltrials.gov/ct2/show/NCT04257578. (Accessed 28 February 2020).
[119] Kite, A Gilead Company, Gilead Sciences. Safety and efficacy of KTE-C19 in adults with refractory aggressive Non-Hodgkin lymphoma, ClinicalTrials.gov; 2015. https://clinicaltrials.gov/ct2/show/NCT02348216. (Accessed 28 February 2020).
[120] Kite, A Gilead Company, Pfizer, Gilead Sciences. Safety and efficacy of axicabtagene ciloleucel in combination with utomilumab in adults with refractory large B-cell lymphoma, ClinicalTrials.gov; 2018. https://clinicaltrials.gov/ct2/show/NCT03704298. (Accessed 28 February 2020).
[121] Kite, A Gilead Company, Gilead Sciences. Efficacy and safety of axicabtagene ciloleucel as first-line therapy in participants with high-risk large B-cell lymphoma, ClinicalTrials.gov; 2020. https://clinicaltrials.gov/ct2/show/NCT03761056. (Accessed 28 February 2020).
[122] Kite, A Gilead Company, Gilead Sciences. A phase 2 multicenter study of axicabtagene ciloleucel in subjects with relapsed/refractory indolent Non-Hodgkin lymphoma, ClinicalTrials.gov; 2017. https://clinicaltrials.gov/ct2/show/NCT03105336. (Accessed 28 February 2020).
[123] Kite, A Gilead Company, Gilead Sciences. Efficacy of axicabtagene ciloleucel compared to standard of care therapy in subjects with relapsed/refractory diffuse large B cell lymphoma, ClinicalTrials.gov; 2018. https://clinicaltrials.gov/ct2/show/NCT03391466. (Accessed 28 February 2020).
[124] Maus MV, Kite, A Gilead Company, Massachusetts General Hospital. Anakinra in Car-T cell mediated neurotoxicity, ClinicalTrials.gov; 2019. https://clinicaltrials.gov/ct2/show/NCT04150913. (Accessed 28 February 2020).
[125] Kite, A Gilead Company, Genentech, Inc., Gilead Sciences. Safety and efficacy of KTE-C19 in combination with atezolizumab in adults with refractory diffuse large B-cell lymphoma (DLBCL) (ZUMA-6), ClinicalTrials.gov; 2016. https://clinicaltrials.gov/ct2/show/NCT02926833. (Accessed 28 February 2020).
[126] Jonsson Comprehensive Cancer Center, National Cancer Institute (NCI). Anakinra in preventing severe chimeric antigen receptor T-cell related encephalopathy syndrome in patients with recurrent or refractory large B-cell lymphoma, ClinicalTrials.gov; 2019. https://clinicaltrials.gov/ct2/show/NCT04205838. (Accessed 28 February 2020).
[127] Gilead Sciences. Safety and efficacy of axicabtagene ciloleucel in combination with either rituximab or lenalidomide in participants with refractory large B-cell lymphoma (ZUMA-14), ClinicalTrials.gov; 2019. https://clinicaltrials.gov/ct2/show/NCT04002401. (Accessed 2 March 2020).
[128] University of Pennsylvania, University of California. Pilot study of autologous T-cells in patients with metastatic pancreatic cancer, ClinicalTrials.gov; 2015. https://clinicaltrials.gov/ct2/show/NCT02465983. (Accessed 2 March 2020).
[129] Novartis Pharmaceuticals. Study of tisagenlecleucel in combination with pembrolizumab in r/r diffuse large B-cell lymphoma patients, ClinicalTrials.gov; 2018. https://clinicaltrials.gov/ct2/show/NCT03630159. (Accessed 2 March 2020).

[130] Abramson Cancer Center of the University of Pennsylvania. Phase I/II study of pembrolizumab in patients failing to respond to or relapsing after anti-CD19 chimeric antigen receptor modified T cell therapy for relapsed or refractory CD19+ lymphomas, ClinicalTrials.gov; 2016. https://clinicaltrials.gov/ct2/show/NCT02650999. (Accessed 2 March 2020).
[131] Novartis Pharmaceuticals. Study of tisagenlecleucel in combination with ibrutinib in r/r diffuse large B-cell lymphoma patients, ClinicalTrials.gov; 2019. https://clinicaltrials.gov/ct2/show/NCT03876028. (Accessed 2 March 2020).
[132] Peter MacCallum Cancer Centre, Australia, Novartis. Clinical trial to assess the efficacy and safety of the combination of tisagenlecleucel and ibrutinib in mantle cell lymphoma, ClinicalTrials.gov; 2020. https://clinicaltrials.gov/ct2/show/NCT04234061. (Accessed 2 March 2020).
[133] Jonsson Comprehensive Cancer Center, National Cancer Institute, Parker Institute for Cancer Immunotherapy. Modified immune cells (CD19/CD20 CAR-T Cells) in treating patients with recurrent or refractory B-cell lymphoma or chronic lymphocytic leukemia, ClinicalTrials.gov; 2019. https://clinicaltrials.gov/ct2/show/NCT04007029. (Accessed 2 March 2020).
[134] M.D. Anderson Cancer Center, National Cancer Institute, Ziopharm Oncology. CD19-specific t-cells in treating patients with advanced lymphoid malignancies, ClinicalTrials.gov; 2015. https://clinicaltrials.gov/ct2/show/NCT02529813. (Accessed 2 March 2020).
[135] Henan Cancer Hospital, The Beijing Pregene Science and Technology Company, Ltd., Yongping Song, Henan Cancer Hospital. The safety and efficacy of CART-19 cells in b-cell acute lymphoblastic leukemia (B-ALL), ClinicalTrials.gov; 2016. https://clinicaltrials.gov/ct2/show/NCT02924753. (Accessed 2 March 2020).
[136] Henan Cancer Hospital, The Beijing Pregene Science and Technology Company, Ltd. The safety and efficacy of CART-19 cells in relapse and refractory patients with CD19+ B-cell lymphoma, ClinicalTrials.gov; 2017. https://clinicaltrials.gov/ct2/show/NCT03101709. (Accessed 2 March 2020).
[137] Children's Hospital of Fudan University, Xiaowen Zhai, Children's Hospital of Fudan University. Study of alloCART-19 cell therapy in pediatric patients with relapsed/refractory B-cell acute lymphoblastic leukemia, ClinicalTrials.gov; 2019. https://clinicaltrials.gov/ct2/show/NCT04173988. (Accessed 2 March 2020).
[138] Masonic Cancer Center, University of Minnesota. MT2017-45: CAR-T cell therapy for heme malignancies, ClinicalTrials.gov; 2018. https://clinicaltrials.gov/ct2/show/NCT03642626. (Accessed 2 March 2020).
[139] Novartis Pharmaceuticals. Tisagenlecleucel vs blinatumomab or inotuzumab for patients with relapsed/refractory B-cell precursor acute lymphoblastic leukemia, ClinicalTrials.gov; 2018. https://clinicaltrials.gov/ct2/show/NCT03628053. (Accessed 2 March 2020).
[140] World Health Organization. Cancer, https://www.who.int/news-room/fact-sheets/detail/cancer; 2018. (Accessed 25 February 2020).
[141] Rennes University Hospital. Impact of Fas/FasL in chemotherapy response in epithelial ovarian carcinoma, ClinicalTrials.gov; 2014. https://clinicaltrials.gov/ct2/show/record/NCT02297958. (Accessed 26 February 2020).
[142] Fundación Salud de los Andes, Universidad Nacional de Colombia, Instituto Colombiano para el Desarrollo de la Ciencia y la Tecnología. Immunogenicity and safety of DCs in breast cancer, ClinicalTrials.gov; 2018. https://clinicaltrials.gov/ct2/show/NCT03450044. (Accessed 2 March 2020).
[143] Changhai Hospital, ImmunoGene Biotechology Co., Ltd, Li Z. Safety and efficacy evaluation of iAPA-DC/CTL combined gemcitabine therapy on advanced pancreatic cancer, ClinicalTrials.gov; 2015. https://clinicaltrials.gov/ct2/show/NCT02529579. (Accessed 2 March 2020).
[144] Pitt JM, André F, Amigorena S, et al. Dendritic cell–derived exosomes for cancer therapy. J Clin Invest 2016;126:1224–32.
[145] Escudier B, Dorval T, Chaput N, et al. Vaccination of metastatic melanoma patients with autologous dendritic cell (DC) derived-exosomes: results of the first phase I clinical trial. J Transl Med 2005;3:10.
[146] Wolfers J, Lozier A, Raposo G, et al. Tumor-derived exosomes are a source of shared tumor rejection antigens for CTL cross-priming. Nat Med 2001;7:297–303.
[147] Dai S, Wei D, Wu Z, et al. Phase I clinical trial of autologous ascites-derived exosomes combined with GM-CSF for colorectal cancer. Mol Ther 2008;16:782–90.
[148] Coombes R. Cancer drug resistance needs urgent attention, says research chief. BMJ 2019;365:l1934. https://doi.org/10.1136/bmj.l1934.

[149] Cohen JT, Chambers JD, Silver MC, Lin P-J, Neumann PJ. Putting the costs and benefits of new gene therapies into perspective, Health Affairs; 2019. https://www.healthaffairs.org/do/10.1377/hblog20190827.553404/full/. (Accessed 26 February 2020).

[150] Silbert S, Yanik GA, Shuman AG. How should we determine the value of CAR T-cell therapy? AMA J Ethics 2019;21:844–51.

[151] Cisplatin. Cost of cisplatin, https://www.cisplatin.org/cost.htm. (Accessed 26 February 2020).

CHAPTER

5

Harnessing metabolism for reinvigorating dysfunctional T cells in cancer

Susana Romero-Garcia, Daniela Alejandra Castro-Flores, and Heriberto Prado-Garcia

Departamento de Enfermedades Cronico-Degenerativas, Instituto Nacional de Enfermedades Respiratorias, "Ismael Cosio Villegas", Ciudad de México, México

Abstract

T cells are essential in the immune response against tumor cells; in addition, immunotherapies such as immune checkpoint blockade (ICB), adoptive T cell therapy, and CAR (chimeric antigen receptor) T cells rely on these cells to trigger the elimination of tumor cells. In order to exert their effector functions, T cells depend on metabolism to obtain energy in the form of ATP, as well as metabolites that are used for macromolecular biosynthesis. However, tumor cells reprogram their metabolism to survive and proliferate, which has deleterious consequences on T cell metabolism. In this chapter, we review some metabolic evasion mechanisms that tumor cells may develop to inhibit anti-tumoral T cell activities. Furthermore, we discuss some potential strategies that may restore T cell effector function through enhancing T cell metabolism.

Abbreviations

Acetyl-CoA	acetyl coenzyme A
AMPK	AMP-activated protein kinase
CAFs	cancer-associated fibroblasts
CAR	chimeric antigen receptor
CPT1	carnitine palmitoyltransferase I
ER	endoplasmic reticulum
ERRα	estrogen-related receptor-alpha
ETC	electron transport chain
$FADH_2$	flavin adenine dinucleotide
FA	fatty acid
FAO	fatty acid oxidation

Immunotherapy in Resistant Cancer:
From the Lab Bench Work to Its Clinical Perspectives
https://doi.org/10.1016/B978-0-12-822028-3.00005-4

HIF-1α	hypoxia-inducible factor 1 alpha
IFN	interferon
IL	interleukin
ICB	immune checkpoint blockade
IRF4	interferon regulatory factor 4
HKII	hexokinase II
LDH	lactate dehydrogenase
MCJ	methylation-controlled J protein
MCT	monocarboxylate transporter
MNC	mononuclear cells
NADH	nicotinamide adenine dinucleotide
OXPHOS	oxidative phosphorylation
PDK1	pyruvate dehydrogenase kinase 1
PGC-1α	peroxisome proliferator-activated receptor gamma coactivator 1-alpha
PEP	phosphoenolpyruvate
PKM2	pyruvate kinase 2
PIP3	phosphatidylinositol-1,4,5-trisphosphate
PI3K	phosphatidylinositol 3-kinase
PIM	proviral integration site for Moloney murine leukemia virus kinase
Raptor	regulatory-associated protein of mTOR
Rheb	ras homolog enriched in brain
S1P	sphingosine 1 phosphate
TCA	tricarboxylic acid cycle
TCR	T cell receptor
TEM	effector memory T cells
TCM	central memory cells
TILs	tumor-infiltrating lymphocytes
TGF-β	transforming growth factor beta

Conflict of interest

No potential conflicts of interest were disclosed by the authors.

Introduction

T cells have the capacity to eliminate tumors by recognizing and responding to tumor antigens. In order to do so, T cells require to undergo a process of activation, proliferation, and differentiation to effector cells. These processes require energy in the form of ATP, building blocks to synthesize proteins, DNA, RNA, lipids, as well as, reducing power in the form of NADPH and NADH. Conversely, tumor cells themselves reprogram their metabolism in order to survive, proliferate, invade surrounding tissue, and metastasize. Tumor metabolic reprogramming changes the microenvironment, thereby, varying nutrient availability, extracellular pH, and oxygen tension levels. Thus, T cells that arrive to the tumor site to recognize and eliminate malignant cells face an adverse environment, where tumor cells have seized and modified the environment. Importantly, lack of nutrients, changes in the pH and O_2 tension level, among others, are factors that might induce dysfunction in T cells.

In this chapter we discuss some of the metabolic evasion mechanisms that tumor cells exert on T cells that induce dysfunction and the inability to respond and eliminate tumor cells. We discuss some strategies for intervening metabolism, which are directed towards enhancing T cell metabolism.

Basics of metabolism

Immune cells require to activate several metabolic pathways to obtain adequate levels of energy in the form of ATP (catabolic metabolism), as well as to synthesize macromolecules from a plethora of intermediates (anabolic metabolism). Both anabolism and catabolism support survival, activation, proliferation, and differentiation. Metabolic pathways are intercrossed, products from one pathway can be synthetic precursors for other metabolic pathways [1]. Some of the most well-known metabolic pathways are briefly described afterward.

Glycolysis. Glycolysis is the metabolic pathway where the cell takes in glucose and processes it to produce two moles of pyruvate, NADH, and ATP per mole of consumed glucose. Glycolytic reactions take place in the cytosol. Cells under low O_2 tensions mainly use glycolysis (anaerobic glycolysis) to obtain energy; remarkably, cells that require to proliferate under normoxia also employ glycolysis, which is known as aerobic glycolysis or Warburg effect. Glycolytic metabolism is crucial to provide substrates for other metabolic pathways, such as TCA (tricarboxylic acid cycle), the pentose phosphate pathway, and fatty acid oxidation (FAO). Glycolysis can provide substrates for other metabolic pathways within the cell [1]. Glucose is incorporated into the cell by means of glucose transporters (Glut); Glut1 is the preponderant glucose transporter in T cells [2]. Under aerobic or anaerobic glycolysis, pyruvate is converted to lactic acid by the enzymes lactate dehydrogenases (LDHs), whereas monocarboxylate transporter 1 (MCT1) and MCT4, respectively, direct the influx and efflux of lactate from the cytosol [3].

The TCA cycle. The TCA cycle (Krebs cycle) is the metabolic pathway that in conjunction with oxidative phosphorylation produces the highest levels of ATP per mol of glucose. Most quiescent and terminally differentiated cells use the TCA cycle; thus, it is a metabolic pathway that is employed for obtaining energy and maintaining longevity. Pyruvate produced from glycolysis or fatty acids oxidation is imported into the mitochondria and converted into acetyl coenzyme A (acetyl-CoA) by pyruvate dehydrogenase (PDH). PDH can be regulated by phosphorylation by pyruvate dehydrogenase kinase 1 (PDK1); this phosphorylation inhibits PDH function, and as a result, pyruvate is redirected into lactate synthesis in the cytosol. This is a mechanism that can regulate acetyl-CoA entry to TCA cycle and favor aerobic glycolysis [4].

Acetyl-CoA is incorporated into the TCA cycle via aldol condensation with oxalacetate to form citrate. NADH and FADH are generated during the TCA cycle; this reducing power can be used by the electron transport chain to support oxidative phosphorylation. In addition, other intermediates from the TCA cycle can be used as substrates for different metabolic pathways. All these reactions of the TCA cycle take place in the mitochondrion.

Oxidative phosphorylation (OXPHOS). This is the process by which ATP is generated by means of the energy obtained from the electron transfer in a stepwise manner, from the reduced NADH and flavin adenine dinucleotide ($FADH_2$) to its final acceptor O_2. OXPHOS, through the electron transport chain, forms a gradient of proton and then pH across the mitochondrial membrane. This gradient activates the ATP synthase, which in turn produces ATP [5].

Fatty acid oxidation (FAO). Also known as β-oxidation, this metabolic pathway transforms fatty acids into diverse metabolites that cells can use for obtaining energy. Among the products that are formed by β-oxidation are NADH, $FADH_2$, and acetyl-CoA. FAO begins in the cytosol, where acetyl-CoA binds covalently to the fatty acid (FA), via an ATP-dependent enzymatic reaction, which generates fatty acid acyl-CoA. Long-chain and medium-chain fatty acids (>6 carbons in the aliphatic tail) are conjugated to carnitine. This reaction is mediated by the carnitine palmitoyltransferase I (CPT1). Then, the FA conjugate is transported into the mitochondria, where carnitine is removed by CPT2. Short-chain fatty acids diffuse passively into the mitochondria, once inside, β-oxidation can take place [1].

The pentose phosphate pathway. This pathway provides pentoses and reducing power in the form of NADPH, which supports survival and proliferation. There are two phases in the pentose phosphate pathway that occur in the cytosol. The first is the oxidative phase, where NADPH is produced, which can be later used for the synthesis of fatty acids and for maintaining an adequate redox status. The second phase is non-oxidative and synthesizes pentoses (5-carbon sugars), which are used for the synthesis of amino acid precursors or nucleotides [1,6].

Fatty acid synthesis. As its names imply, this pathway generates lipids that are necessary for cell growth, organelle synthesis, as well as for the synthesis of metabolites used in other metabolic pathways. Several metabolites from glycolysis, the pentose phosphate pathway, and the TCA cycle are used for the fatty acid synthesis. Some enzymes that are essential for fatty acid synthesis are fatty acid synthase (FASN) and ACC (acetyl-CoA carboxylase), which are regulated by SREB (sterol regulatory element-binding protein). Most reactions for the synthesis of fatty acids occur in the cytosol [1].

Amino acid metabolic pathways. Amino acids are essential not just as building blocks for protein synthesis, as they additionally serve as substrates for different metabolic pathways involved in proliferation and cell growth. Amino acids can be used in the synthesis of fatty acids, as well as other biosynthetic processes, such as the synthesis de novo of purines and pyrimidines, where glutamine and aspartate are needed. Glutamine can be incorporated into the TCA cycle for providing ATP in proliferating cells, such as actively proliferating T cells, which are highly dependent on this amino acid [1]; glutaminolysis also provides substrates for the synthesis of amino acids, lipids, and nucleic acids [7].

The metabolism of T cells during activation

During the course of its life, a T cell might have different metabolic requirements, which depend on the cell itself (state of activation/differentiation) as well as microenvironmental conditions.

T cells that have not been activated by its cognate antigen (naive T cells) are quiescent; thus, their metabolism is directed to use predominantly OXPHOS to generate ATP, as quiescent cell metabolic requirements are low.

When T cells are activated by signals from the TCR, costimulatory receptors, and cytokines, several transduction signals are activated that lead to the activation of metabolic pathways. Activation of metabolism is one of the first steps that a T cell must follow after TCR engagement, which is crucial for a full development of effector functions. Activation of T cells induces aerobic glycolysis within minutes; this early event is independent of CD28 signaling; nevertheless, for a sustained glycolysis, CD28 ligation is needed. TCR-induced glycolysis is initiated by activation of ZAP-70 and Lck kinases, which in turn phosphorylates pyruvate dehydrogenase kinase 1 (PDK1); this event is independent of glucose flux into the cell. PDK1 activity inhibits the entry of pyruvate to TCA cycle and thus favors aerobic glycolysis (see Fig. 2). This process has been observed to be essential for the production of cytokines (IFN-γ, IL-2, and TNF-α), albeit other effector functions, such as proliferation and cytolytic function, are not affected [4].

Glycolysis is maintained by the interaction with costimulatory molecules, being the best characterized the CD28 molecule. CD28 interaction with its ligands (CD80/CD86) activates phosphatidiylinositol 3-kinase (PI3K) and the serine-threonine kinase Akt (see Figs. 1A and 2). The latter in turn favors the localization of Glut1 on the cell membrane, thus increasing glucose uptake [8]. Akt additionally increases the expression of the glycolytic enzymes hexokinase and phosphofructokinase and is essential for the activation of the mammalian target of rapamycin (mTOR). There are two complexes that form mTOR, mTORC1, and mTORC2, and the activation of these complexes is crucial to satisfy the needs of the T cell for activation, proliferation, differentiation, and long-term effector functions. mTOR activation induces the expression of the transcriptional regulators c-Myc, HIF-1α (hypoxia-inducible factor 1 alpha), estrogen-related receptor-alpha (ERRα), as well as IRF4 (interferon regulatory factor 4). HIF-1α is a transcriptional factor that is expressed under hypoxia and regulates glycolysis in this condition; nevertheless, it can be induced by mTORC1 in a non-oxygen-dependent pathway to modulate aerobic glycolysis [9]. c-Myc has been long known to regulate cell cycle entrance; it also regulates glycolysis and glutaminolysis. c-Myc induces the expression of some of the enzymes and transporters involved in these pathways in an independent way from HIF-1α [10]. ERRα, which is associated with the proliferation of breast carcinomas, also modulates glycolysis and induces mitochondrial biogenesis via peroxisome proliferator-activated receptor gamma coactivator 1-alpha (PGC-1α) in CD4+ T cells [11]. IRF4 is a transcription factor whose activation depends on the strength of TCR ligation. IRF4 supports several processes in CD8+ T cells, such as survival, proliferation, and differentiation [12] (see Fig. 1A). This transcription factor also increases the expression the receptors associated with T-cell exhaustion [13]. IRF4 mediates its function by increasing molecules required for the aerobic glycolysis, such as HIF-1α, glucose transporters, hexokinase 2, and aldolases [14].

On the contrary, the AMP-activated protein kinase (AMPK) is a regulator that senses changes in the AMP/ATP ratio. By inhibiting mTOR activity, AMPK dampens anabolism, inhibiting lipid and protein biosynthesis, as well as cell proliferation. AMPK increases energy production during metabolic stress [15]. Two ways of inhibiting mTOR by AMPK are through the activation via phosphorylation of the GTP-binding protein Rheb (Ras homolog enriched in brain), which inhibits mTOR activity; in the other way, AMPK phosphorylates the regulatory protein Raptor (regulatory-associated protein of mTOR) on S722/792 sites leading to mTOR inhibition [7]. Under low glucose levels, AMPK promotes oxidative metabolism and allows the T cells to respond under this condition. Thus, AMPK is equally important for long-term responses in vivo [16] (see Fig. 1B).

Following T cell activation, energy and biosynthetic precursors are needed to support clonal expansion and effector functions; thus, T cells undergo metabolic reprogramming to glycolysis and inhibition of OXPHOS for energy production. As pyruvate does not enter

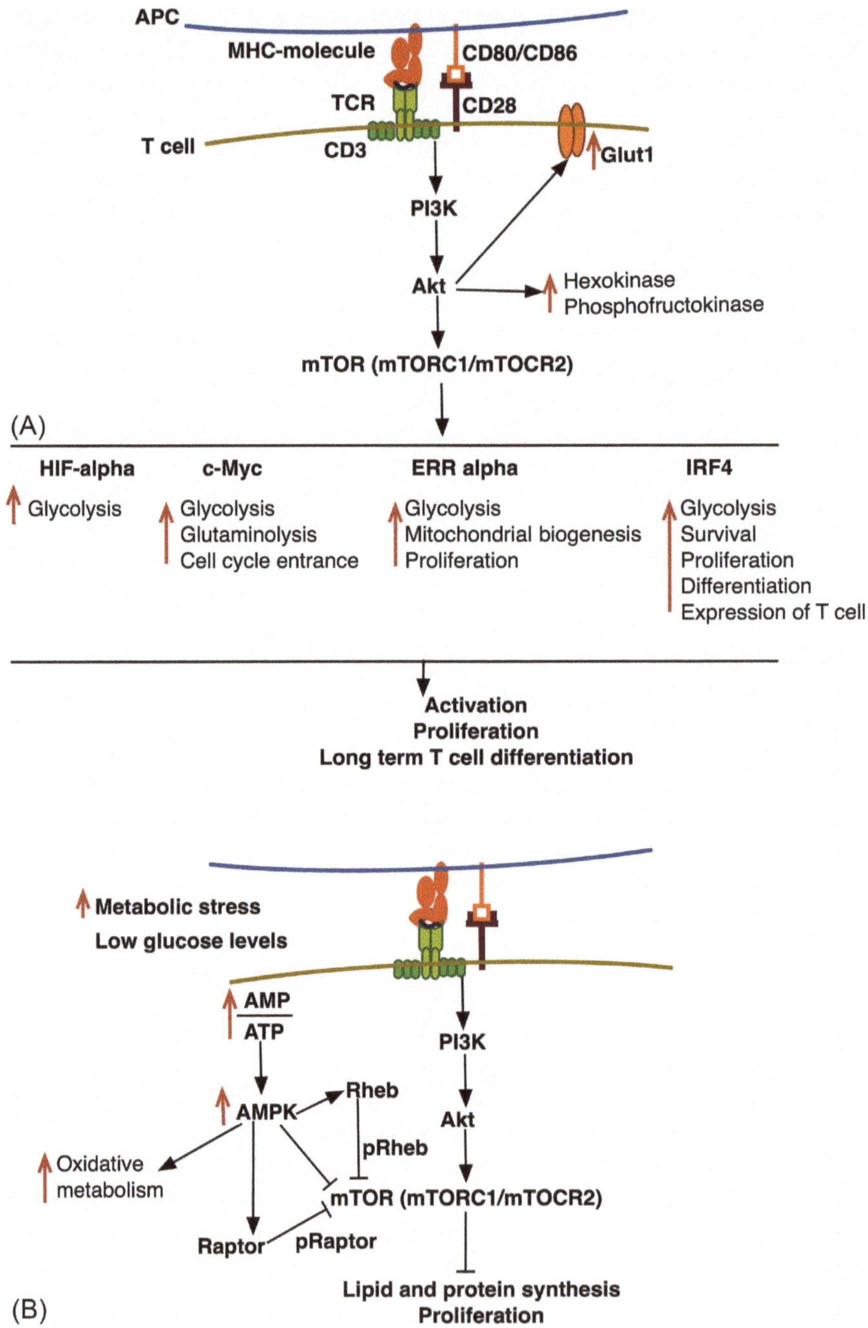

FIG. 1 (A) Metabolic pathways associated with T cell activation that support energy production and biosynthetic precursors, in order that T cells can exit the quiescent state, proliferate, and differentiate. (B) Metabolic stress, such as low glucose levels, inhibits T cell activation via AMPK, which consequently inhibits mTOR.

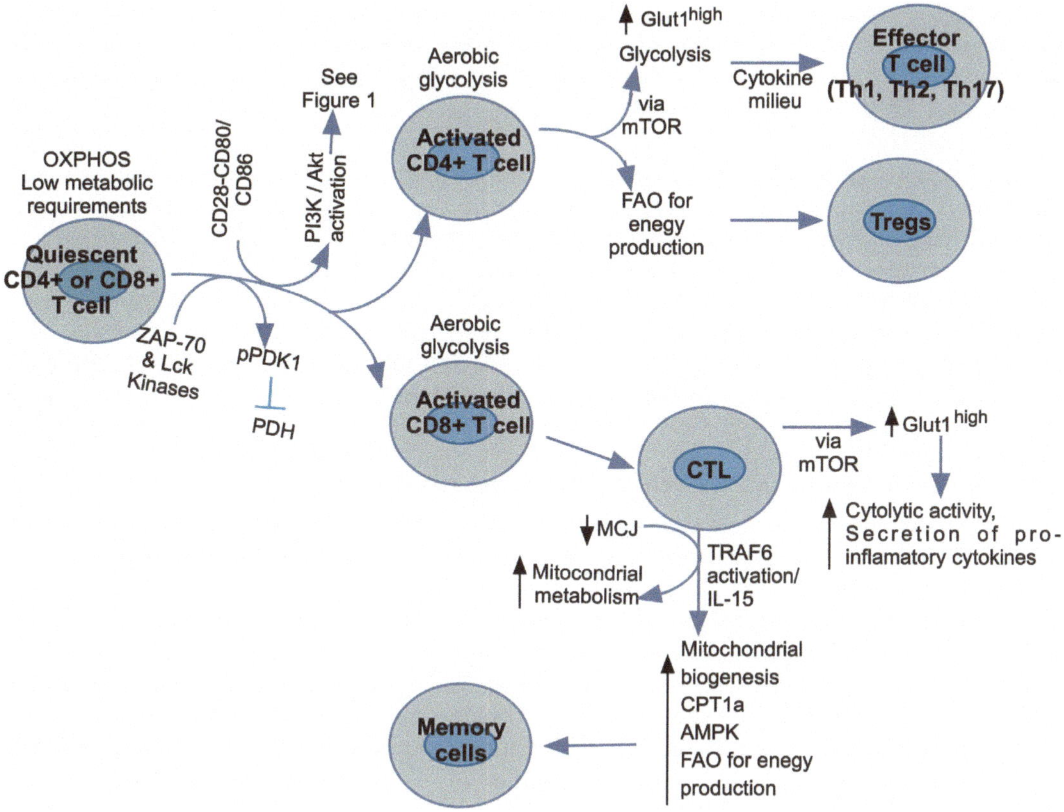

FIG. 2 Metabolic profile of T cells during their different maturation stages.

the TCA cycle, there is an increased amount of lactate that is secreted [8]; however, high levels of lactate in the microenvironment might have deleterious effects on T cells [17].

Once activated, and depending on the cytokine milieu and costimulatory signals, CD4+ T cells differentiate into effector T cells (e.g., Th1, Th2, Th17) or induced regulatory T cells (iTregs). CD8+ T cells differentiate to effector T cells (cytotoxic T lymphocytes, CTLs), which have cytolytic activity, as well as the potential to secrete proinflammatory cytokines (see Fig. 2). Different effector functions of T cells have been shown to be associated with the activation of mTOR (Almeida 2016), because, among other molecules, mTOR regulates the expression of Glut1. Both CD4+ and CD8+ T cells expressing high surface levels of Glut1 ($Glut1^{high}$) show enhanced proliferation and produce higher amounts of cytokines (e.g., IFN-γ) with respect to their $Glut1^{low}$ counterpart [2]. Metabolic requirements vary from each specialized subset, whereas Th1, Th2, and Th17 subsets present a highly glycolytic metabolism and express high levels of Glut1 upon activation, regulatory T cells (both natural occurring and iTregs) depend on FAO for obtaining energy (see Fig. 2). Likewise, FAs inhibit cytokine production by Th1, Th2, and Th17 cells, whereas Tregs retain their suppressive capacity [18].

Effector CD8+ T cells depend on glycolysis to upregulate effector cytokines such as IFN-γ; accordingly, under glucose deprivation this cytokine is downregulated in mouse models.

Other effector molecules (TNF-α, perforin, and granzymes B and C), but not IL-2, are also downregulated by glucose deprivation [19]. Of note, in humans glucose deprivation does not affect IFN-γ production by CD8+ T cells, albeit there is a reduction in cell viability and proliferation [20]. Even so, many effector functions of CD8+ T cells depend on glucose availability [21].

T cells equally require amino acids to sustain their metabolic necessities. L-Arginine is essential for T cell responses, as lack of this amino acid has been shown to downregulate CD3ζ chain, which consequently affects several pathways in T cell activation and effector functions [22]. Another essential amino acid is glutamine, because, like tumor cells, T cells are highly dependent on this amino acid [23,24]. T cells import glutamine via ERK activation in order to proliferate and produce cytokines (IL-2 and IFN-γ) (see Fig. 2). Costimulation via CD28 upregulates glutamine transporters, as well as the required enzymes to incorporate glutamine into the TCA cycle [25]. Glutaminolysis is important to replenish the TCA cycle, which makes mitochondria essential for T cell activation and effector functions.

The role of mitochondria in T cell metabolic reprogramming

Mitochondria are crucial in T cell activation, proliferation, and even differentiation, not just because they produce ATP, but rather for other metabolic processes. Besides cell respiration, mitochondria are involved in glutaminolysis, the metabolism of serine, β-oxidation, Fe^{2+} metabolism, as well as for Ca^{2+}-dependent activation [26–29]. Furthermore, mitochondrial production of ROS is needed to support T cell signaling [30]. Mitochondria are located in proximity during the formation of the immunological synapse, which highlights their role at the initiation of T cell activation [31]. Nonetheless, mitochondrial content is different in T cell subpopulations, CD4+ T cells have a higher mitochondrial content than CD8+ T cells. Accordingly, CD4+ T cells show a higher oxidative metabolism upon activation, whereas CD8+ T cells show a more glycolytic metabolism that promotes a higher proliferation rate in these cells compared with CD4+ T cells [32].

Besides providing energy and substrates, mitochondria participate in different processes of T cell activation and differentiation. TCR-engagement and costimulation result in an increase of mitochondrial mass and membrane potential and synthesis of mitochondrial DNA to promote mitochondrial biogenesis (see Fig. 1) [33,34]. The stimulation of CD4+ T cells via CD3/CD28 promotes activation of the TCA cycle after 4-h post-stimulation, while mitochondrial mass is greatly increased after 24 h. T cell activation induces mitochondrial enzymes involved in folate-mediated one-carbon metabolism, which is essential for T cell survival. This pathway synthesizes one-carbon compounds required to synthesize purine, thymidylate, and NADPH [25]. To support T cell activation, mitochondria remodel their cristae. This process has been shown to be primordial in CD8+ T cells for developing protective immunity against tumors in vivo and it requires of CD28 costimulation [35]. Activation of mitochondrial biogenesis is dependent on mTORC1 and is crucial for T cells to exit quiescence phase and enter in G1-phase of the cell cycle. T cell activation and subsequent proliferation depend on TCA cycle and mitochondrial biogenesis through overlapping pathways that are distinct from glycolysis [34]. This has been shown using a model in CD4+ T cells where deletion of transcription factor A mitochondrial (TFAM), an essential molecule for inducing mitochondrial biogenesis, reduces the proliferative capacity compared with wild type CD4+ T cells [36]. Also, defects in COX10 (a molecule required in synthesis of ETC complex IV) reduces T cell proliferation and promotes apoptosis [34]. Thus, T cells need functional mitochondrial so that they can be activated, proliferate, and differentiate to effector cells [37].

Memory CD8+ T cells mainly use FAO as a source of energy; these cells increase both glycolysis and OXPHOS following activation, as opposed to naive T cells [26]. Memory T cells present a higher mitochondrial content, increased spare respiratory capacity, and higher levels of proteins involved in OXPHOS [38]. CD8+ T cell memory differentiation depends on the activation of TNF receptor-associated factor 6 (TRAF6), which in turn activates AMPK and thus promotes FAO [26]. In addition, mitochondrial biogenesis and expression of CPT1a are promoted by IL-15, a cytokine that is critical for promoting memory T cell differentiation [39]. MCJ (methylation-controlled J protein) acts as a negative regulator of ETC, whose loss causes sustained OXPHOS and increased cytokine secretion (IL-2 and IFN-γ). MCJ reduces mitochondrial metabolism during the transition from effector to rested effector T cells and subsequent differentiation to memory T cells (see Fig. 2). Thus, mitochondrial metabolism is essential to define the reprogramming of CD8+ T cells to memory or effector subsets [40].

Metabolic alterations observed in cancer-associated T cells

Glucose depletion and lactate accumulation

Tumor cells consume large amounts of glucose to sustain their metabolism; as a result, glucose levels within the tumor microenvironment are reduced [41]. The depletion of glucose has consequences on the effector function of T cells. Early studies in vitro demonstrated that altering glucose consumption with a glucose analogue decreases T cell proliferation and reduces the expression of IFN-γ and TNF-α, while IL-2 production is relatively unaffected [21]. In a sarcoma mouse model, Chang et al. proved that tumor cells compete for glucose with T cells, which impairs IFN-γ production and glycolytic metabolism in CD8+ T cells [42]. Accordingly, high levels of Glut1 and LDH5 expressed on renal cell carcinoma cells are associated with low infiltrations of CD8+ and granzyme-B+ T cells, as evidenced by immunohistochemistry [43]. This study proposed that the induction of the glycolytic metabolism in renal carcinoma cells negatively impacts the anti-tumoral response. Another evidence that reveals how tumoral glycolytic metabolism inhibits T cell anti-tumoral response comes from a study in hepatocellular carcinoma. The extracellular matrix metalloproteinase inducer (CD147), a chaperone of MCT1 and MCT4, enhances glucose metabolism via the PI3K/Akt/mTOR pathway in hepatocellular carcinoma cells, which is associated with a reduction in the infiltration of CD8+ T cells. CD147 has been reported to be upregulated in several types of tumors [44].

Some studies have shown that CD8+ T cells infiltrating tumors are defective on glycolysis. CD8+ TILs infiltrating human renal cell carcinoma have defects in both glucose uptake and glycolysis in vitro, even though these cells exhibit normal levels of Glut1 compared to CD8+ T cells from healthy donors [45]. In lung cancer patients, memory CD8+ T cells infiltrating the pleural compartment (pleural effusion CD8+ T cells) fail to increase the expression of Glut1 and consequently do not consume glucose after stimulation ex vivo via CD3/CD28 under hypoxia. This phenomenon is not observed in memory CD8+ T cells from peripheral blood, indicating that memory T cells are rendered dysfunctional by the tumor microenvironment and are unable to respond in situ against tumor cells under hypoxic environments. Of note, CD8+ T cells maintain this reduced glycolytic activity even in the presence of media with supra-physiological concentrations of glucose [46]. These studies suggest that there is an intrinsic alteration on glycolysis that impairs CD8+ T cells function.

Using the B16cOVA melanoma mouse model, Gemta et al. showed that CD8+ TILs have low glycolytic activity. Contrary to pleural effusion CD8+ T cells or CD8+ TILs from renal cell carcinoma, CD8+ TILs from melanoma do not have reduced the expression of Glut1 nor inefficient glucose consumption. Instead, these cells are defective in the enzymatic activity of enolase 1; consequently, low levels of phosphoenolpyruvate are produced, thus inhibiting aerobic glycolysis [47]. Taken together, these studies indicate that tumor cells might induce glycolysis dysfunction by different mechanisms and at different levels.

CD4+ T cells are also affected by glucose deprivation. Ascitic fluid from ovarian cancer patients induces ER stress and suppresses the expression of Glut1 and glucose uptake in CD4+ T cells via activation of IRE1α-XBP1. Of note, IRE1α-XBP1 also induces defects in mitochondrial respiration and limits glutamine influx [48]. Using a melanoma mouse model, Ho et al. showed that tumor cells reduce glucose availability of intratumoral Th1 cells, while augmenting the production of TGF-β by T cells. Glucose-deprivation reduces the levels of phosphoenolpyruvate, which increases the activity of calcium transporter sarco/ER Ca2 + -ATPase (SERCA) and consequently impairs TCR-induced Ca2+ flux [49]. The increase in tumor glycolysis has been demonstrated to promote resistance to adoptive T cell therapy (ATC) in melanoma patients [50]; thus, the glycolytic metabolism of tumor cells is a factor to consider the design of T cell-based therapies.

Tumors not only consume glucose in detriment of effector T cell functions, but they also secrete lactate into the microenvironment, which also have deleterious effects on T cells. Lactate secretion via lactate-proton symport is one of the mechanisms that lower pH (<6.9), thus creating acidic conditions within the microenvironment. Acidic pH reduces the cytotoxicity of CD8+ T cells by decreasing IFN-γ and granzyme B levels in vitro [51]. Lactate serum levels positively correlate with tumor burden. Lactate inhibits proliferation and cytokine production (IFN-γ and IL-2) on CD8+ T cells in vitro. Lactate reduces the levels of intracellular perforin and granzyme B, which represses the cytolytic activity of CD8+ T cells [17]. These observations have been corroborated using a mouse melanoma model. Brand et al. showed that tumors expressing high levels of LDHA exhibit low numbers of CD8+ T cells that secrete low levels of cytokines (IFN-γ and IL-2) and sensitize them to apoptosis [52]. Of note, tumor cells are not the only source of lactate as cancer-associated fibroblasts (CAFs) also secrete high quantities of this metabolite. In a model in vitro of prostate cancer, lactate secreted by CAFs reduces the number of Th1 cells, concomitantly increasing the proportion of Tregs identified as FOXP3+CD25+CD4+ T cells [53].

Effect of immunosuppressive cytokines and metabolites on T cell metabolism

Immunosuppressive cytokines have been long known to inhibit effector functions of the anti-tumoral immune response; nonetheless, until recently a direct effect on T cell metabolism has begun to be described. TGF-β has been long known as an immunosuppressive cytokine that contributes to tumor survival by inhibiting T cell responses. This cytokine impairs the respiratory capacity of effector/memory CD4+ T cells [54], which show a more oxidative metabolism compared with CD8+ T cells. Dimeloe et al. demonstrated that TGF-β inhibits ATP synthase activity and reduces IFN-γ production by effector/memory CD4+ T cells; nonetheless, no effects were observed in the secretion of TNF-α, IL-17, and IL-4, albeit there was a slight reduction in IL-10 secretion [54]. Thus, TGF-β also contributes to inhibit CD4+ T cell responses by inhibiting mitochondrial metabolism.

Even before the emergence of the field of immunometabolism some enzymes and products of the metabolism of tumor cells or immune cells had been previously identified as suppressors of the T cell response. Tumor cells, as well as M2-like macrophages and myeloid-derived suppressor cells, increase the expression of IDO (indoleamine-2,3-dioxygenase) and Arginase 1 (Arg1) [55–57]. These enzymes break down tryptophan and arginine, respectively, consequently reducing the levels of these amino acids, thus inhibiting T cell function. In addition, the degradation of tryptophan by IDO produces kynurenine, which is a metabolite that suppresses T cell function by downregulating CD3ζ chain and sensitizes to T cell apoptosis [58].

Metabolic T cell exhaustion in cancer

T cell exhaustion is a stage of differentiation characterized by reduced effector functions, as well as sustained and constitutive expression of co-inhibitory receptors. Exhausted T cells have been shown to present a transcriptional state that is different from functional effector or memory T cells [13]. It has been suggested that tumors promote T cell exhaustion because tumor cells chronically stimulate T cells under an immunosuppressive microenvironment. Many studies have identified the mechanisms involved in the induction of exhaustion; among them, the alterations in metabolism and mitochondrial function appear to play a central role. PD-1 signaling, together with persisting activation of mTOR, induces metabolic dysfunction in T cells by reducing glucose uptake, suppressing respiration, and dysregulating mitochondrial function [59].

In a transplantable melanoma mouse model, Zhang et al. showed that CD8+ TILs increase the expression of PD-1 and LAG-3 (two markers of exhaustion), as well as reduce the expression of IFN-γ, granzyme B, and perforin. Interestingly, TILs appear to be exhausted regardless of antigen specificity. LAG-3, but not PD-1, increased expression is induced by hypoxia via HIF-1α upregulation in CD8+ TILs. Thus, hypoxia might have a role in T cell exhaustion. However, in this study, the co-expression of these markers was not demonstrated. Zhang et al. showed in this study that PD-1 expression was not necessarily linked to T cell exhaustion. Thus, it is likely that hypoxia and low levels of glucose send opposing metabolic signals, which lead to T cell dysfunction [60].

Siska et al. reported that B cell leukemia increases the expression of co-inhibitory receptors PD-1 and TIM-3 in T cells, which was associated with a reduced expression of Glut1, HKII, and glucose uptake. The authors found that the dysfunction in glycolysis was caused by defects in Akt/mTORC1 signaling. Although the authors attribute this metabolic dysfunction to T cell exhaustion, other markers associated with this state were not increased. Thus, metabolic dysfunction probably was caused by a dysregulated signaling induced by the expression of PD-L1 on lymphoma cells [61].

We previously indicated that pleural effusion CD8+ T cells show reduced Glut1 expression and diminished glucose consumption under hypoxia [46]. To evaluate glucose uptake in PD-1+ CD8+ T cells, we studied CD8+ T cells infiltrating the pleural compartment from lung cancer patients, who were admitted to the National Institute of Respiratory Diseases "Ismael Cosio Villegas." Pleural effusion was obtained by thoracocentesis used for routine diagnostic procedures. None of the patients received any kind of therapy against cancer, nor they had clinical signs of infection prior to taking pleural effusion samples. As memory cells are predominantly infiltrating the pleural compartment [62], we focused on this subset (see Fig. 3A and B). Under hypoxia (2% O_2) and after 24 h of stimulation with anti-CD3 antibody, T cells increased the expression of PD-1, as well as glucose uptake (2NBDG+); nonetheless, the percentage of 2NBDG+/PD-1+ cells were higher after costimulation via CD28 in memory (CD45RA–CD27+) CD8+ T cells from healthy donors (see Fig. 3C). Of note, PD-1 has been shown to be expressed by activated T cells from healthy subjects [63]. A similar behavior was observed in memory CD8+ T cells from peripheral blood obtained from cancer patients. Memory CD8+ T cells infiltrating the pleural compartment from non-malignant diseases (tuberculosis and pneumonia), which were included as an additional comparison group, also showed a high percentage of 2NBDG+/PD-1+ cells (see Fig. 3B and C). Thus, there was a higher proportion memory CD8+ T cells that consumed glucose after stimulation via CD3/CD28 compared with CD3 stimulation because CD28 activates Akt [8], which increases glycolysis, while expression of PD-1 was linked to T cell activation. Conversely, in memory CD8+ T cells infiltrating the pleural compartment from lung cancer patients, the frequencies of 2NBDG+/PD-1+ cells were lower after CD3/CD28 stimulation compared with CD3 stimulation alone, whereas the percentages of PD-1+ memory CD8+ T cells that did not consume glucose (2NBDG−/PD-1+cells) tended to be higher after CD3/CD28 stimulation (see Fig. 3C). Although up to 80% of cells expressed PD-1 after stimulation, it is unlikely that altered glucose consumption is caused by exhaustion because CD8+ T cells do not co-express the classic markers of exhaustion TIM-3 and LAG-3 [64]. Deregulated PD-1 expression might be responsible for the lower glucose consumption observed in memory CD8+ T cells infiltrating malignant pleural effusions that were stimulated under hypoxia.

Mitochondrial defects in T cells induced by the tumor microenvironment

In a melanoma mouse model, Scharping et al. showed that TILs, in particular CD8+ T cells, lose mitochondrial functions. This phenomenon was characterized by reduced mitochondrial mass, decreased ability to take up glucose, and defective oxygen consumption. CD8+ T cells from adenocarcinoma and Lewis lung carcinoma exhibited similar alterations [65]. This loss of mitochondrial function is particular of tumor-specific T cells and correlates with the expression of inhibitory receptors. However, PD-1 blockade does not rescue T cells from mitochondrial dysfunction. Of note, the downregulation of transcriptional coactivator PGC-1α represses mitochondrial biogenesis and is responsible for mitochondrial defects in TILs [65]. Also, short-term hypoxia (16 h) and glucose depletion have effects in mitochondria. In the transplantable melanoma mouse model of Zhang et al., CD8+ T cells downregulate transcripts involved in TCA cycle (IDH3a and MDH2), ROS detoxification (SOD1), and electron

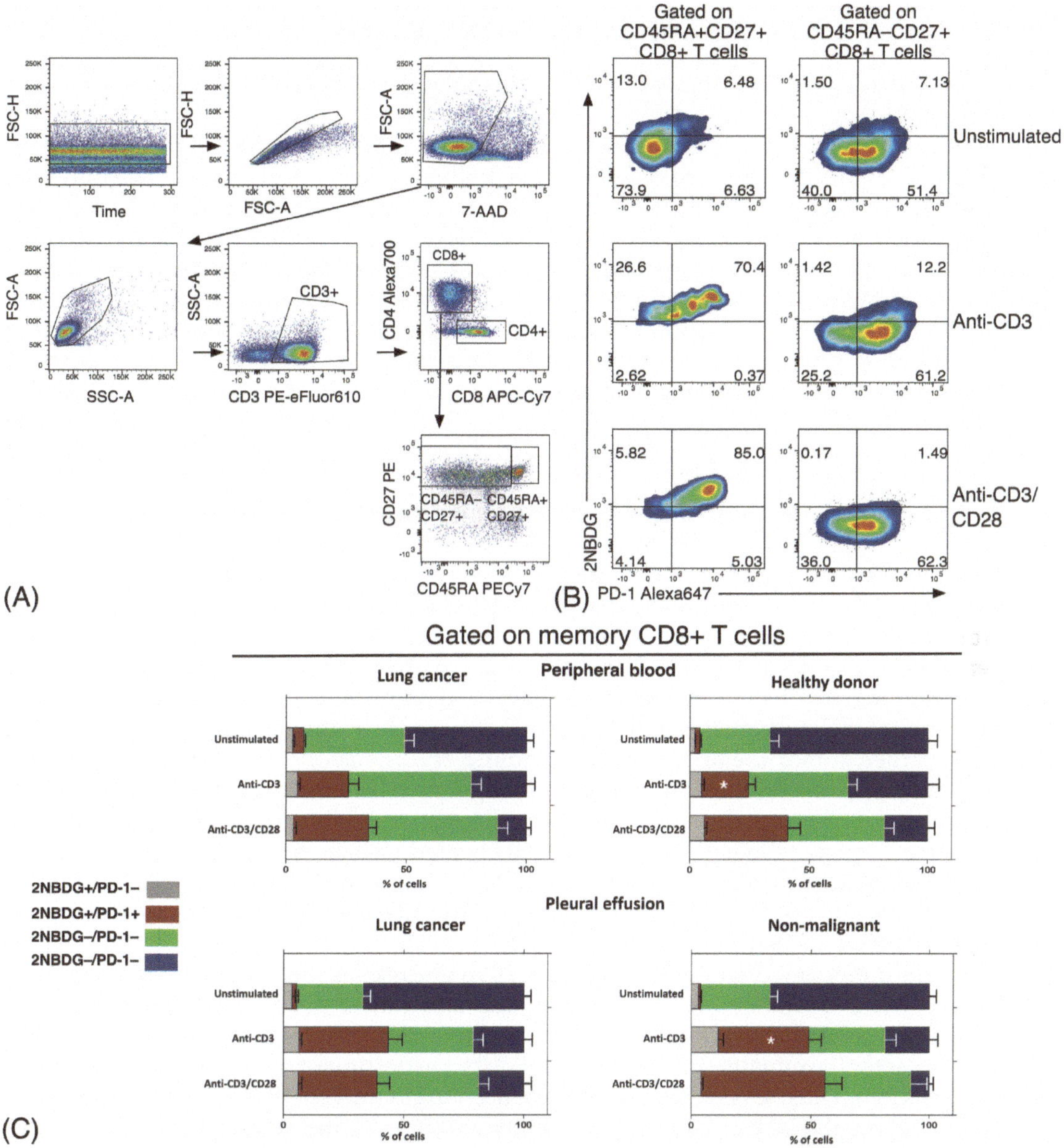

FIG. 3 CD8+ T cells infiltrating pleural effusion from lung cancer patients do not consume glucose after CD3/CD28 stimulation under hypoxia. MNC were stimulated with anti-CD3 or anti-CD3/CD28 antibodies under hypoxia for 48, and then, 2NBDG uptake assay and immunophenotype were done as reported in [46]. (A) Representative flow cytometric strategy. FSC-H vs time and FSC-A vs FSC-H plots were done to exclude clogs and select singlets respectively. Then, viable cells (7-aminoactinomycin-negative cells) were gated and FSC vs SSC dot-plot graphs were done to select lymphocytes. From a gate of CD3+ vs SSC-A, a CD4+ vs CD8+ graph plot was done and CD8+ cells were gated. Finally, the analysis of PD-1 expression and 2NBDG uptake was performed on CD45RA+CD27+ (naive) and CD45RA–CD27+ (memory cells) that were gated from CD3+CD8+ T cells. (B) Representative analysis of PD-1 and 2NBDG (glucose uptake) in memory and naive CD8+ T cells infiltrating pleural compartment from a lung cancer patient. Quadrants were set according to fluorescence minus one control. Naive CD8+ T cells were 2NBDG+/PD-1+ in a high proportion of the population after anti-CD3 stimulation, which was further increased when cells were stimulated with anti-CD3/CD28. Conversely, memory CD8+ T cells did not increase glucose uptake even when they were stimulated with both stimuli (anti-CD3 and anti-CD28). (C) The analysis of 2NBDG and PD-1 on CD45RA–CD27+ (memory cells) CD8+ T cells were under hypoxia for 48 h. Pleural effusion and peripheral blood from lung cancer patients ($n = 14$), patients with non-malignant diseases ($n = 9$), and peripheral blood from healthy donors (HD, $n = 11$) were evaluated. Bars depict mean ± standard error. *$P < .05$ respect to anti-CD3/CD28 stimulation.

transport chain (COX5b), but increase the expression of transcripts of molecules involved in FA uptake [60].

However, alterations in mitochondria might be induced through different mechanisms. In a renal cell carcinoma mouse model, CD8+ TILs maintain their mitochondrial mass, and the PD-1 pathway appears to be involved in repressing the production of mitochondrial ROS and downregulation of PGC-1α [66]. On the contrary, Siska et al. reported that CD8+ TILs also maintain their mitochondrial mass and the expression of PGC-1α appears to be normal. Nonetheless, mitochondria are smaller and rounder compared to mitochondria from control CD8+ T cells, which suggest that mitochondria are fragmented. CD8+ TILs also show mitochondrial hyperpolarization and produce high levels of mitochondrial ROS. The authors suggest that high mitochondrial membrane polarization may identify anti-tumoral T cells albeit with limited activity [45].

Targeting metabolism for restoring anti-tumoral T cell responses

It took decades of research, but now immunotherapy is considered a feasible strategy for treating cancer patients. Immunotherapy includes the use of blocking agents against co-inhibitory molecules expressed on T cells (known as immune checkpoint blockade, ICB), transfer of naturally occurring TILs, or the use of genetically engineered T cells (chimeric antigen receptor, CAR T cells). However, the employ of immunotherapy has had some drawbacks because of the mechanisms that tumor cells present to evade the immune response. As discussed earlier, tumor metabolic reprogramming and metabolic T cell dysfunction have negative repercussions on the elimination of malignant cells. In consequence, strategies for intervening metabolism are directed towards two branches: inhibiting tumor cell metabolism or enhancing T cell metabolism. The first is well covered elsewhere [67]. With respect to the latter, one challenge is that T cells and tumor cells share similarities in their cellular and mitochondrial metabolism. Thus, the strategy of targeting alterations that are apparently tumor specific might have deleterious effects on T cells.

T cells and tumor cells compete for nutrients like glucose, as indicated earlier. Moreover, T cell glycolysis is dysfunctional in tumor models and cancer patients; however, restoring glycolysis is problematic, because induction of high glycolytic activity in CD8 + T cells may favor effector (terminally differentiated cells), but in detriment of formation of long-lived memory cells [68]. Thus, several proposals have tried to circumvent these circumstances. One strategy to bypass the lack of glucose availability and boost anti-tumoral TILs activity is by increasing intracellular levels of phosphoenolpyruvate. Ho et al. overexpressed phosphoenolpyruvate carboxykinase (PCK1) in CD4+ T cells; this enzyme converts oxalacetate into phosphoenolpyruvate. CD4+ T cells overexpressing PCK1 secrete higher quantities of IFN-γ and express higher levels of CD40L, which leads to maturation of tumor-associated macrophages and suppresses melanoma growth [49]. Another strategy is the use of acetate as an alternative carbon source. This is because acetate is required for optimal function of memory CD8+ T cells. Qiu et al. showed that, under glucose-restricted conditions, acetate promotes chromatin accessibility and enhances IFN-γ transcription in T cells. Furthermore, acetyl-CoA synthetase expression in CD8+ T cells favors tumor clearance in a lymphoma mouse model [69].

Inhibition of glucose metabolism by targeting glycogen synthase-3β (GSK-3β) has been applied in CAR T cells, as GSK-3β favors glycolytic metabolism. Sabatino et al. reported a procedure to generate clinic-grade CAR T cells directed towards CD19 with a stem cell memory phenotype. The authors activated naive precursors in the presence of IL-7, IL-21, and a GSK-3β pharmacological inhibitor. These CAR T cells show an increased ability to produce cytokines and degranulate; also they downregulate glycolytic enzymes as well as Glut1 and increase their spare respiratory capacity [70].

Another approach is direct inhibition of mTOR and consequently reducing glycolysis in T cells. Chatterjee et al. showed that by inhibiting proviral integration site for Moloney murine leukemia virus (PIM) kinases, which promote mTORC1 activity, T cells lower glycolysis as evidenced by reduced glucose uptake, as well as glycolysis-associated genes. This strategy has the advantage of an increased differentiation towards central memory T cell phenotype accompanied with a decreased expression in CD38 and PD-1 molecules. Importantly, adoptive transfer of T cells in combination with anti-PD-1 and pan-PIM inhibitor increases survival of mice implanted with B16-F10 melanoma [71].

Metabolic reprogramming of CD8+ T cells, so that they increase FA catabolism through FAO, has been proposed as a mean to increase the activity of CD8+ T cells that are under glucose-restricted conditions [60,72,73]. In this regard, sphingosine 1 phosphate (S1P) is a bioactive lipid molecule that, besides favoring growth, proliferation as well as survival of tumors, inhibits T cell effector functions and favors Treg differentiation. Inhibition of the enzyme that synthesizes S1P (sphingosine kinase 1, SphK1) increases anti-tumoral T cell activity against B16-F10 melanoma. Genetic and pharmacological inhibition of Sphk1 enhances T cell migration, IFN-γ, and IL-17 secretion, and memory T cell formation by increasing OXPHOS and lipolysis [73]. Thus, pharmacological inhibition of S1P synthesis is a feasible way of inhibiting tumor growth and concomitantly boosting anti-tumoral responses.

The anti-diabetic drug metformin has been shown to promote the differentiation to memory T cells, thus increasing anti-tumoral activity [26,74]. One of the mechanisms of action of metformin is through AMPK activation, which restores mitochondrial FAO. Metformin has been shown to promote tumor rejection via participation of CD8+ T cells in tumor-bearing mice. Metformin effect is through AMPK activation and protects CD8+ T cells from apoptosis, favoring differentiation to memory T cells and their migration into tumors [75]. Treatment with metformin increases the frequencies of memory stem and central memory CD8+ T cells in lung cancer patients compared with those who do not take the drug. Metformin also increases the expression of transcription factor Eomesodermin (Eomes), which is involved in the generation and maintenance of memory T cells. Interestingly, metformin also downregulates the expression of PD-1 and TIM-3, while promoting the expression of the activation marker CD69 [74]. Thus, co-administration of metformin in patients receiving immunotherapy might enhance the anti-tumoral response.

Currently, ICB has great relevance in the treatment of several types of cancers. To enhance its effectivity, several combinatorial schemes have been proposed, among them is the use of molecules that modulate the metabolism. CTLA-4 (cytotoxic T lymphocyte-associated antigen-4) is a co-inhibitory molecule inducible after T cell activation and is involved in the induction and maintenance of tolerance. The blockade of this molecule has been shown to increase T cell responses and control tumor growth [72,76]. Pedicord et al. showed that combining anti-CTLA-4 with rapamycin, an inhibitor of mTOR improves CD8+ T cell recall responses and increases the frequencies of cytokine-producing cells. Interestingly, combination but not the treatments alone increased mitochondrial biogenesis and spare capacity in CD8+ T cells. The combination of anti-CTLA-4 with rapamycin reduced tumor mass and increased survival in mice with EL4 lymphoma [72].

In a colon carcinoma mouse model, PD-1 blockage increases in tumor-responding CD8+ T cells mitochondrial mass, mitochondrial membrane potential, mitochondrial superoxide, and ROS production. In addition, the combination of drugs known to activate PGC-1a (oltipraz and bezafibrate) with anti-PD-L1 therapy has a synergistic anti-tumoral activity [66]. Thus, in some tumors, mitochondrial activation of T cells might synergize with the blockage of PD-1/PD-L1 pathway to enhance anti-tumoral responses [65,66].

Metformin has also been shown to target PD-L1 molecule. Cha et al. showed that in breast cancer cells, treatment with metformin decreases the expression of PD-L1 on the cytoplasmic membrane. Metformin activates AMPK that phosphorylates serine 195 of PD-L1, which induces an abnormal glycosylation and blocks translocation to Golgi, thus promoting degradation of PD-L1. The reduction of PD-L1 expression caused by metformin reduces tumor growth and increases the infiltration of CD8+ T cells that synthesize granzyme B in mice with 4 T1 breast cancer. These findings were corroborated in breast cancer patients treated with metformin, as patients that responded to metformin treatment, showed AMPK phosphorylation, and exhibited decreased levels of PD-L1 in tumor tissues. In addition, the combination of metformin with anti-CTLA-4 reduces tumor growth in a breast cancer mouse model [77].

Although the use of the inhibitory checkpoints is one of the strategies that has recalled the attention of oncologists, another approach involves costimulation of T cells. For instance, CD137 (4-1BB) is a costimulatory receptor expressed on activated T cells, and the stimulation via this molecule has been shown to increase the anti-tumoral immunity when combined with other therapies. In a melanoma mouse model, Menk et al. showed that costimulation by CD137 increases mitochondrial mass, by increasing mitochondrial fusion and biogenesis. This mitochondrial "priming" increases the anti-tumoral activity of adoptive T cell therapy [78]. CD137 has also been shown to enhance survival and increased formation of memory T cells, when signaling domains for this molecule are included in the CAR architecture of CD19-specific CAR T cells. CD137 domains enhance respiratory capacity, mitochondrial biogenesis, and FAO. Therefore, it is feasible to engineer CAR T cells so that they can reprogram their metabolism [79].

Conclusion

Cancer therapy has evolved since the incorporation of immune-based treatments. These therapies have achieved a great success in terms of tumor reduction and overall survival of cancer patients, in particular the use of ICB. However, as many tumors are still refractory to this kind of therapy and there is the risk of the emergence of tumor resistance, metabolic interventions might provide an attractive back up to immunotherapy. Hence, understanding both T cell and tumor cell metabolisms is mandatory to develop rational strategies that inhibit tumor cells and boost T cell anti-tumoral functions.

References

[1] O'Neill LA, Kishton RJ, Rathmell J. A guide to immunometabolism for immunologists. Nat Rev Immunol 2016;16:553–65.

[2] Cretenet G, Clerc I, Matias M, Loisel S, Craveiro M, Oburoglu L, Kinet S, Mongellaz C, Dardalhon V, Taylor N. Cell surface Glut1 levels distinguish human CD4 and CD8 T lymphocyte subsets with distinct effector functions. Sci Rep 2016;6:24129.

[3] Romero-Garcia S, Moreno-Altamirano MM, Prado-Garcia H, Sanchez-Garcia FJ. Lactate contribution to the tumor microenvironment: mechanisms, effects on immune cells and therapeutic relevance. Front Immunol 2016;7:52.

[4] Menk AV, Scharping NE, Moreci RS, Zeng X, Guy C, Salvatore S, Bae H, Xie J, Young HA, Wendell SG, Delgoffe GM. Early TCR Signaling induces rapid aerobic glycolysis enabling distinct acute T cell effector functions. Cell Rep 2018;22:1509–21.

[5] Beckermann KE, Dudzinski SO, Rathmell JC. Dysfunctional T cell metabolism in the tumor microenvironment. Cytokine Growth Factor Rev 2017;35:7–14.

[6] Cole L, Kramer PR. Chapter 1.3: Sugars, fatty acids, and energy biochemistry. In: Cole L, Kramer PR, editors. Human physiology, biochemistry and basic medicine. Boston: Academic Press; 2016. p. 17–30.

[7] Chang CH, Pearce EL. Emerging concepts of T cell metabolism as a target of immunotherapy. Nat Immunol 2016;17:364–8.

[8] Jacobs SR, Herman CE, Maciver NJ, Wofford JA, Wieman HL, Hammen JJ, Rathmell JC. Glucose uptake is limiting in T cell activation and requires CD28-mediated Akt-dependent and independent pathways. J Immunol 2008;180:4476–86.

[9] MacIver NJ, Michalek RD, Rathmell JC. Metabolic regulation of T lymphocytes. Annu Rev Immunol 2013;31:259–83.

[10] Wang R, Dillon CP, Shi LZ, Milasta S, Carter R, Finkelstein D, McCormick LL, Fitzgerald P, Chi H, Munger J, Green DR. The transcription factor Myc controls metabolic reprogramming upon T lymphocyte activation. Immunity 2011;35:871–82.

[11] Michalek RD, Gerriets VA, Nichols AG, Inoue M, Kazmin D, Chang CY, Dwyer MA, Nelson ER, Pollizzi KN, Ilkayeva O, Giguere V, Zuercher WJ, Powell JD, Shinohara ML, McDonnell DP, Rathmell JC. Estrogen-related receptor-alpha is a metabolic regulator of effector T-cell activation and differentiation. Proc Natl Acad Sci U S A 2011;108:18348–53.

[12] Yao S, Buzo BF, Pham D, Jiang L, Taparowsky EJ, Kaplan MH, Sun J. Interferon regulatory factor 4 sustains CD8 (+) T cell expansion and effector differentiation. Immunity 2013;39:833–45.

[13] Prado-Garcia H, Romero-Garcia S. The role of exhaustion in tumor-induced T-cell dysfunction in cancer. In: - Rezaei N, editor. Cancer immunology: a translational medicine context. Cham: Springer International Publishing; 2020. p. 117–32.

[14] Man K, Miasari M, Shi W, Xin A, Henstridge DC, Preston S, Pellegrini M, Belz GT, Smyth GK, Febbraio MA, Nutt SL, Kallies A. The transcription factor IRF4 is essential for TCR affinity-mediated metabolic programming and clonal expansion of T cells. Nat Immunol 2013;14:1155–65.

[15] Faubert B, Boily G, Izreig S, Griss T, Samborska B, Dong Z, Dupuy F, Chambers C, Fuerth BJ, Viollet B, Mamer OA, Avizonis D, DeBerardinis RJ, Siegel PM, Jones RG. AMPK is a negative regulator of the Warburg effect and suppresses tumor growth in vivo. Cell Metab 2013;17:113–24.

[16] Blagih J, Coulombe F, Vincent Emma E, Dupuy F, Galicia-Vázquez G, Yurchenko E, Raissi Thomas C, van der Windt Gerritje JW, Viollet B, Pearce Erika L, Pelletier J, Piccirillo Ciriaco A, Krawczyk Connie M, Divangahi M, Jones RG. The energy sensor AMPK regulates T cell metabolic adaptation and effector responses in vivo. Immunity 2015;42:41–54.

[17] Fischer K, Hoffmann P, Voelkl S, Meidenbauer N, Ammer J, Edinger M, Gottfried E, Schwarz S, Rothe G, Hoves S, Renner K, Timischl B, Mackensen A, Kunz-Schughart L, Andreesen R, Krause SW, Kreutz M. Inhibitory effect of tumor cell-derived lactic acid on human T cells. Blood 2007;109:3812–9.

[18] Michalek RD, Gerriets VA, Jacobs SR, Macintyre AN, MacIver NJ, Mason EF, Sullivan SA, Nichols AG, Rathmell JC. Cutting edge: distinct glycolytic and lipid oxidative metabolic programs are essential for effector and regulatory CD4+ T cell subsets. J Immunol 2011;186:3299–303.

[19] Cham CM, Gajewski TF. Glucose availability regulates IFN-gamma production and p70S6 kinase activation in CD8+ effector T cells. J Immunol 2005;174:4670–7.

[20] Renner K, Geiselhoringer AL, Fante M, Bruss C, Farber S, Schonhammer G, Peter K, Singer K, Andreesen R, Hoffmann P, Oefner P, Herr W, Kreutz M. Metabolic plasticity of human T cells: preserved cytokine production under glucose deprivation or mitochondrial restriction, but 2-deoxy-glucose affects effector functions. Eur J Immunol 2015;45:2504–16.

[21] Cham CM, Driessens G, O'Keefe JP, Gajewski TF. Glucose deprivation inhibits multiple key gene expression events and effector functions in CD8+ T cells. Eur J Immunol 2008;38:2438–50.

[22] Choi BS, Martinez-Falero IC, Corset C, Munder M, Modolell M, Müller I, Kropf P. Differential impact of L-arginine deprivation on the activation and effector functions of T cells and macrophages. J Leukoc Biol 2008;85:268–77.

[23] Carpenter L, Halestrap AP. The kinetics, substrate and inhibitor specificity of the lactate transporter of Ehrlich-Lettre tumour cells studied with the intracellular pH indicator BCECF. Biochem J 1994;304(Pt 3):751–60.

[24] Yaqoob P, Calder PC. Glutamine requirement of proliferating T lymphocytes. Nutrition 1997;13:646–51.

[25] Carr EL, Kelman A, Wu GS, Gopaul R, Senkevitch E, Aghvanyan A, Turay AM, Frauwirth KA. Glutamine uptake and metabolism are coordinately regulated by ERK/MAPK during T lymphocyte activation. J Immunol 2010;185:1037–44.

[26] Pearce EL, Walsh MC, Cejas PJ, Harms GM, Shen H, Wang LS, Jones RG, Choi Y. Enhancing CD8 T-cell memory by modulating fatty acid metabolism. Nature 2009;460:103–7.

[27] Ron-Harel N, Santos D, Ghergurovich JM, Sage PT, Reddy A, Lovitch SB, Dephoure N, Satterstrom FK, Sheffer M, Spinelli JB, Gygi S, Rabinowitz JD, Sharpe AH, Haigis MC. Mitochondrial biogenesis and proteome remodeling promote one-carbon metabolism for T cell activation. Cell Metab 2016;24:104–17.

[28] Ma EH, Bantug G, Griss T, Condotta S, Johnson RM, Samborska B, Mainolfi N, Suri V, Guak H, Balmer ML, Verway MJ, Raissi TC, Tsui H, Boukhaled G, Henriques da Costa S, Frezza C, Krawczyk CM, Friedman A, Manfredi M, Richer MJ, Hess C, Jones RG. Serine is an essential metabolite for effector t cell expansion. Cell Metab 2017;25:345–57.

[29] Mehta MM, Weinberg SE, Chandel NS. Mitochondrial control of immunity: beyond ATP. Nat Rev Immunol 2017;17:608–20.

[30] Sena Laura A, Li S, Jairaman A, Prakriya M, Ezponda T, Hildeman David A, Wang C-R, Schumacker Paul T, Licht Jonathan D, Perlman H, Bryce Paul J, Chandel NS. Mitochondria are required for antigen-specific T cell activation through reactive oxygen species Signaling. Immunity 2013;38:225–36.

[31] Quintana A, Schwindling C, Wenning AS, Becherer U, Rettig J, Schwarz EC, Hoth M. T cell activation requires mitochondrial translocation to the immunological synapse. Proc Natl Acad Sci U S A 2007;104:14418–23.

[32] Cao Y, Rathmell JC, Macintyre AN. Metabolic reprogramming towards aerobic glycolysis correlates with greater proliferative ability and resistance to metabolic inhibition in CD8 versus CD4 T cells. PLoS ONE 2014;9(8): e104104.

[33] D'Souza AD, Parikh N, Kaech SM, Shadel GS. Convergence of multiple signaling pathways is required to coordinately up-regulate mtDNA and mitochondrial biogenesis during T cell activation. Mitochondrion 2007;7:374–85.

[34] Tan H, Yang K, Li Y, Shaw TI, Wang Y, Blanco DB, Wang X, Cho JH, Wang H, Rankin S, Guy C, Peng J, Chi H. Integrative proteomics and phosphoproteomics profiling reveals dynamic Signaling networks and bioenergetics pathways underlying T cell activation. Immunity 2017;46:488–503.

[35] Geltink RI, O'Sullivan D, Corrado M, Bremser A, Buck MD, Buescher JM, Firat E, Zhu X, Niedermann G, Caputa G, Kelly B, Warthorst U, Rensing-Ehl A, Kyle RL, Vandersarren L, Curtis JD, Patterson AE, Lawless S, Grzes K, Qiu J, Sanin DE, Kretz O, Huber TB, Janssens S, Lambrecht BN, Rambold AS, Pearce EJ, Pearce EL. Mitochondrial Priming by CD28. Cell 2017;171: 385–97.e11.

[36] Baixauli F, Acin-Perez R, Villarroya-Beltri C, Mazzeo C, Nunez-Andrade N, Gabande-Rodriguez E, Ledesma MD, Blazquez A, Martin MA, Falcon-Perez JM, Redondo JM, Enriquez JA, Mittelbrunn M. Mitochondrial respiration controls lysosomal function during inflammatory T cell responses. Cell Metab 2015;22:485–98.

[37] Prado-Garcia H, Sandoval-Martinez R, Romero-Garcia S. T-cell metabolism and its dysfunction induced by cancer. In: Rezaei N, editor. Cancer immunology. Cham: Springer International Publishing; 2020. p. 107–16.

[38] Buck Michael D, O'Sullivan D, Klein Geltink Ramon I, Curtis Jonathan D, Chang C-H, Sanin David E, Qiu J, Kretz O, Braas D, van der Windt Gerritje JW, Chen Q, Huang Stanley C-C, O'Neill Christina M, Edelson Brian T, Pearce Edward J, Sesaki H, Huber Tobias B, Rambold Angelika S, Pearce Erika L. Mitochondrial dynamics controls t cell fate through metabolic programming. Cell 2016;166:63–76.

[39] van der Windt G, Everts B, Chang C-H, Curtis JD, Freitas TC, Amiel E, Pearce EJ, Pearce EL. Mitochondrial respiratory capacity is a critical regulator of CD8+ T cell memory development. Immunity 2012;36.

[40] Champagne DP, Hatle KM, Fortner KA, D'Alessandro A, Thornton TM, Yang R, Torralba D, Tomas-Cortazar J, Jun YW, Ahn KH, Hansen KC, Haynes L, Anguita J, Rincon M. Fine-tuning of CD8(+) T cell mitochondrial metabolism by the respiratory chain repressor MCJ dictates protection to influenza virus. Immunity 2016;44:1299–311.

[41] Romero-Garcia S, Lopez-Gonzalez JS, Baez-Viveros JL, Aguilar-Cazares D, Prado-Garcia H. Tumor cell metabolism: an integral view. Cancer Biol Ther 2011;12:939–48.

[42] Chang C-H, Qiu J, O'Sullivan D, Buck MD, Noguchi T, Curtis JD, Chen Q, Gindin M, Gubin MM, van der Windt G, Tonc E, Schreiber RD, Pearce EJ, Pearce EL. Metabolic competition in the tumor microenvironment is a driver of cancer progression. Cell 2015;162:1229–41.

[43] Singer K, Kastenberger M, Gottfried E, Hammerschmied CG, Buttner M, Aigner M, Seliger B, Walter B, Schlosser H, Hartmann A, Andreesen R, Mackensen A, Kreutz M. Warburg phenotype in renal cell carcinoma: high expression of glucose-transporter 1 (GLUT-1) correlates with low CD8(+) T-cell infiltration in the tumor. Int J Cancer 2011;128:2085–95.

[44] Li X, Zhang Y, Ma W, Fu Q, Liu J, Yin G, Chen P, Dai D, Chen W, Qi L, Yu X, Xu W. Enhanced glucose metabolism mediated by CD147 contributes to immunosuppression in hepatocellular carcinoma. Cancer Immunol Immunother 2020;1–14.

[45] Siska PJ, Beckermann KE, Mason FM, Andrejeva G, Greenplate AR, Sendor AB, Chiang YJ, Corona AL, Gemta LF, Vincent BG, Wang RC, Kim B, Hong J, Chen CL, Bullock TN, Irish JM, Rathmell WK, Rathmell JC. Mitochondrial dysregulation and glycolytic insufficiency functionally impair CD8 T cells infiltrating human renal cell carcinoma. JCI Insight 2017;2:e93411.

[46] Prado-Garcia H, Romero-Garcia S, Castro-Flores DA, Rumbo-Nava U. Deficient glucose uptake is linked to impaired Glut1 expression upon CD3/CD28 stimulation in memory T cells from pleural effusions secondary to lung cancer. Scand J Immunol 2019;90:e12802.

[47] Gemta LF, Siska PJ, Nelson ME, Gao X, Liu X, Locasale JW, Yagita H, Slingluff Jr. CL, Hoehn KL, Rathmell JC, Bullock TNJ. Impaired enolase 1 glycolytic activity restrains effector functions of tumor-infiltrating CD8(+) T cells. Sci Immunol 2019;4: eaap9520.

[48] Song M, Sandoval TA, Chae CS, Chopra S, Tan C, Rutkowski MR, Raundhal M, Chaurio RA, Payne KK, Konrad C, Bettigole SE, Shin HR, Crowley MJP, Cerliani JP, Kossenkov AV, Motorykin I, Zhang S, Manfredi G, Zamarin D, Holcomb K, Rodriguez PC, Rabinovich GA, Conejo-Garcia JR, Glimcher LH, Cubillos-Ruiz JR. IRE1alpha-XBP1 controls T cell function in ovarian cancer by regulating mitochondrial activity. Nature 2018;562:423–8.

[49] Ho P-C, Bihuniak Jessica D, Macintyre Andrew N, Staron M, Liu X, Amezquita R, Tsui Y-C, Cui G, Micevic G, Perales Jose C, Kleinstein Steven H, Abel ED, Insogna Karl L, Feske S, Locasale Jason W, Bosenberg Marcus W, Rathmell Jeffrey C, Kaech SM. Phosphoenolpyruvate is a metabolic checkpoint of anti-tumor T cell responses. Cell 2015;162.

[50] Cascone T, McKenzie JA, Mbofung RM, Punt S, Wang Z, Xu C, Williams LJ, Wang Z, Bristow CA, Carugo A, Peoples MD, Li L, Karpinets T, Huang L, Malu S, Creasy C, Leahey SE, Chen J, Chen Y, Pelicano H, Bernatchez C, Gopal YNV, Heffernan TP, Hu J, Wang J, Amaria RN, Garraway LA, Huang P, Yang P, Wistuba II, Woodman SE, Roszik J, Davis RE, Davies MA, Heymach JV, Hwu P, Peng W. Increased tumor glycolysis characterizes immune resistance to adoptive T cell therapy. Cell Metab 2018;27:977–87 e4.

[51] Nakagawa Y, Negishi Y, Shimizu M, Takahashi M, Ichikawa M, Takahashi H. Effects of extracellular pH and hypoxia on the function and development of antigen-specific cytotoxic T lymphocytes. Immunol Lett 2015;167:72–86.

[52] Brand A, Singer K, Koehl Gudrun E, Kolitzus M, Schoenhammer G, Thiel A, Matos C, Bruss C, Klobuch S, Peter K, Kastenberger M, Bogdan C, Schleicher U, Mackensen A, Ullrich E, Fichtner-Feigl S, Kesselring R, Mack M, Ritter U, Schmid M, Blank C, Dettmer K, Oefner Peter J, Hoffmann P, Walenta S, Geissler Edward K, Pouyssegur J, Villunger A, Steven A, Seliger B, Schreml S, Haferkamp S, Kohl E, Karrer S, Berneburg M, Herr W, Mueller-Klieser W, Renner K, Kreutz M. LDHA-associated lactic acid production blunts tumor immunosurveillance by T and NK cells. Cell Metab 2016;24:657–71.

[53] Comito G, Iscaro A, Bacci M, Morandi A, Ippolito L, Parri M, Montagnani I, Raspollini MR, Serni S, Simeoni L, Giannoni E, Chiarugi P. Lactate modulates CD4+ T-cell polarization and induces an immunosuppressive environment, which sustains prostate carcinoma progression via TLR8/miR21 axis. Oncogene 2019;38:3681–95.

[54] Dimeloe S, Gubser P, Loeliger J, Frick C, Develioglu L, Fischer M, Marquardsen F, Bantug GR, Thommen D, Lecoultre Y, Zippelius A, Langenkamp A, Hess C. Tumor-derived TGF-β inhibits mitochondrial respiration to suppress IFN-γ production by human CD4+ T cells. Sci Signal 2019;12:eaav3334.
[55] Rodriguez PC, Zea AH, DeSalvo J, Culotta KS, Zabaleta J, Quiceno DG, Ochoa JB, Ochoa AC. L-arginine consumption by macrophages modulates the expression of CD3 zeta chain in T lymphocytes. J Immunol 2003;171:1232–9.
[56] Uyttenhove C, Pilotte L, Theate I, Stroobant V, Colau D, Parmentier N, Boon T, Van den Eynde BJ. Evidence for a tumoral immune resistance mechanism based on tryptophan degradation by indoleamine 2,3-dioxygenase. Nat Med 2003;9:1269–74.
[57] Rodriguez PC, Quiceno DG, Zabaleta J, Ortiz B, Zea AH, Piazuelo MB, Delgado A, Correa P, Brayer J, Sotomayor EM, Antonia S, Ochoa JB, Ochoa AC. Arginase I production in the tumor microenvironment by mature myeloid cells inhibits T-cell receptor expression and antigen-specific T-cell responses. Cancer Res 2004;64:5839–49.
[58] Lee GK, Park HJ, Macleod M, Chandler P, Munn DH, Mellor AL. Tryptophan deprivation sensitizes activated T cells to apoptosis prior to cell division. Immunology 2002;107:452–60.
[59] Bengsch B, Johnson AL, Kurachi M, Odorizzi PM, Pauken KE, Attanasio J, Stelekati E, McLane LM, Paley MA, Delgoffe GM, Wherry EJ. Bioenergetic insufficiencies due to metabolic alterations regulated by the inhibitory receptor PD-1 are an early driver of CD8(+) T cell exhaustion. Immunity 2016;45:358–73.
[60] Zhang Y, Kurupati R, Liu L, Zhou XY, Zhang G, Hudaihed A, Filisio F, Giles-Davis W, Xu X, Karakousis GC, Schuchter LM, Xu W, Amaravadi R, Xiao M, Sadek N, Krepler C, Herlyn M, Freeman GJ, Rabinowitz JD, Ertl HCJ. Enhancing CD8+ T cell fatty acid catabolism within a metabolically challenging tumor microenvironment increases the efficacy of melanoma immunotherapy. Cancer Cell 2017;32: 377–91.e9.
[61] Siska PJ, van der Windt GJ, Kishton RJ, Cohen S, Eisner W, MacIver NJ, Kater AP, Weinberg JB, Rathmell JC. Suppression of Glut1 and glucose metabolism by decreased Akt/mTORC1 signaling drives T cell impairment in B cell leukemia. J Immunol 2016;197:2532–40.
[62] Prado-Garcia H, Aguilar-Cazares D, Flores-Vergara H, Mandoki JJ, Lopez-Gonzalez JS. Effector, memory and naive CD8+ T cells in peripheral blood and pleural effusion from lung adenocarcinoma patients. Lung Cancer 2005;47:361–71.
[63] Legat A, Speiser DE, Pircher H, Zehn D, Fuertes Marraco SA. Inhibitory receptor expression depends more dominantly on differentiation and activation than "exhaustion" of human CD8 T cells. Front Immunol 2013;4:455.
[64] Prado-Garcia H, Romero-Garcia S, Puerto-Aquino A, Rumbo-Nava U. The PD-L1/PD-1 pathway promotes dysfunction, but not "exhaustion", in tumor-responding T cells from pleural effusions in lung cancer patients. Cancer Immunol Immunother 2017;66:765–76.
[65] Scharping NE, Menk AV, Moreci RS, Whetstone RD, Dadey RE, Watkins SC, Ferris RL, Delgoffe GM. The tumor microenvironment represses T cell mitochondrial biogenesis to drive intratumoral T cell metabolic insufficiency and dysfunction. Immunity 2016;45:374–88.
[66] Chamoto K, Chowdhury PS, Kumar A, Sonomura K, Matsuda F, Fagarasan S, Honjo T. Mitochondrial activation chemicals synergize with surface receptor PD-1 blockade for T cell-dependent antitumor activity. Proc Natl Acad Sci U S A 2017;114:E761–70.
[67] Luengo A, Gui DY, Vander Heiden MG. Targeting metabolism for cancer therapy. Cell Chem Biol 2017;24:1161–80.
[68] Sukumar M, Liu J, Ji Y, Subramanian M, Crompton JG, Yu Z, Roychoudhuri R, Palmer DC, Muranski P, Karoly ED, Mohney RP, Klebanoff CA, Lal A, Finkel T, Restifo NP, Gattinoni L. Inhibiting glycolytic metabolism enhances CD8+ T cell memory and antitumor function. J Clin Invest 2013;123:4479–88.
[69] Qiu J, Villa M, Sanin DE, Buck MD, O'Sullivan D, Ching R, Matsushita M, Grzes KM, Winkler F, Chang CH, Curtis JD, Kyle RL, Van Teijlingen BN, Corrado M, Haessler F, Alfei F, Edwards-Hicks J, Maggi Jr. LB, Zehn D, Egawa T, Bengsch B, Klein Geltink RI, Jenuwein T, Pearce EJ, Pearce EL. Acetate promotes T cell effector function during glucose restriction. Cell Rep 2019;27:2063–74 e5.
[70] Sabatino M, Hu J, Sommariva M, Gautam S, Fellowes V, Hocker JD, Dougherty S, Qin H, Klebanoff CA, Fry TJ, Gress RE, Kochenderfer JN, Stroncek DF, Ji Y, Gattinoni L. Generation of clinical-grade CD19-specific CAR-modified CD8+ memory stem cells for the treatment of human B-cell malignancies. Blood 2016;128:519–28.
[71] Chatterjee S, Chakraborty P, Daenthanasanmak A, Iamsawat S, Andrejeva G, Luevano LA, Wolf M, Baliga U, Krieg C, Beeson CC, Mehrotra M, Hill EG, Rathmell JC, Yu XZ, Kraft AS, Mehrotra S. Targeting PIM kinase with PD1 inhibition improves immunotherapeutic antitumor T-cell response. Clin Cancer Res 2019;25:1036–49.

[72] Pedicord VA, Cross JR, Montalvo-Ortiz W, Miller ML, Allison JP. Friends not foes: CTLA-4 blockade and mTOR inhibition cooperate during CD8+ T cell priming to promote memory formation and metabolic readiness. J Immunol 2015;194:2089–98.

[73] Chakraborty P, Vaena SG, Thyagarajan K, Chatterjee S, Al-Khami A, Selvam SP, Nguyen H, Kang I, Wyatt MW, Baliga U, Hedley Z, Ngang RN, Guo B, Beeson GC, Husain S, Paulos CM, Beeson CC, Zilliox MJ, Hill EG, Mehrotra M, Yu XZ, Ogretmen B, Mehrotra S. Pro-survival lipid Sphingosine-1-phosphate metabolically programs T cells to limit anti-tumor activity. Cell Rep 2019;28:1879–93 e7.

[74] Zhang Z, Li F, Tian Y, Cao L, Gao Q, Zhang C, Zhang K, Shen C, Ping Y, Maimela NR, Wang L, Zhang B, Zhang Y. Metformin enhances the antitumor activity of CD8(+) T lymphocytes via the AMPK-miR-107-Eomes-PD-1 pathway. J Immunol 2020;204:2575–88.

[75] Eikawa S, Nishida M, Mizukami S, Yamazaki C, Nakayama E, Udono H. Immune-mediated antitumor effect by type 2 diabetes drug, metformin. Proc Natl Acad Sci U S A 2015;112:1809–14.

[76] Prado-Garcia H, Romero-Garcia S. The role of exhaustion in tumor-induced T-cell dysfunction in cancer. In: - Rezaei N, editor. Cancer immunology. Cham: Springer International Publishing; 2020. p. 117–32.

[77] Cha JH, Yang WH, Xia W, Wei Y, Chan LC, Lim SO, Li CW, Kim T, Chang SS, Lee HH, Hsu JL, Wang HL, Kuo CW, Chang WC, Hadad S, Purdie CA, McCoy AM, Cai S, Tu Y, Litton JK, Mittendorf EA, Moulder SL, Symmans WF, Thompson AM, Piwnica-Worms H, Chen CH, Khoo KH, Hung MC. Metformin promotes antitumor immunity via endoplasmic-reticulum-associated degradation of PD-L1. Mol Cell 2018;71:606–20 e7.

[78] Menk AV, Scharping NE, Rivadeneira DB, Calderon MJ, Watson MJ, Dunstane D, Watkins SC, Delgoffe GM. 4-1BB costimulation induces T cell mitochondrial function and biogenesis enabling cancer immunotherapeutic responses4-1BB metabolically supports T cell function. J Exp Med 2018;215:1091–100.

[79] Kawalekar OU, O'Connor RS, Fraietta JA, Guo L, McGettigan SE, Posey AD, Patel PR, Guedan S, Scholler J, Keith B, Snyder NW, Blair IA, Milone MC, June CH. Distinct signaling of coreceptors regulates specific metabolism pathways and impacts memory development in CAR T cells. Immunity 2016;44:380–90.

CHAPTER

6

The IgM as a tool for recognition of early tumoral antigens

Pedro Ostoa-Saloma

Departamento de Inmunología, Instituto de Investigaciones Biomédicas, Universidad Nacional Autónoma de México, Ciudad de México, México

Abstract

Breast cancer is the most common cancer in women in the world, therefore, research aimed at finding ways of early detection maintains great importance in this disease. Tools that contribute to the recognition of cancer warning signs can lead to early diagnosis in a healthy population to identify people with the disease at an initial stage when symptoms are not yet apparent. Early detection will increase the survival rate of patients and, therefore, an effective treatment option to fight the disease. The innate immune system recognizes and eliminates cancer cells. This chapter describes the advantages of IgM, in their innate form called natural IgM, to identify tumor antigens in order to find a candidate to be considered as an early diagnostic tool for breast cancer. Most research on biomarkers in breast cancer has focused on the use of IgG, taking advantage of its properties even as a therapeutic tool. However, there is currently no diagnostic method for early detection of breast cancer. The chapter will also try to answer the following questions: In a precancerous condition, is it possible to identify and differentiate the antigens recognized by both IgM and IgG and determine which immunoglobulin is best for identifying antigens in early stages? How would it be the recognition of IgM towards tumor antigens in the middle of a genetic background associated with resistance and/or susceptibility? By measuring the pattern of IgM recognition of tumor antigens, would it be possible to differentiate which individuals are more susceptible and which less? And therefore, by means of IgM, antigens associated with susceptibility and/or resistance to breast cancer could be recognized.

Abbreviations

aa	amino acids
BCR	B cell receptor
BPA	bisphenol A
Cμ	constant region
CD	cluster of differentiation
IgG	immunoglobulin G
IgM	immunoglobulin M
MBL	mannose-binding lectin

Immunotherapy in Resistant Cancer:
From the Lab Bench Work to Its Clinical Perspectives
https://doi.org/10.1016/B978-0-12-822028-3.00002-9

mIgM	membrane-bound IgM
MZ	marginal zone
NK	natural killer
sIgM	secretable IgM
TAAs	tumor-associated antigens
TACAs	tumor-associated carbohydrate antigens
TLR	toll-like receptors
Vμ	variable region

Conflict of interest

No potential conflicts of interest were disclosed by the authors.

Introduction

In 1909, Paul Ehrlich postulated that the immune system not only kills pathogenic bacteria but also suppresses the growth of carcinomas with great frequency, providing antibodies against malignant cells [1]. Fifty years later, Frank Macfarlane Burnet and Lewis Thomas reviewed the topic of natural immune protection against cancer. Burnet considered that tumor-cell-specific neo-antigens could cause an effective immune reaction that eliminates the development of cancer cells and defined the concept of immunovigilance [2,3]. That is, the transformed cells are killed by a process consisting of an immediate immune response that provides antibodies against malignant cells, as well as an inherited secondary immune response derived from B cells [4,5].

We currently know that immunovigilance corresponds to innate immunity since this is the first line of defense and stimulates the adaptive immune response [6]. The innate immune system is based on TLR, which do not recognize individual-specific structures, but rather specific molecular patterns associated with pathogens. These specific patterns are conserved and repetitive structures, such as carbohydrates in glycoproteins and glycolipids that are independently expressed by mutational events [7] and are detected independently of T cells [8]. It was not until the 1990s that this theory was more robust with studies in knockout models, validating the existence of immunovigilance in induced tumors. Within the immunovigilance theory, a process called immuno-editing was proposed, which consists of three phases: (1) elimination: where both the innate and adaptive systems recognize and eliminate tumor cells, (2) balance: where a population of malignant cells is eliminated and another portion of the malignant cells resists the effector mechanisms of the immune system, and (3) escape: where tumor cells evade the mechanisms of the immune system [9–11].

Innate immunity is the body's first line of defense. Included in the important cellular components involved in the immuno-editing process are NK lymphocytes, which are lymphoid cells that come from the same precursor to T lymphocytes but do not express the CD3 marker nor the rearrangement for T-cell receptor expression.

Another element of innate immunity is IgM. IgM is the first immunoglobulin isotype that is expressed in naïve B lymphocytes. IgM is made up of μ heavy chains (56–60 kDa) of 576 aa. The variable region (Vμ) contains about 124 aa. The constant region is made up of four domains (Cμ1–Cμ4) and contains 452 aa. It is expressed as a membrane-bound monomer in

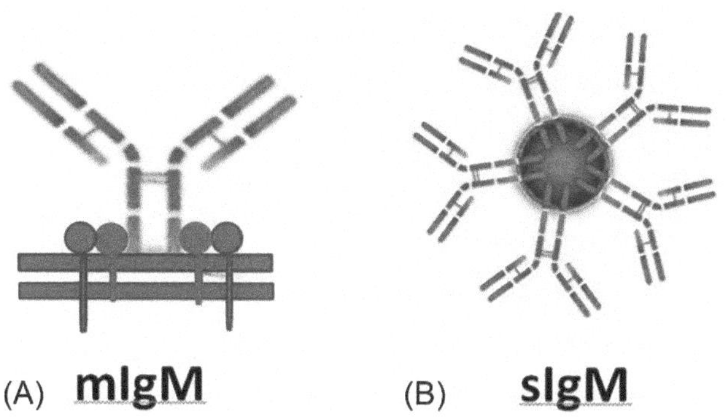

FIG. 1 Schematic representation of (A) membrane-bound IgM (mIgM) and (B) secretory IgM (sIgM).

the B lymphocyte (called mIgM), which is linked to two glycoproteins Igα (CD79α) and Ig (CD79) as a central part of the BCR (Fig. 1A); likewise, its secretable form is found in the form of a pentamer (called sIgM) whose molecular weight is approximately 970 kDa. Each monomer is linked by disulfide bonds, and each pentamer has a J-link or chain (Fig. 1B). In its soluble form, it has ten antigen-binding sites, so it has a high avidity but a low affinity. It is very effective in activating complement through the classic C1q pathway when the antigen is on the surface of an invading pathogen, senescent cells, cellular debris, or precancerous, or cancerous cells since only one molecule of IgM is needed to activate it, followed by neutralization and opsonization (Fig. 1) [12].

There are two types of IgMs that depend on the subtype of the B-lymphocyte population that generates them. The follicular B2-lymphocyte or B-lymphocyte subtype ($IgM^{low}IgD^{high}CD21^{high}CD23^{high}$) produces adaptive IgM, which is produced in response to an antigenic challenge. The recognized antigen depends on the response by Th lymphocytes, which activate B2 lymphocytes, making it a type of secondary T-dependent response. Immunization-induced IgM differs from natural IgM in its structure at antigen-binding centers, affinity, repertoire specificity, and the spectrum of its functions [13]. Adaptive IgM antibodies comprise a relatively small fraction of circulating molecules, are monoreactive, have a higher affinity ($Kd = 10^{-7}$–10^{-11} mol^{-1}), and their variable regions contain point mutations that evidence a somatic hyper-mutation process [14,15].

The other subtype is the one that comes from B1 or CD5+ cells and B cells from the MZ, which do not require affinity maturation to provide early protection [16]. B-1 cells possess the $B220^{low}IgM^{hi}CD23^{low}/CD43^{+}IgD^{low}$ phenotype. These cells exhibit characteristics of activated cells and are larger in cytoplasmic size and complexity than B2 cells [17]. B-1 cells produce natural antibodies, which are produced without the need for a previous antigenic stimulus, making it an independent primary T response [18]. This natural IgM preferentially binds to conserved structures like cell surface glycoproteins and glycolipids.

Natural IgM has lambda chains, unlike adaptive IgM, which has kappa chains [19]. Natural IgM is the first antibody that is synthesized in the newborn, so it represents the body's first line of defense in humoral immunity. There are high levels of IgM in the first years of the body's life, which decrease with age but stay present [20–22]. Natural IgM circulates in healthy individuals even in the absence of exogenous antigenic stimulation or antigen-driven selection [23,24]. The level of natural IgM in the serum of newborns and in animals growing

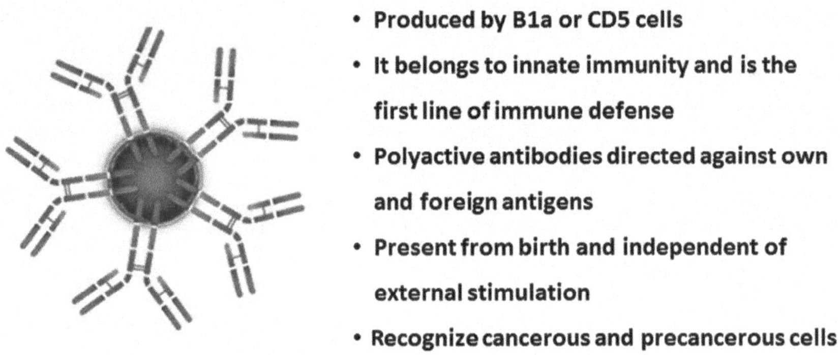

FIG. 2 Main characteristics of natural IgM antibodies.

under sterile conditions and with an antigen-free diet does not differ from that of normal animals [12]; natural IgM is also found in humans [25]. Natural IgM plays an important role in primary defense mechanisms [26–28] since it is associated with the recognition and elimination of precancerous and cancer cells [29–31], preferentially binding to post-transcriptionally modified antigens on the cell surface that are tumor specific as carbohydrate epitopes [32–34]. The carbohydrate epitopes recognized by natural IgM are stably expressed in a variety of tumors and at various precursor stages. Unlike single-chain peptide-based epitopes, glyco-epitopes share structural homologies beyond the boundaries of protein families. Glyco-epitopes are cross-reactive and are preferred as targets for natural IgM antibodies [1].

Natural IgM antibodies are germline encoded and not affinity matured. More than 80% of natural IgM antibodies are expressed by the VH genes of the VH3 gene family [15] and have a relatively low affinity (Kd $= 10^{-4}$–10^{-7} mol^{-1}) [14]. The efficacy of the antigen-antibody interaction is enhanced by the relative potency of IgM in the involvement of the complement pathway; unlike IgG, a single IgM molecule can bind C1q and activate complement [35].

Fig. 2 shows the different functions of natural and adaptive IgM against transformed cells, identified in experiments with murine models. The behavior of IgM and IgG throughout the development of a breast cancer tumor is also shown. Natural and adaptive IgM remains constant from the time the cell transformation begins until the tumor is established. IgG is present only in the first stage of adaptive immunity, but when the tumor is established, IgG is immunodeficient, but IgM is maintained.

In innate immunity, natural IgM protects the organism using different strategies, such as: (1) complement activation by the classic route in collaboration with C1q [36], where IgM binds strongly to complement factor C1q and activates the complement cascade [37]; (2) direct neutralization [38–40], and (3) elimination of apoptotic cells by phagocytosis [41] due to the binding of IgM to MBL, which in turn binds to apoptotic cells [42].

Natural IgM and breast cancer

Breast cancer is a heterogeneous disease with tumors that express a variety of aberrant proteins [43]. Natural IgM can recognize TAA that has undergone post-translational modifications. It also mediates the destruction of tumor tissue by recognizing TACAs [44].

The presence of post-translational modifications, such as glycosylation, phosphorylation, oxidation, and proteolysis, can induce the generation of a new epitope. Proteins in modified cells are poorly localized, mutated, or misfolded, or their expression is aberrant, and they are associated with carcinogenic processes, such as cell cycle progression, signal transduction activation, proliferation, and apoptosis [45,46]. Cell surface glycans that are secreted into the serum by malignant cells provide a mechanism for monitoring tumor burden. Many malignant cells, but not normal cells, overexpress CD20, ECFR, and HER2, allowing these proteins to be used as a diagnostic tool, but not for early diagnosis [47]. The natural IgM against TAA found in the serum of patients with breast cancer can be easily detected, in addition to being inherently stable and persistent in the serum for a relatively long period, because they generally do not undergo proteolysis, like other polypeptides [46].

Currently, there are blood tests that identify high-level tumor antigens in patients with metastatic disease, but these are too insensitive for use in the early diagnosis of breast cancer [43]. TAAs modulate the transmembrane signaling that is required for tumor-cell proliferation, invasion, and metastasis [48]. Natural IgM against TAA can be used as an early signal of breast cancer in vivo and allow earlier detection than current methods. Furthermore, natural IgM is detected in the asymptomatic stage of cancer up to 5 years before the onset of the disease [49]. Some monoclonal IgM antibodies have been isolated from patient tumors. Table 1 shows examples of IgM antibodies used against breast cancer tumor antigens.

The idea of using natural IgM as an immunodiagnostic tool for the early detection of breast cancer arose from the need to develop a diagnostic system that would make it possible to differentiate the population with the highest risk of developing breast cancer (beyond hereditary cancer) to have a closer follow-up of this population since current tools, such as self-examination and mammography, have a sensitivity much lower than immunological sensitivity.

Based on the literature, it is clear that most of the research has focused on the use of IgG for the diagnosis of breast cancer through the establishment of biomarkers. It should be said that, to date, there is no reliable diagnostic method for detecting breast cancer in its early stages.

The first question was whether, in a cancer condition, antigens recognized by both IgM and IgG could be assessed and differentiated in a murine cancer model. Fig. 3 shows that IgM has a greater range of antigenic recognition and acts earlier than IgG antibodies, which indicated that the use of IgM was feasible, given its potential, in developing a serological analysis for the early diagnosis of cancer [57].

Another result of this work was that the mice showed great variability in the recognition of tumor antigens. Apparently, there was no recognition of common antigens among the evaluated individuals. Natural IgM antibodies are known to be encoded by the germline and do not show maturation in their affinity (as happens with IgG), that is, their variability depends on the variability that occurs in parental genetic recombination. This information led to the question of how IgM recognition of tumor antigens would be towards the same tumor in the middle of a different genetic background. It has been reported that there is a susceptibility to breast cancer based on the genetic background of mice [58] (something similar happens in humans). Therefore, it was planned to apply tumor cells to murine representatives of a resistant strain, a sensitive strain, and an intermediate strain. The serum of the different individuals was obtained and using two-dimensional immunoblots, and the pattern of recognition towards the antigens of the applied tumor was resolved, that is, the same tumor for all the individuals. Could

TABLE 1 IgM antibodies used against breast cancer tumor antigens.

Antibody	Target	Description	Reference
FC-2.15	Carbohydrate motifs of different glycoproteins	Murine monoclonal antibody. It specifically recognizes CaMa cells and the motifs of different glycoproteins and is capable of mediating complement lysis in vitro. It induces anti-tumor response	[50]
SC-1	Decay acceleration factor-B-specific carbohydrate epitope (DAF or CD55)	Natural antibody. It induces apoptosis by receptor cross-linking, both in vitro and in experimental systems in vivo	[51]
PAM-1	CFR-1 (cysteine-rich fibroblast growth factor receptor)	Monoclonal antibody. It blocks growth factor receptors, such as EGFR or FGFR overexpressed in malignant cells, leading to starvation and cell death	[33]
SAM-6	Variant of GRP78 with a molecular weight of 82 kDa overexpressed in tumor cells	Natural antibody. It is internalized through endocytosis to the tumor cell and is responsible for a lethal accumulation of oxidized lipoproteins followed by apoptosis	[52]
3EL.2	Serum breast antigen (SBA)	Murine monoclonal antibody. The identification of SBA in blood could be used as a diagnosis, as it is abnormally high in patients with breast cancer	[53]
IgM anti-P10s	P10s recognize glyco-sphingolipids (GSL), such as GD2, GM2, and the Lewis Y antigen (LeY)	It mediates complement-dependent cytotoxicity. It is more cytotoxic to tumor cells than other antibodies	[54]
IgM anti-MUC1	Polymorphic epithelial mucin (PEM or MUC1 in epitopes CA 15.3 and CA 27.29) and MUC16 (CA 125)	Anti-MUC1. It contributes to the destruction of MUC1-expressing tumor cells, whether circulating or established	[55]
IgM anti-CEA	Carcinoembryogenic antigen (CEA)	CEA has been found in patients with ductal carcinoma in situ, so it may be an early marker of the tumorigenic process	[56]

FIG. 3 Number, total sum, average, and standard deviation of antigens recognized by natural IgM through 2D immunoblot using preimmune serum from female and male mice and having a 4T1 cell lysate as antigenic background.

Female mice		
Spots	IgM	IgG
N° Mouse		
1	86	0
2	39	0
3	65	0
4	45	4
5	98	4
6	89	7
7	45	2
9	35	6
10	61	7
∑ Spots per serum	563	183
Average	62.6	20.33
SD	23.6	16.73

one, by measuring the pattern of IgM recognition of tumor antigens, differentiate which mouse was susceptible and which was resistant?

At first glance, it was not easy to discern that the patterns were different, but how could a recognition pattern of a two-dimensional immunoblot be measured? In other words, is it possible to transform the recognition pattern into a mathematical language to make it quantitative? This question led to the development of the concept of "immunological signature." The idea of the immunological signature is to transform the pattern of IgM recognition of a 2D immunoblot into a digital language. The procedure was as follows: once the coordinates of the recognized antigens within each 2D immunoblot image were calculated, the images were divided into 10 columns × 10 rows. In each grid, matrices were established, assigning a score of 0 if there was no antigen in the cell and 1 if there were one or more antigens. The matrix was converted into a vector, placing row N immediately after its predecessor. Therefore, instead of a 10 × 10 matrix, a vector with 100 locations containing values of 0 and 1 was generated (Fig. 4). This vector was used as input to a Python script to perform a full grouping of links with the cluster package. For this analysis, the metric in which the distance between two points is the sum of the absolute differences in their Cartesian coordinates was selected. The resulting hierarchical grouping is presented as a dendrogram [59].

With this available methodology, we returned to the problem of two-dimensional patterns obtained from mice with different genetic backgrounds. Different immunological signatures were obtained both from individuals that were sensitive and from individuals that were less sensitive to breast cancer. The generated dendrogram almost perfectly grouped resistant individuals on the one hand and sensitive individuals on the other. The intermediate individuals were grouped on one side or the other interchangeably. Therefore, it was found that through patterns of antigenic recognition by IgM, mice that genetically have different susceptibilities to breast cancer can be differentiated and grouped (Fig. 5) [60]. The analysis also led to the recognition of the responsible antigens that made the difference between individuals being on the sensitive or the resistant side. This result raised the possibility that, by means of IgM, antigens associated with susceptibility and/or resistance to breast cancer could be

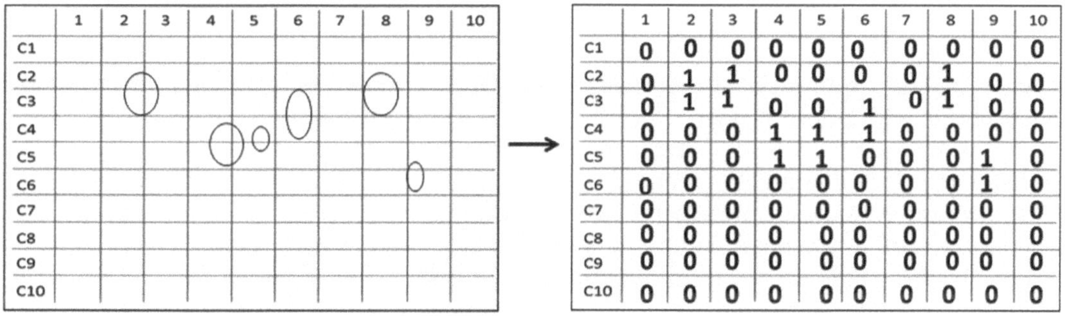

FIG. 4 Graphic explanation of how an immunological signature is obtained.

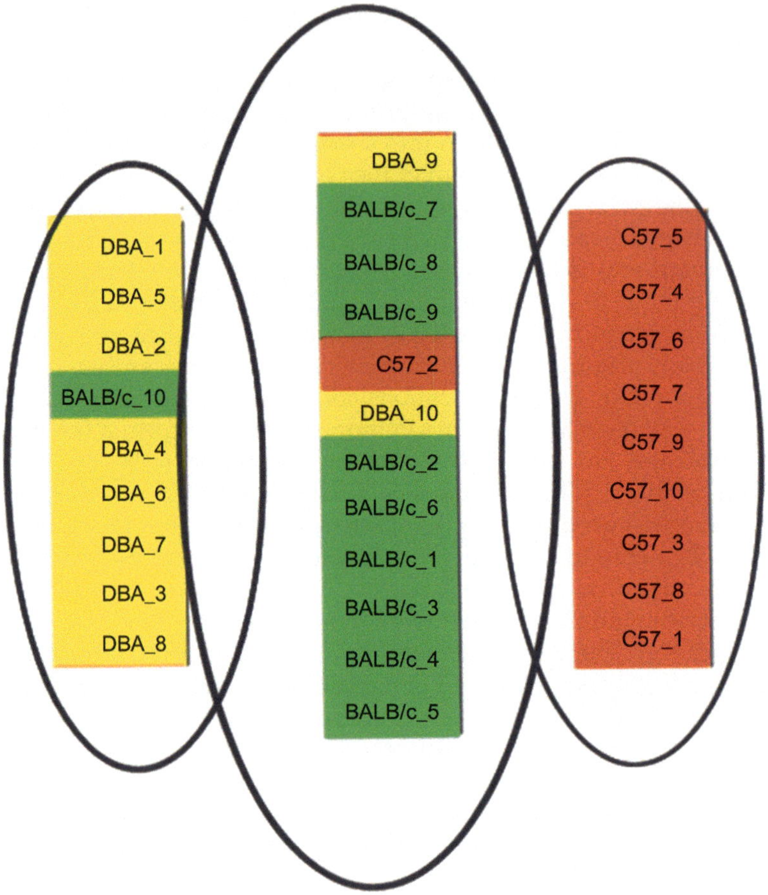

FIG. 5 Diagram showing the grouping of different mouse strains with different susceptibility to cancer according to their immunological signatures: strain DBA/2J (susceptible), C57BL/6J (resistant), and Balb C (external group) (see ref. [60]).

recognized. The application of this methodology is currently being carried out using sera from patients with breast cancer.

Continuing with the idea of using IgM for an immunodiagnostic test, the question arose as to whether the lack of a reliable method of early diagnosis was due to environmental effects on the immune system, specifically the IgM humoral immune response. One of the chemicals with which humans are most widely in contact is BPA. Serum was obtained from mice that each received a single neonatal administration of bisphenol A to assess the pattern of IgM recognition on tumor antigens in the absence and presence of BPA. Serum was obtained from the mice to which single neonatal administrations of bisphenol A were applied. The pattern of IgM recognition on tumor antigens in the absence and presence of BPA was evaluated and found to be affected by the presence of BPA. This means that individuals treated with BPA fail to recognize antigens that are normally recognized by all individuals, meaning that in the presence of BPA, there is less identification of antigens or the identification of antigens common to all mice simply disappears. Therefore, for the development of an immunodiagnostic method, the involvement of environmental factors on the immune system must be taken into account [61].

Conclusion

When analyzing the 2D immunoblot images, it was observed that there is a great variability of the IgM-mediated humoral immune response towards experimental breast cancer in mice.

There is a recognition of tumor antigens by natural IgM before the detectable appearance of the tumor and it does not decrease with time, so it can be considered as an important tool for early diagnosis of tumors.

The variation in the distribution pattern of spots recognized by IgM in 2D immunoblot images can be expressed as an immunological signature, which opens the possibility of correlating a particular pattern with resistance or susceptibility.

Antigenic recognition mediated by natural IgM towards the early appearance of an experimental tumor in mice was greater than the recognition of IgG.

Acknowledgments

This research was funded by Project Grant IT200120 from Programa de Apoyo a Proyectos de Innovación Tecnológica (PAPIIT), Dirección General de Asuntos del Personal Académico (DGAPA), Universidad Nacional Autónoma de Mexico (UNAM), to Pedro Ostoa Saloma.

References

[1] Vollmers HP, Brandlein S. Natural IgM antibodies: from parias to parvenus. Histol Histopathol 2006;21:1355–66.
[2] Burnet FM. Immunological surveillance in neoplasia. Transplant Rev 1971;7:3–25.
[3] Dunn GP, Old LJ, Schreiber RD. The immunobiology of cancer immunosurveillance and immunoediting. Immunity 2004;21:137–48.
[4] Milner EC, Anolik J, Cappione A, Sanz I. Human innate B cells: a link between host defense and autoimmunity? Springer Semin Immunopathol 2005;26:433–52.
[5] Karin M, Lawrence T, Nizet V. Innate immunity gone awry: linking microbial infections to chronic inflammation and cancer. Cell 2006;124:823–35.
[6] Hoebe K, Janssen E, Beutler B. The interface between innate and adaptive immunity. Nat Immunol 2004;5:971–4.
[7] Janeway Jr. CA. Pillars article: approaching the asymptote? Evolution and revolution in immunology. Cold Spring Harb Symp Quant Biol. 1989;54:1-13. J Immunol 2013;191:4475–87.
[8] Vollmers HP, Brandlein S. Natural antibodies and cancer. N Biotechnol 2009;25:294–8.
[9] Kim R, Emi M, Tanabe K. Cancer immunoediting from immune surveillance to immune escape. Immunology 2007;121:1–14.
[10] Topfer K, Kempe S, Muller N, Schmitz M, Bachmann M, Cartellieri M, Schackert G, Temme A. Tumor evasion from T cell surveillance. J Biomed Biotechnol 2011;2011:918471.
[11] Ravelli A, Roviello G, Cretella D, Cavazzoni A, Biondi A, Cappelletti MR, Zanotti L, Ferrero G, Ungari M, Zanconati F, Bottini A, Alfieri R, Petronini PG, Generali D. Tumor-infiltrating lymphocytes and breast cancer: beyond the prognostic and predictive utility. Tumour Biol 2017;39:1010428317695023. https://doi.org/10.1177/1010428317695023.
[12] Klimovich VB. IgM and its receptors: structural and functional aspects. Biochemistry (Mosc) 2011;76:534–49.

[13] Baumgarth N, Chen J, Herman OC, Jager GC, Herzenberg LA. The role of B-1 and B-2 cells in immune protection from influenza virus infection. Curr Top Microbiol Immunol 2000;252:163–9.
[14] Zhou ZH, Tzioufas AG, Notkins AL. Properties and function of polyreactive antibodies and polyreactive antigen-binding B cells. J Autoimmun 2007;29:219–28.
[15] Notkins AL. Polyreactivity of antibody molecules. Trends Immunol 2004;25:174–9.
[16] Vesely MD, Kershaw MH, Schreiber RD, Smyth MJ. Natural innate and adaptive immunity to cancer. Annu Rev Immunol 2011;29:235–71.
[17] Merino M, Gruppi A. Origen y desarrollo de linfocitos B1: Una población celular involucrada en defensa y autoinmunidad. Medicina (B Aires) 2006;66:165–72.
[18] Vas J, Gronwall C, Silverman GJ. Fundamental roles of the innate-like repertoire of natural antibodies in immune homeostasis. Front Immunol 2013;4:4.
[19] Vollmers HP, Brandlein S. Natural IgM antibodies: the orphaned molecules in immune surveillance. Adv Drug Deliv Rev 2006;58:755–65.
[20] Ebrahimnezhad S, Jazayeri M, Hassanian S, Avan A. Current status and prospective regarding the therapeutic potential of natural antibodies in cancer therapy. J Cell Physiol 2017;232:2649–52.
[21] Griffin DO, Holodick NE, Rothstein TL. Human B1 cells in umbilical cord and adult peripheral blood express the novel phenotype CD20+ CD27+ CD43+ CD70. J Exp Med 2011;208:67–80.
[22] Lobo PI. Role of natural autoantibodies and natural IgM anti-leucocyte autoantibodies in health and disease. Front Immunol 2016;7:198.
[23] Madi A, Hecht I, Bransburg-Zabary S, Merbl Y, Pick A, Zucker-Toledano M, Quintana FJ, Tauber AI, Cohen IR, Ben-Jacob E. Organization of the autoantibody repertoire in healthy newborns and adults revealed by system level informatics of antigen microarray data. Proc Natl Acad Sci U S A 2009;106:14484–9.
[24] Merbl Y, Zucker-Toledano M, Quintana FJ, Cohen IR. Newborn humans manifest autoantibodies to defined self molecules detected by antigen microarray informatics. J Clin Invest 2007;117:712–8.
[25] Schettino EW, Chai SK, Kasaian MT, Schroeder Jr. HW, Casali P. VHDJH gene sequences and antigen reactivity of monoclonal antibodies produced by human B-1 cells: evidence for somatic selection. J Immunol 1997;158:2477–89.
[26] Bohn J. Are natural antibodies involved in tumour defence? Immunol Lett 1999;69:317–20.
[27] Boes M, Schmidt T, Linkemann K, Beaudette BC, Marshak-Rothstein A, Chen J. Accelerated development of IgG autoantibodies and autoimmune disease in the absence of secreted IgM. Proc Natl Acad Sci U S A 2000;97:1184–9.
[28] Brandlein S, Pohle T, Ruoff N, Wozniak E, Muller-Hermelink HK, Vollmers HP. Natural IgM antibodies and immunosurveillance mechanisms against epithelial cancer cells in humans. Cancer Res 2003;63:7995–8005.
[29] Vollmers HP, Brandlein S. The "early birds": natural IgM antibodies and immune surveillance. Histol Histopathol 2005;20:927–37.
[30] Brandlein S, Eck M, Strobel P, Wozniak E, Muller-Hermelink HK, Hensel F, Vollmers HP. PAM-1, a natural human IgM antibody as new tool for detection of breast and prostate precursors. Hum Antibodies 2004;13:97–104.
[31] Vollmers H, Brändlein S. Death by stress: antibody induced apoptosis. Methods Find Exp Clin Pharmacol 2005;27:185–91.
[32] Hensel F, Hermann R, Schubert C, Abe N, Schmidt K, Franke A, Shevchenko A, Mann M, Muller-Hermelink HK, Vollmers HP. Characterization of glycosylphosphatidylinositol-linked molecule CD55/decay-accelerating factor as the receptor for antibody SC-1-induced apoptosis. Cancer Res 1999;59:5299–306.
[33] Rauschert N, Brandlein S, Holzinger E, Hensel F, Muller-Hermelink HK, Vollmers HP. A new tumor-specific variant of GRP78 as target for antibody-based therapy. Lab Invest 2008;88:375–86.
[34] Kobata A, Amano J. Altered glycosylation of proteins produced by malignant cells, and application for the diagnosis and immunotherapy of tumours. Immunol Cell Biol 2005;83:429–39.
[35] Manson JJ, Mauri C, Ehrenstein MR. Natural serum IgM maintains immunological homeostasis and prevents autoimmunity. Springer Semin Immunopathol 2005;26:425–32.
[36] Ogden CA, Kowalewski R, Peng Y, Montenegro V, Elkon KB. IGM is required for efficient complement mediated phagocytosis of apoptotic cells in vivo. Autoimmunity 2005;38:259–64.
[37] Quartier P, Potter PK, Ehrenstein MR, Walport MJ, Botto M. Predominant role of IgM-dependent activation of the classical pathway in the clearance of dying cells by murine bone marrow-derived macrophages in vitro. Eur J Immunol 2005;35:252–60.
[38] Ochsenbein AF, Fehr T, Lutz C, Suter M, Brombacher F, Hengartner H, Zinkernagel RM. Control of early viral and bacterial distribution and disease by natural antibodies. Science 1999;286:2156–9.

[39] Baumgarth N, Herman OC, Jager GC, Brown LE, Herzenberg LA, Chen J. B-1 and B-2 cell-derived immunoglobulin M antibodies are nonredundant components of the protective response to influenza virus infection. J Exp Med 2000;192:271–80.
[40] Jayasekera JP, Moseman EA, Carroll MC. Natural antibody and complement mediate neutralization of influenza virus in the absence of prior immunity. J Virol 2007;81:3487–94.
[41] Chen Y, Khanna S, Goodyear CS, Park YB, Raz E, Thiel S, Gronwall C, Vas J, Boyle DL, Corr M, Kono DH, Silverman GJ. Regulation of dendritic cells and macrophages by an anti-apoptotic cell natural antibody that suppresses TLR responses and inhibits inflammatory arthritis. J Immunol 2009;183:1346–59.
[42] Nauta AJ, Raaschou-Jensen N, Roos A, Daha MR, Madsen HO, Borrias-Essers MC, Ryder LP, Koch C, Garred P. Mannose-binding lectin engagement with late apoptotic and necrotic cells. Eur J Immunol 2003;33:2853–63.
[43] Molina R, Duffy MJ, Aronsson AC, Lamerz R, Stieber P, Van Dalen A, et al. Tumour markers in breast cancer–EGTM recommendations. European Group on Tumor Markers. Anticancer Res 1999;19:2803–5.
[44] Monzavi-Karbassi B, Hennings LJ, Artaud C, Liu T, Jousheghany F, Pashov A, Murali R, Hutchins LF, Kieber-Emmons T. Preclinical studies of carbohydrate mimetic peptide vaccines for breast cancer and melanoma. Vaccine 2007;25:3022–31.
[45] Anderson KS, Ramachandran N, Wong J, Raphael JV, Hainsworth E, Demirkan G, Cramer D, Aronzon D, Hodi FS, Harris L, Logvinenko T, LaBaer J. Application of protein microarrays for multiplexed detection of antibodies to tumor antigens in breast cancer. J Proteome Res 2008;7:1490–9.
[46] Tan HT, Low J, Lim SG, Chung MC. Serum autoantibodies as biomarkers for early cancer detection. FEBS J 2009;276:6880–904.
[47] Kobata A, Amano J. Altered glycosylation of proteins produced by malignant cells, and application for the diagnosis and immunotherapy of tumours. Immunol Cell Biol 2005;83:429–39.
[48] Shishido SN, Varahan S, Yuan K, Li X, Fleming SD. Humoral innate immune response and disease. Clin Immunol 2012;144:142–58.
[49] Fernandez MF. Autoantibodies in breast cancer sera: candidate biomarkers and reporters of tumorigenesis. Cancer Lett 2005;230:187–98.
[50] Capurro M, Bover L, Portela P, Livingston P, Mordoh J. FC-2.15, a monoclonal antibody active against human breast cancer, specifically recognizes Lewis(x) hapten. Cancer Immunol Immunother CII 1998;45:334–49.
[51] Liu C, Tseng L, Su J, Chang K, Chu P, Tai W, Shiau C, Chen K. Novel sorafenib analogues induce apoptosis through SHP-1 dependent STAT3 inactivation in human breast cancer cells. Breast Cancer Res 2013;15:R63.
[52] Pohle T, Brändlein S, Ruoff N, Bra S, Mu H, Vollmers H. Lipoptosis: tumor-specific cell death by antibody-induced intracellular lipid accumulation. Cancer Res 2004;64:3900–6.
[53] Verring A, Clouth A, Ziolkowski P, Oremek G. Clinical usefulness of cancer markers in primary breast cancer. Int Scholar Res Netw ISRN Pathol 2011; ID 817618:1–4.
[54] Ragupathi G, Liu N, Musselli C, Powell S, Lloyd K, Livingston P. Antibodies against tumor cell glycolipids and proteins, but not mucins, mediate complement-dependent cytotoxicity. J Immunol 2005;174:5706–12.
[55] Von Mensdorff-Pouilly S, Verstraeten A, Kenemans P, Snijdewint F, Kok A, Van Kamp G, Paul M, Van Diest P, Meijer S, Hilgers J. Survival in early breast cancer patients is favorably influenced by a natural humoral immune response to polymorphic epithelial mucin. J Clin Oncol 2000;18:574–83.
[56] Conry R, Allen K, Lee S, Moore S, Shaw D, LoBuglio A. Human autoantibodies to carcinoembryonic antigen (CEA) induced by a vaccinia-CEA vaccine. Clin Cancer Res 2000;6:34–41.
[57] Díaz-Zaragoza M, Hernández R, Ostoa-Saloma P. 2D immunoblots show differential response of mouse IgG and IgM antibodies to antigens of mammary carcinoma 4 T1 cells. Cancer Cell Int 2014;14:9–16.
[58] Hoag W. Spontaneous cancer in mice. Ann N Y Acad Sci 1963;108:805–31.
[59] Diaz-Zaragoza M, Hernández-Avila R, Govezensky T, Mendoza L, Meneses-Ruiz M, Ostoa-Saloma P. Comparison patterns of 4 T1 antigens recognized by humoral immune response mediated by IgG and IgM antibodies in female and male mice with breast cancer using 2D-immnunoblots. Immunobiology 2015;220:1050–8.
[60] Díaz-Zaragoza M, Hernández-Ávila R, Ostoa-Saloma P. Recognition of tumor antigens from 4T1 cells by natural IgM in three strains of mice with different susceptibilities to spontaneous breast cancer. Oncology Lett 2017;13:271–4.
[61] Hernández-Avila R, Palacios-Arreola M, Nava-Castro K, Morales-Montor J, Ostoa-Saloma P. Neonatal bisphenol A exposure affects the IgM humoral immune response to 4T1 breast carcinoma cells in mice. Int J Environ Res Publ Health 2019;16:1784.

CHAPTER

7

The inflammation during colorectal cancer: A friend or a foe?

Itzel Medina-Andrade[a,b,*], *Jonadab E. Olguín*[a,b,*], *Tonathiu Rodríguez*[b], *and Luis I. Terrazas*[a,b]

[a]Laboratorio Nacional en Salud: Diagnóstico Molecular y Efecto Ambiental en Enfermedades Crónico-degenerativas, Facultad de Estudios Superiores Iztacala, Universidad Nacional Autónoma de México, Tlalnepantla, México [b]Unidad de Biomedicina, Facultad de Estudios Superiores Iztacala, Universidad Nacional Autónoma de México, Tlalnepantla, México

Abstract

In the past two decades, a close relationship between inflammation and carcinogenesis has been described. In fact, since 2011, the inflammation has been considered as one of the ten hallmarks of cancer. However, natural selection did not evolve inflammation to cause troubles and diseases to our body. Both types of inflammation, acute and chronic, are natural events orchestrated by the immune system to generate protection against infectious and noninfectious insults, including colorectal cancer (CRC). Strong evidence in humans and some animal models of CRC indicates that the absence of molecules associated with inflammation induces an increased number of tumors and a reduction in apoptosis, suggesting that inflammation is involved in protection when the tumor has been established. However, another line of evidence, also well supported, suggests that the inflammatory processes during cancer development, especially in CRC, are very undesirable because it accelerates the tumor growth in many cases. Therefore, many questions arise regarding the relationship between inflammation and CRC: What are the steps and decisions taken by the immune system to guide the inflammation to favor CRC development? Or maybe, is the inflammation a process involved in protection against CRC establishment? The balance between the inflammation, their causes and checkpoints is extremely important during CRC development, and there are studies looking to solve these questions. In this chapter, we try to establish some of the stronger proofs suggesting a major role for inflammation in the protection during early CRC development. The understanding of the complex relationship between inflammation and cancer may help to develop combinatory therapies (drugs and immunomodulators) to blockade the possible collateral damage induced by chronic inflammation that favors CRC elimination.

[*]These authors contributed equally to this work.

Immunotherapy in Resistant Cancer: From the Lab Bench Work to Its Clinical Perspectives
https://doi.org/10.1016/B978-0-12-822028-3.00003-0

Abbreviations

15d-PGJ2	15-deoxy-Δ12,14-prostaglandin J2
5-FU	5-fluorouracil
AOM	azoxymethane
APC	adenomatous polyposis coli
CAC	colitis-associated colorectal cancer
CARD	caspase recruitment domain
CCL2	chemokine (C—C motif) ligand 2
DCs	dendritic cells
CEA	carcinoembryonic antigen
C-myc	c-myc gene
CRC	colorectal cancer
CTLA-4	cytotoxic T lymphocyte antigen
CXCL1	chemokine (C—X—C motif) ligand 1
CXCL5	C—X—C motif chemokine 5
Cyclin D1	cyclin d1 protein
DPD	enzyme dihydropyrimidine dehydrogenase
DSS	dextran sulfate sodium
Eb1	end-binding protein 1
EGF	epidermal growth factor
EGFR	epidermal growth factor receptor
Foxp3	forkhead box P3
G-CSF	granulocyte colony-stimulating factor
GM-CSF	granulocyte-macrophage colony-stimulating factor
GSK-3β	glycogen synthase kinase-3 β
hCG	human chorionic gonadotropin
HLA-DR	human leukocyte antigen
IBD	inflammatory bowel disease
ID2	inhibitor of DNA binding 2
IFN-γ	interferon-gamma
IFN-γR	interferon-gamma receptor
IgA	immunoglobulin A
IGF1	insulin-like growth factor 1
IL-1	interleukin-1
IL-12	interleukin-12
IL-12p70	interleukin-12p70
IL-12Rb2	interleukin-12 receptor beta2
IL-13	interleukin-13
IL-17	interleukin-17
IL-18	interleukin-18
IL-1R	interleukin-1 receptor
IL-1R1	interleukin-1 receptor type 1
IL-1α	interleukin-1-alpha
IL-1β	interleukin-1-beta
IL-22	interleukin-22
IL-23	interleukin-23
IL-4	interleukin-4
IL4Rα	interleukin-4 receptor α
IL-6	interleukin-6
iNOS	inducible nitric oxide synthase
IQGAP1	IQ motif containing GTPase-activating protein 1
ITF2	immunoglobulin transcription factor 2
K-ras	Kirsten rat sarcoma

LV	leucovorin
M1	macrophages 1
M2	macrophages 2
mAbs	monoclonal antibodies
MMP4	matrix metalloproteinase 4
MMP7	matrix metalloproteinase 7
mRNA	messenger RNA
MUC2	mucin 2
NF-κB	nuclear factor kappa-light-chain-enhancer of activated B cells
NK	natural killer
NOD	nucleotide-binding oligomerization domain-like receptors
NO	nitric oxide
PD1	programmed death 1
PDL1	programmed death-ligand 1
PGD2	prostaglandin D2
PG-F2α	prostaglandin F2α
Pla2g5	phospholipase A2 group V
PPAR-γ	peroxisome proliferator-activated receptor gamma
PTEN	phosphatidylinositol-3,4,5-trisphosphate 3-phosphatase
Reg3b	regenerating islet-derived protein 3-beta
RNI	reactive nitrogen intermediates
RORC	RAR-related orphan receptor C
Rorγt	RAR-related orphan receptor gamma
ROS	reactive oxygen species
Smo	smoothened, frizzled class receptor
SNP	single nucleotide polymorphisms
SPM	specialized pro-resolving mediators
STAT1	signal transducer and activator of transcription 1
STAT3	signal transducer and activator of transcription 3
STAT4	signal transducer and activator of transcription 4
STAT6	signal transducer and activator of transcription 6
TAMs	tumor-associated macrophages
T-bet	transcription factor T-bet
T-box21	T-box transcription factor 21
TGF-β	transforming growth factor beta 1
Th1	T helper cells 1
Th2	T helper cells 2
Th17	T helper cells 17
TLRs	toll-like receptor
TLR-2	toll-like receptor-2
TLR-4	toll-like receptor-4
TNFR	tumor necrosis factor receptor
TNF-α	tumor necrosis factor alpha
TNF-β	tumor necrosis factor beta
TNM staging	classification of malignant tumors
Treg	regulatory T cell
TS	enzyme thymidylate synthase
VEGF	vascular endothelial growth factor
WHO	World Health Organization
WT	wild type

Conflict of interest

No potential conflicts of interest were disclosed by the authors.

Introduction

Since 2002, colorectal cancer (CRC) has prevailed as the third most common type of cancer worldwide, and currently, it is the second type of cancer causing more fatalities around the world (Globocan 2018). The world regions that remain with a higher number of patients suffering from CCR are Northern Europe, North America, and Oceania.

The development of CRC has different origins: genetic (hereditary), sporadic, and inflammatory-mediated as in colitis-associated colon cancer (CAC). Only 5% of CRC cases are hereditary, while 75% are sporadic, being associated with environmental factors. The remaining 20% are associated with dysregulated inflammatory responses in the colon [1]. Patients diagnosed with intestinal inflammatory diseases such as ulcerative colitis or Crohn's disease have an increased probability to develop colon cancer.

In general, the treatment for CRC has been almost unspecific for many years; it depends on the stage of cancer development and includes surgery, ablation, embolization, radiation, chemotherapy, and recently [2] immunotherapy. CRC is becoming a major public healthcare problem, and it is necessary to know how to reduce its incidence and mortality, with strategies less aggressive than surgery, radiation, and chemotherapy. By understanding how this type of cancer initiates its development and why our own body is unable to recognize the first steps of this problem, we can fight against it.

The inflammatory response has been described as one of the major checkpoints involved in the development of CRC [2]. However, strong proofs in some models of CRC have demonstrated that the absence of molecules associated with inflammation favors an increase in the number of tumors together with reduced apoptosis, suggesting that inflammation is involved in protection [3,4]. It is important to mention that inflammation is a strong process requiring some steps including vascular changes, recruitment of immune cells, and release of cytokines [5], which is regularly triggered by infectious agents or tissue injury. After the inflammation performs its basic function to clear infectious and noninfectious damage, the antiinflammatory process that reduces the impact and damage caused by a strong inflammatory response is performed, inducing both surveillance and tissue-repairing mechanisms [5]. However, an exacerbated or uncontrolled anti-inflammatory response has been associated with immunosuppression that reduces the ability of the immune system to fight cancer; some tumors, such as colon cancer, take advantage of this suppression to develop and grow.

Here, we describe inflammation as a physiological process in the colon, the major role of the type of inflammatory immune response being part of the protection mechanism when the tumor has been established as well as the critical role of both excessive anti-inflammatory and immunosuppressive responses being responsible for the ineffective protective role of the immune response during CRC development. Finally, we discuss which side of the coin has a major role in the development of CRC, whether the inflammatory or the anti-inflammatory arm.

Acute inflammation has a physiological role in the colon

Colon is a part of the gastrointestinal tract. It serves to reabsorb water, electrolytes, and some nutrients that were not digested and absorbed in the small intestine. The colon epithelium is simple, cylindrical, and continually renewed by immature cells that arise from invaginations known as Lieberkühn crypts, which, in turn, also shelter goblet cells and enteroendocrine cells [6]. Histologically, the colon is composed of several layers: the submucosa, the muscular mucosa, the lamina propria, and the mucosa (Fig. 1). The lamina propria is the region sheltering both innate immune cells, such as macrophages, eosinophils, dendritic cells (DCs), and adaptive immune cells, such as CD4+ and CD8+ T lymphocytes [6]. The mucosa is the region that shapes the intestinal lumen favoring constant interaction with the colon's microbiota, which is made up of trillions of different species of bacteria [7]. The interaction between the bacteria and the cells of the immune system in the colon generates an environment of symbiosis between both populations (Fig. 1). On the one hand, the microbiota contributes to protection by preventing the growth of pathogens, and on the other hand, the immune system allows the microbiota to establish itself [5]. However, when the colonic mucosa suffers any physical, chemical, or biological damage, as in the case of CRC, the microbiota can interact with innate immune receptors, such as Toll-like receptors (TLRs), which are located on the surface of immune cells in the lamina propria, such as macrophages and DCs. Also, the invasion by pathogens in the colon and damaged mucosa is recognized by TLRs. The recognition of either the pathogen or the microbiota by TLRs induces a rapid activation of cells of the innate immune response toward an inflammatory profile. For example, after such activation, macrophages acquire an enhanced antimicrobial activity mediated by the secretion of molecules such as nitric oxide (NO) and reactive oxygen species

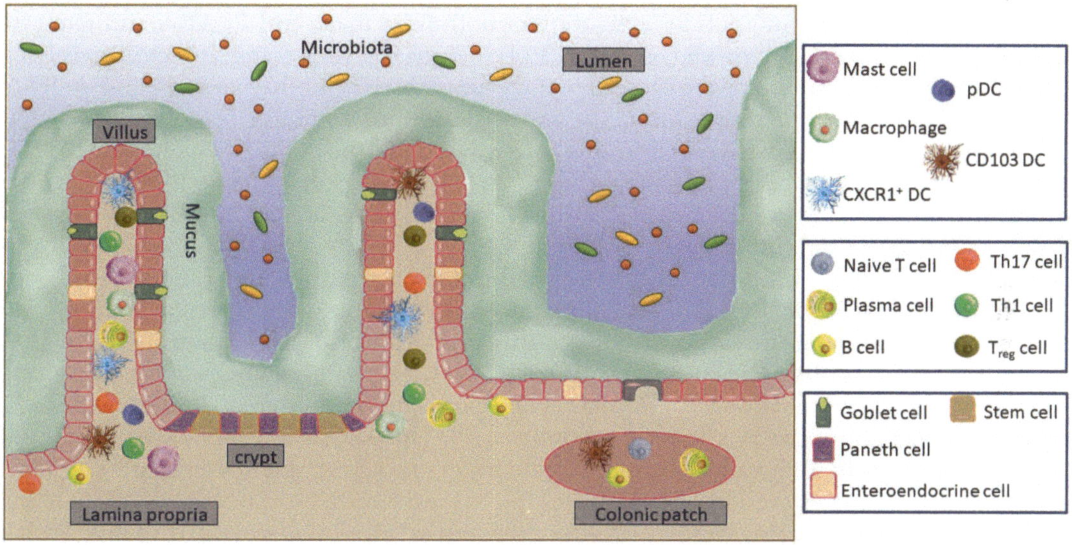

FIG. 1 General scheme of colon and immune system cells. Colon is composed by a mucus layer, a simple layer of cylindrical epithelium, and a region known as lamina propria. Mucus layer is the outermost region that faces intestinal lumen and fulfills the function of a protective barrier against microbiota. In turn, simple cylindrical epithelium forms several tubular glands known as Lieberkühn crypts. These crypts are integrated by various cell types such as stem cells present at the crypts' base, which give rise to all other cell types such as goblet cells that produce mucus present in abundant amounts in colon, paneth cells that are producers of antimicrobial peptides, and enteroendocrine cells. These cell types are found in smaller amounts in colon than in small intestine. Likewise, in lamina propria lymphoid tissue-harboring immune cells such as B cells and T lymphocytes can be found. These tissues are called colon patches. On the other hand, other types of immune cells such as macrophages, dendritic cells (DCs), mast cells, and lymphocytes can be found outside colon patches.

(ROS), which are essential for the destruction of pathogens. Additionally, these inflammatory macrophages, also called classically activated macrophages or M1, are characterized by the production and secretion of pro-inflammatory cytokines, such as tumor necrosis factor (TNF)-α, IL-1β, and IL-12 [8]. In turn, inflammatory cytokines and chemokines activate and recruit more cells of the immune system, while ROS and NO induce oxidative damage in proteins, lipids, and the DNA of microbiota or pathogenic bacteria as well as in healthy neighboring tissues and in cells near to the site of the inflammation. This inflammatory microenvironment supports the recruitment of naive T lymphocytes that activate them preferentially toward a Th1-type profile, which will further intensify the inflammatory and cytotoxic responses through the production of IFN-γ by CD4+ and CD8+ T lymphocytes [9,10]. Once the inflammatory response has been established and its purpose has been fulfilled, which is the elimination of its cause, the next step is the resolution of the inflammation. The resolution of the inflammation consists in limiting inflammatory signals and promoting mucosal repair through diverse soluble mediators such as lipid mediators (prostaglandins), specialized pro-resolving mediators (SPM; lipoxins, resolvins, protectins, and maresins), and protein mediators (anti-inflammatory cytokines and growth factors) [11]. The expression of arginase-1 whose function is to metabolize L-arginine into L-ornithine also helps in tissue repair by inducing de novo synthesis of collagen [12].

Prostaglandins are molecules of a lipid nature derived from arachidonic acid, which have been described for their regulatory effects on inflammation. Prostaglandins PGD2 and 15d-PGJ2 can suppress inflammation by activation of both IkB kinase and PPAR-γ, inducing the inhibition of the nuclear factor-κB (NF-κB) [13,14]. NF-κB is a transcription factor mainly associated with the promotion of the secretion of inflammatory cytokines such as TNF-α, IL-1β, IL-12, and IL-18, while PPAR-γ inhibits the expression of inflammatory cytokines, which in turn promotes the differentiation of immune cells toward anti-inflammatory phenotypes [11]. Also, it has been suggested that prostaglandin F2α (PG-F2α) contributes to the resolution of chronic inflammation [15]. Like prostaglandins, SPM are molecules responsible for the regulation of inflammation, limiting excessive infiltration of leukocytes and production of inflammatory cytokines. Likewise, the anti-inflammatory cytokines IL-4, IL-13, IL-10, and TGF-β and the growth factors EGF, VEGF, and IGF1 contribute to the restoration of both the architecture and the function of colon's mucosa. All these anti-inflammatory signals can switch the T lymphocyte profile toward a Th2 profile. This phenomenon is called "phenotypic plasticity" of immune cells [16]. Th2 cells, neutrophils, and mast cells can produce IL-4 and IL-13, activating the WNT signal pathway [17]. This signal induces tissue repair and epithelial regeneration through β-catenin activation, which regulates cell proliferation-associated genes such as cyclin D1 and c-myc [18]. Macrophages, lymphocytes, and intestinal cells produce TGF-β, favoring cellular proliferation and extracellular matrix formation [19]. In fact, it has been shown that the blockade of TGF-β reduces the migration of damaged intestinal mucosa [20]. Thus, TGF-β not only is involved in tissue repairing, but also has a role in cellular migration after intestinal tissue damage. Both macrophages and neutrophils phagocytize cellular debris from damaged tissue and secrete growth factors such as TGF-β, IGF-1, VEGF, and MMP [21,22], which favor the remodeling of the extracellular matrix, angiogenesis, and proliferation of colon epithelial progenitors [23]. In studies using GM-CSF, its capacity to repair harmed colon mucosa was shown in a model of colitis induced by dextran sulfate sodium (DSS), by favoring the infiltration of monocytes and macrophages in the lamina propria [24]. Another cellular type having a major role in the anti-inflammatory process in the colon mucosa is regulatory T (Treg) cells (see below for details). Likewise, they regulate the activation of macrophages and neutrophils, inducing the M2 and N2 profile, respectively [25], favoring strong anti-inflammatory events that may end in immunosuppression [25]. Finally, in addition to the active participation of immune cells during mucosal repair, intestinal epithelial cells are also capable of stimulating signals that promote cell proliferation and tissue repair, through the activation of the transcription factor STAT3. For example, in a study of acute colitis, it was shown that STAT3 regulates genes associated with wound healing (*smo*), proliferation (*myc, Reg3b, Pla2g5*), and apoptosis regulation (*mcl1 and surviving*); therefore, STAT3 is implicated in mucosal wound healing [26]. Thus, in the colon, the inflammation is a physiological process of rapid response to damage. However, in cases of prolonged inflammatory response in the colon, it may result in an uncontrolled chronic inflammation that may be closely associated with the development of diseases such as ulcerative colitis and colon cancer.

Do we have evidence that chronic inflammation is promoting both CAC and CRC?

In the past decade, it has been established that inflammation precedes the onset of CAC. However, inflammation is highly unlikely to initiate the development of CRC. CAC and CRC differ in the timing and order of mutations in specific genes as well as in the participation of the immune response. But in both CAC and CRC, the appearance of mutations is a requirement to initiate tumor development. In CAC, the hallmark of chronic inflammation is its ability to cause mutations [27]. In chronic inflammation, there is a constant infiltration of immune cells into the colon, such as macrophages and neutrophils, which are one of the main sources of ROS and reactive nitrogen intermediates (RNI) [27], which are highly reactive and induce damage to the DNA of intestinal epithelial cells. This promotes the acquisition of mutations in p53 gene and poli adenomatous polyp (APC) gene, in early and late stages, respectively [28–30]. The p53 gene is located on the chromosome 17p, and its activation induces cell cycle arrest, apoptosis, and senescence in the presence of cellular stress, such as DNA damage, hypoxia, and activation of oncogenes such as *C-myc* and *Ras* [31,32]. The p53 mutation prevents cells from being able to repair the DNA damage generated by ROS and RNI and from being sensitive to apoptosis stimuli and cell cycle arrest signals, which favors the accumulation of mutations. Furthermore, APC is a tumor suppressor gene that regulates the transcription of numerous genes involved in cell proliferation (WNT/β-catenin pathway, cyclin D, C-myc), regulation of the cytoskeleton (EB1, IQGAP1), and cell adhesion (E-cadherin) [33]. Studies have shown that the specific loss of APC in stem cells results in progressive neoplastic growth [34]. On the other hand, the development of CRC is initiated by mutations, in early stages in the APC gene and in late stages in the p53 gene. In fact, the percentage of mutation in the APC gene in this type of cancer is approximately 80% [35]. Subsequently, the appearance of consecutive mutations in genes, such as those of the DNA error repair mechanism, K-ras, and GSK-3 (glycogen synthase kinase 3, a protein that controls β-catenin activation) [7], culminates in the formation of a tumor. Likewise, different studies have suggested that once the tumor has been established, cells of the immune response may promote an inflammatory microenvironment [7,36,37]. For example, in a CRC model in Apcmin/+ mice lacking one or two iNOS alleles, it was observed that these mice develop significantly fewer colon tumors compared to Apcmin/+ iNOS+ / + controls [38]. However, in one study, the type, density, and location of immune cells in patients with CRC were described to have a prognostic value. For example, there is a positive correlation between the presence of Th1 (inflammatory) polarization markers and memory and cytotoxic T cells and, a low incidence of tumor recurrence [39]. Furthermore, the involvement of cells in the immune response also differs between CAC and CRC. In an analysis of tumor specimens from 24 patients with CAC and 48 patients with CRC, it was reported that patients with CAC had significantly lower levels of CD3+, CD8+, and CD3+Foxp3+ cells and a high expression of PDL1+ (programmed death-ligand 1) associated with cell anergy, while patients with CAC who displayed a higher expression of CD3+CD8+ and CD3+Foxp3+ cells had a better overall survival. In contrast, patients harboring CRC had a higher CD3+Foxp3+ cell infiltration and a lower CD3+CD8+ cell infiltration as well as a low PDL1+ expression, while CRC patients with a high CD3+CD8+, Foxp3+, and PDL1+ cell expression had a better overall survival [40]. Therefore, CAC and CRC differ in the appearance of mutations in specific genes, participation of immune cells, and appearance of different inflammatory microenvironments. However, the role of inflammation as a promoter of tumor development has become controversial.

The paradoxical role of inflammation between CAC and CRC

As mentioned earlier, CAC is a subtype of colon cancer. Its development has been partially associated with inflammatory bowel diseases (IBDs) such as ulcerative colitis and Crohn's disease (Fig. 2). It has been estimated that the risk of developing CAC when diagnosed with an IBD is 2% after 10 years, 8% after 20 years, and 18% after 30 years [41]. Being critical, the percentage of association between both the diagnosis of IBD and the risk of developing CAC is low, but it is widely accepted. And even patients diagnosed with any of these inflammatory diseases are under constant medical surveillance [41,42]. However, it has been suggested that this type of epidemiological studies may have unreliable conclusions, due to the heterogeneity of the criteria among studies. There are relevant factors that are partially considered: severity and duration of inflammation, IBD extension, age of onset, sex, family history of sporadic CRC, follow-up time, drug treatments, exposure to any surgical intervention [42], or even coexistence of primary sclerosis cholangitis, which is a rare cholestatic liver disease characterized by progressive fibroinflammatory destruction of the intra- and extrahepatic biliary ducts [43]. Therefore, the risk of developing CAC in patients with inflammatory diseases does not fall exclusively on the inflammation, but also falls on both environmental and genetic factors such as the family record of CRC [44]. In addition to this, the association between inflammatory bowel diseases and CAC has begun to change, since more recent studies have given evidence showing that the cumulative incidence of developing CAC in patients with colitis is approximately 0% at 10 years, 2.5% at 20 years, 7.6% at 30 years, 10.8% at 40 years, and 13.5% at 45 years since the diagnosis of colitis [45,46]. All this evidence suggests that the association between IBD and the risk of developing CAC has a lower incidence rate than previously suggested. Therefore, these data support the beneficial role of the inflammatory response in avoiding CRC development.

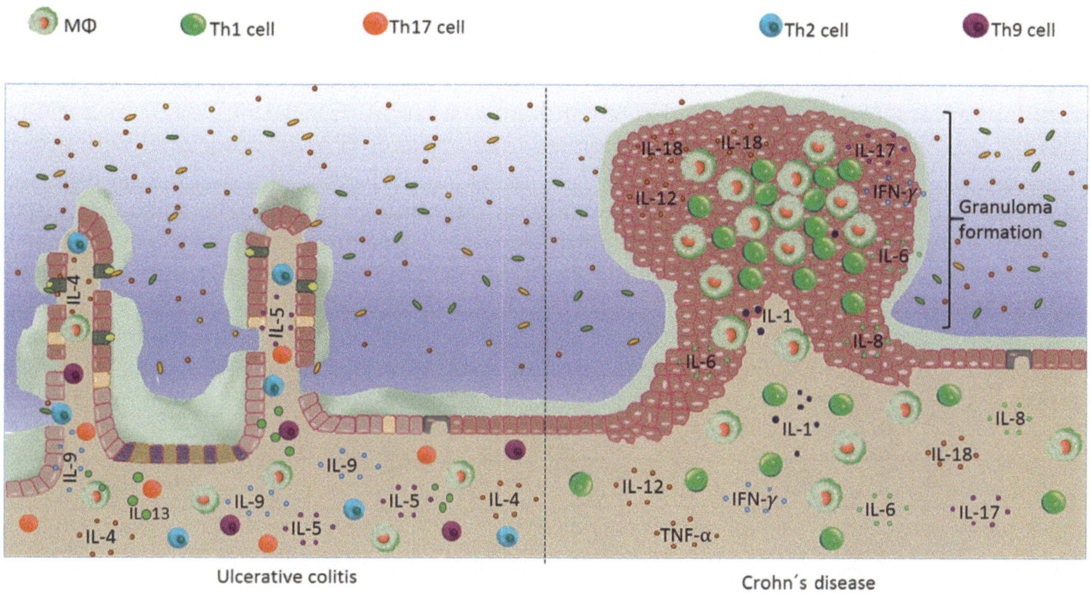

FIG. 2 Immune response during inflammatory bowel diseases (IBDs). Ulcerative colitis (UC) and Crohn's disease (CD) are the two main types of IBDs characterized by an imbalance in homeostasis between immune system cells. UC is characterized by ulcerations that are limited to mucus layer of the colon. These ulcerations start in rectum and can spread through entire colon. The cells infiltrated during UC are cells of a Th9 immune response which produce IL-9, cells of a Th2 immune response (anti-inflammatory) which produce IL-13, IL-4, IL-5, and cells of a Th17 immune response. The production of these cytokines can induce apoptosis and alterations in tight junction proteins in intestinal epithelial cells. Furthermore, CD lesions can occur in any region along the digestive tract. These lesions are characterized by the formation of fistulas, ulcers, and granulomas in mucus layer colon. The presence of antigen-presenting cells such as macrophages infiltrated into mucus layer, favoring the production of cytokines such as IL-12 and IL-18, which activate lymphocytes to a Th1 (inflammatory) profile. Activation of Th1 profile produces secretion of inflammatory cytokines such as IFN-γ, TNF-α, IL-1β, and IL-6, which favor chronic intestinal inflammation during CD.

Are either inflammatory bowel diseases or Th17 inflammatory pathway associated with CAC and CRC development?

There is growing evidence for an alternative kind of inflammation having a major role in the pathology and pathogenesis during CAC development. Different studies support the hypothesis of the role of IL-17 and its signaling pathway in the development of CAC. IL-17 cytokine family members (IL-17A, IL-17F, IL-17E, and IL-17C) have a role during pathogenic diseases, but recently, there has been an increased association between IL-17 cytokines and allergic, autoimmune and, relevant to the purposes of this chapter, chronic inflammatory diseases [47]. CD4+ naive T cells migrating to the secondary lymphoid organs acquire a Th17 phenotype by their exposure to IL-1β, IL-6, TGF-β, IL-21, and IL-23 cytokines in the microenvironment [47]. These CD4+Th17+ cells express their own master transcription factor Rorγt. As in CD4+ cells, IL-17A and IL-17E are also produced by CD8+, T-CD4+, and innate tissue-resident cells. Particularly in the intestinal mucosa, Th17 cells have a protective role, inducing the expression of molecules such as β-defensins, S100 protein, and lipocalin-2 that stimulate epithelial barrier functions limiting the bacterial growth [47]. A hallmark of the Th17-mediated inflammation is the recruitment of neutrophils induced by granulocyte colony-stimulating factor (G-CSF) and also chemokines such as CXCL1, CXCL5, and CCL2 [47,48]; also, Th17 cells recruit tumor-associated macrophages (TAMs) associated with tumor development during CAC [49,50] (Fig. 3). It is important to highlight that the inflammatory response of Th17 is not related to Th1-mediated inflammatory response because the origin, milieu to induce the CD4+ cell differentiation, necessary cytokines, master transcription factors involved, and the final effector functions are different between these two subpopulations of T helper cells.

Some cumulative and overwhelming evidence points out that IL-17 has a role in the pathogenesis of CAC and CRC. For example, a dense infiltration of CD4+IL-17+ cells in human samples of CRC, ulcerative colitis, and CAC has been shown [51]. In line with these

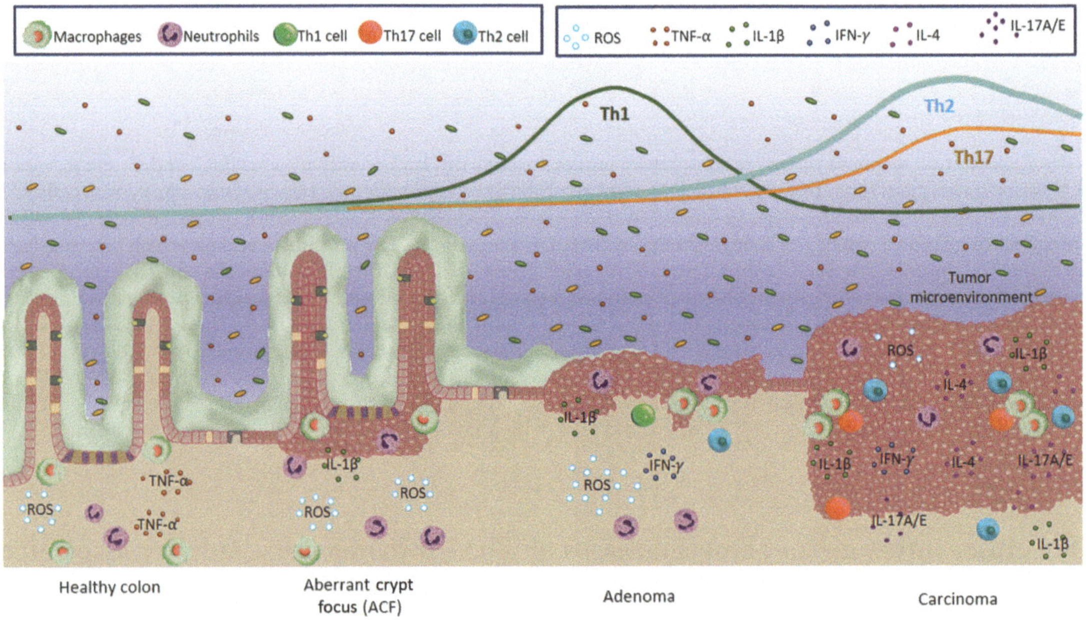

FIG. 3 The transition of the immune response during the development of tumors in CAC. In colon, a wide variety of immune system cells are infiltrated to maintain intestinal homeostasis. However, in IBDs the infiltration of inflammatory or Th1 cells such as M1 macrophages and neutrophils produces large amounts of reactive oxygen species (ROS), which generate mutations in genes such as p53. The p53 mutation prevents cells from being sensitive to apoptosis stimuli and cell cycle arrest signals, leading to an increased proliferation of cells with accumulated mutations. This in turn leads to the generation of aberrant crypt foci (ACF), which are early neoplastic lesions that later can lead to the formation of adenomas that are considered the precursors to colon cancer. The appearance of preneoplastic lesions can lead to dysbiosis of the microbiota that further intensifies the inflammatory response. Once the inflammatory response has generated enough mutations to trigger tumor formation, a transition occurs between the Th1 inflammatory response and the Th2 anti-inflammatory and Th17 inflammatory response. That is, the intensity of the Th1 response gradually begins to decrease and becomes transitory, while the intensity of the Th2 and Th17 responses begins to increase and becomes permanent or chronic. The establishment of a Th2 and Th17 response in tumor microenvironment is exploited by tumor cells to grow, proliferate, migrate, and invade other tissues.

observations, in the same study, a complementary testing on a mouse model of AOM-DSS CAC showed that Rorγt−/− mice displayed signs of intense colitis, but developed fewer tumors [51], indicating that IL-17 signaling was necessary for colon tumor development. Supporting this idea, the Th1 immune response profile is associated with a better outcome than a Th17 profile in patients coursing with CAC [52] (Figs. 3 and 4). Moreover, in Apcmin+/− mice, it was shown that infection with the human *Bacteroides fragilis* triggered a rapid development of colitis and, therefore, tumor formation. This outcome was related to IL-17 because the injection of anti-IL-17 blocking antibody avoided the colon tumor formation [49]. Another independent work demonstrated that both IL-23 and IL-17 signaling promoted tumor development in the polyposis APC−/− mouse model. Here, the IL-23 was locally produced by tumor-recruited myeloid cells that probably were activated by microbial products [53]. Interestingly, IL-17 signaling on transformed enterocytes (a type of epithelial intestinal cell) promotes CRC development [49]. Additionally, in a different work, it was shown that endogenous IL-17E also promotes colonic tumor development [52]. In line with

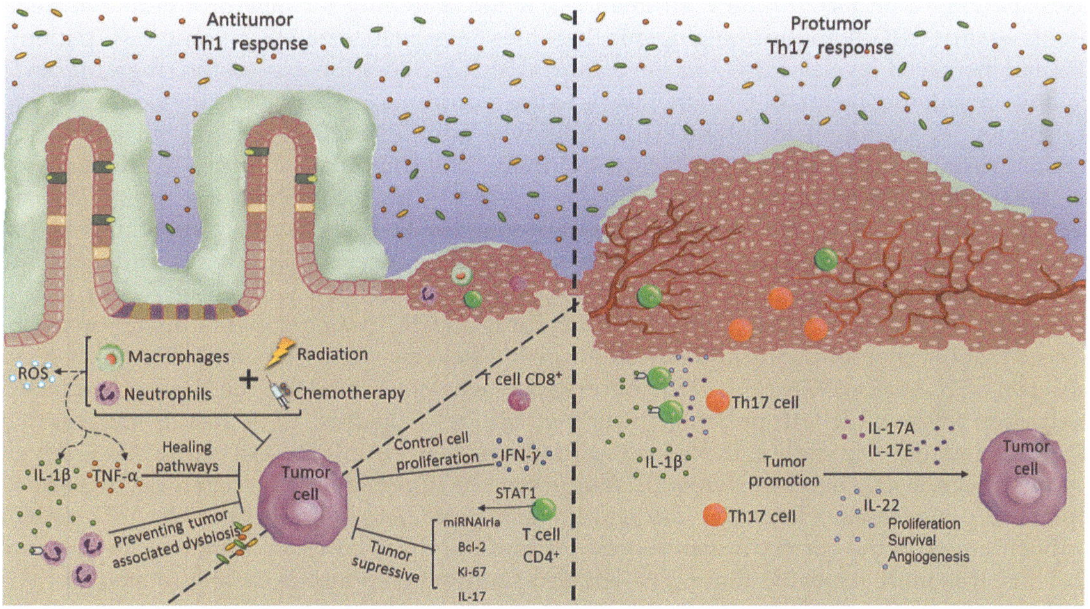

FIG. 4 The inflammatory anti-tumor and inflammatory pro-tumor microenvironment. Th1 inflammation has typically been associated as a pro-tumor response. However, reactive oxygen species (ROS) and the presence of enzyme inducible nitric oxide synthase (iNOS) during treatment with chemotherapy or radiation increase cancer cells' death. Likewise, the production of TNF-α by macrophages promotes the repair of mucus layer during IBDs, while IL-1 receptor expression on neutrophils is capable of controlling microbiota invasion into the colon, thereby preventing dysbiosis associated with tumor development. In turn, the activation of transcription factor STAT1 promotes an anti-tumor response by inactivating miRNA181a, Bcl-2, ki-67, and IL-17, while IFN-γ production by activated $CD4^+$ and $CD8^+$ lymphocytes is capable of controlling cell proliferation and stimulating apoptosis in cancer cells. On the other hand, immune response polarization toward a Th17 profile and the secretion of IL-17A and IL-17E in tumor microenvironment can favor proliferation, survival, and angiogenesis in tumor cells. However, IL-1 receptor expression on lymphocytes stimulates the secretion of IL-17 and IL-22, which further contributes to pro-tumor microenvironment.

these results, the specific absence of IL-1R in T cells decreased tumor-elicited inflammation dependent on IL-17 and IL-22, thereby reducing CRC progression, supporting the idea that IL-1 is involved in Th17 and CRC development [54] (Fig. 4). In a retrospective study analyzing 125 frozen colorectal tumor samples from patients with CRC, it was shown that genes such as RORC and IL-17A were associated with a Th17 profile and these patients had a poor prognosis. Moreover, patients with increased expression levels of genes such as T-bet, IL-12Rb2, and STAT4 that are associated with a Th1 profile had a prolonged disease-free survival [55]. With these results, the authors suggest that, in the early, first, or second stage of CRC, a premature inflammatory response dominated by a Th17 profile predicts a rapid progression to incurable metastatic disease [55]. In another study with samples of both human colorectal adenomas and colorectal carcinomas, increased levels of IL-17 mRNA were found in adenomas samples, but much higher levels were found in carcinoma samples. This higher IL-17 expression was associated with increased dysplasia severity, and these observations were confirmed by immunohistochemistry [56]. Thus, IL-17 may promote inflammation and colon tumorigenesis as well as poor prognosis and may become a target of treatment in colon cancer patients (Figs. 3 and 4). In line with this idea, while searching for new or alternative treatments against CRC, some anti-angiogenic therapies have been tested [57]. Some CRC patients display increased levels of IL-17 in the tumor, which induces the expression of granulocyte colony-stimulating factor (G-CSF) that leads to the recruitment of immature myeloid cells into the tumor. A treatment to inhibit Th17 cells was administered together with antibodies against vascular endothelial growth factor (VEGF); this combination improved the efficacy of the single anti-VEGF treatment [57]. Thus, patients with colorectal adenomas display higher levels of IL-17 in all stages throughout carcinoma; IL-17 induces production of factors associated with tumor proliferation, survival, and angiogenesis [52] (Fig. 4). In fact, some polymorphisms in IL-17A, IL-17E, and IL-23 genes are associated with an increased risk of CRC [52]. For example, in one study with 102 Tunisian CRC patients in treatment, it was observed that the chemotherapy and radiotherapy inefficiency was associated with mutated genotypes of IL-17A G197A SNP. This observation was confirmed by the fluorescent-based restriction fragment length polymorphism method [58]. Together, these results suggest that Th17-mediated inflammation is playing a different role in both IBDs and during CAC development (Figs. 2–4). Therefore, while IBDs appear to be poorly and less related to CAC development, it is necessary to clarify the real incidence of CAC derived from the IBD patient population, with the correct inflammatory response and the correct parameters to measure it. Thus, it is more probable that a Th17-biased immune response is closely related to CAC and CRC development in murine models and human samples than previously estimated.

A paradox: protective role of inflammatory cytokines, transcription factors, and immune cells during CRC

Several investigations have associated the immune response as a promoter of the development of CAC [59] because during inflammation, both macrophages and neutrophils produce large amounts of ROS (Fig. 3), which in turn can interact with the DNA of intestinal epithelial cells, generating mutations in tumor suppressor genes and, as a consequence, activating proto-oncogenes (*C-myc, Ras, PTEN*) that lead to the formation of a tumor [60,61]. However,

ROS has also been associated with possible antioncogenic properties since in several types of cancer, they are able to decrease the resistance of tumor cells to conventional therapeutic agents [62–64]. In fact, studies using different CRC-derived cell lines have shown that gene transfer of iNOS (which is an enzyme that produces the majority of reactive nitrogen intermediates during inflammation) can be a possible radiosensitizer, given that this gene transference is capable of increasing the apoptosis rate in cancer cells during radiotherapy [65]. In addition to this, cancer cells are more sensitive to cisplatin treatment in the presence of NO [66]. In murine models of chronic inflammation, mice deficient in iNOS are more susceptible to carcinogenic processes developing more polyps [67,68]. All these results suggest the protective role of a specific inflammatory response with iNOS as a weapon against established tumor; at this point, it is important to highlight that one of the most important NO producers are inflammatory (also called M1) macrophages.

In the same way, the inflammatory cytokine TNF-α produced during the early stages of inflammation has been described in the origin, development, survival, and promotion of tumor growth since it promotes the activation of the Wnt/β-catenin pathway [69], ending in either CAC or CRC [70,71]. However, in a DSS-induced experimental colitis model, TNF-α decreases tumor damage by promoting healing through the Wnt/β-catenin signaling pathway [72–74]. This regulates genes implicated in stem cell differentiation in the intestine (ID2, ITF2), cell proliferation (C-myc, CCND1), and cell migration (MMP7, MMP4) [75]. Likewise, it has been suggested that the systemic production of TNF-α protects against the spontaneous development of colitis and, consequently, of CAC development [76]. In addition, in sporadic CRC models (Apcmin/+ mice), genetic elimination of TNF-α does not decrease the tumor incidence [77]. In the case of the inflammatory cytokine IL-1β, it has been associated as a promoter of tumor development since it is capable of favoring angiogenesis, Th17-mediated inflammation, cancer cells proliferation, and suppression of the immune response during CRC and CAC [78–80] (Fig. 4). However, like TNF-α, IL-1β has a contrasting role since its exogenous injection has been implied in tumor regression [81]. In a different and interesting approach, in a murine model of CAC, it was described that the colon tumor burden decreased with the inhibition of NLRP3 inflammasome, while the tumor burden increased when low levels of IL-1β and IL-18 at the tumor site were observed [82]. Also, it has been described that the absence of the IL-1 receptor in myeloid cells, particularly neutrophils, is capable of increasing the intensity of CRC, while the genetic elimination of the IL-1 receptor (IL-1R1) only in epithelial cells decreases tumorigenesis in an APC model of CRC. Therefore, signaling through the IL-1 pathway causes specific responses depending on the type of cell analyzed, which determines both the intensity of the inflammation in the tumor microenvironment, and either the progression or reduction of CRC [54] (Fig. 4). IFN-γ is a major inflammatory cytokine that is capable of intensifying both the inflammatory and cytotoxic immune responses. IFN-γ is produced by CD4+, CD8+, and NK cells in response to stimuli generated in primed macrophages and neutrophils through IL-12 and IL-18 [83]. IFN-γ is involved in the suppression of IL-4 signaling and, therefore, of Th2 differentiation [83]. IFN-γ has a major role in the induction of apoptosis, regulation of cell proliferation and cell cycle, and in the activation of transcriptional factors [84,85]. Their signaling downstream involves type 1 and type 2 IFN-γ receptors (IFN-γR1 or IFN-γR2), which are directly linked to the signal transducer and activator of transcription 1 (STAT1), which in turn induces the expression of T-bet (T-box21), the master transcription factor controlling IFN-γ expression in T cells [85,86]. But what is the role of IFN-γ during CRC and CAC? We and others have shown that, in a murine model of CAC, IFN-γ-deficient mice (IFN-γ−/−) developed both a larger number of tumors and higher levels of anti-inflammatory cytokines than wild-type (WT) mice [87]. In Apcmin/+ IFN-γ−/− murine models, it was shown that the absence of IFN-γ induces a higher number of adenomas; consequently, approximately 50% of mice developed adenocarcinoma [87]. In vitro, knockdown of IFN-γR1 in the HT-29 colon cell line favors cell proliferation and colony formation, which is the first step to develop adenomas [87]. Moreover, loss of IFN-γ during CRC promotes the development of tumors and their progression toward adenocarcinoma [87]. In addition, in ovarian cancer cell lines and in tumor cells of ovarian tumors, IFN-γ had anti-proliferative activity through apoptosis induction [88]. Interestingly, it was shown that genetic variations in either IFN-γ or IFN-γR subunits are associated with CRC development, inducing a worse prognosis [87]. Also patients with IBD have distinct patterns of IFN-γ methylation [89], which is probably implicated to a pathological condition. All these results that implicate iNOS, IL-1β, TNF-α, and IFN-γ begin to support the hypothesis for a protective role of a Th1-mediated inflammatory response during CAC and CRC in both situations—when the tumor is already established and when the anti-inflammatory and suppressive immune response is increasing (Fig. 4).

Inflammatory transcription factors involved in protection

Similar to inflammatory cytokines, transcription factors associated with the inflammatory immune response play an important role during the development of CAC and CCR. For example, STAT1 and STAT4 are transcription factors promoting the inflammatory immune response and are directly associated with the production of interferons. In murine models, particularly during the early stages of the onset of CAC, it has been suggested that STAT1 plays a tumor suppressive role, through the control of cell proliferation and the promotion of apoptosis [4]. In addition, STAT1 has been shown to be able to suppress the growth of the LoVo and SW480 colon cell lines, by negatively regulating the miRNA-181a, which is associated with a poor prognosis in patients with CRC [90] (Fig. 4). Conversely, STAT6 is a transcription factor involved in promoting the anti-inflammatory response; it is activated by the binding of IL-4 and IL-13 to its receptor [3]. Both STAT6 and STAT1 are mutually excluding molecules. Interestingly and underlining the role of anti-inflammation in the development of CRC, it has been described that STAT6 is capable of negatively regulating the inflammatory cytokines involved in the development of CRC [3]. However, in several studies, it has been demonstrated that STAT6 is a tumor promoter. For example, in a murine model of CAC, it has been observed that the absence of STAT6 promotes a decrease in the number of tumors, generating a significant delay in tumor progression [3]. Also, STAT6 has been described as a promoter of metastasis during CRC since it favors the migration, invasion, proliferation, and resistance to apoptosis [3]. Both the cytokine IL-4 and its receptor, which are the triggers of the STAT6 pathway, contribute to the growth of tumors favoring the formation of aberrant crypts foci and cell proliferation during CAC [91]. On the other hand, IL-13-mediated activation of STAT6 favors the epithelial-mesenchymal transition, thereby increasing the tumor development of CRC [92]. As we mentioned above, T-bet transcription factor controls IFN-γ expression in T cells. Interestingly, therapy using anti-PD1 antibody in metastatic CRC patients with microsatellite instability has been shown to be effective [93]; thus, the blockade of this critical immune checkpoint generated a remarkable inhibition of CRC growth that was associated with higher number of T-bet+ infiltrating lymphocytes, which correlated with higher IFN-γ levels [93]. In an orthotopic murine model using MC38 colon adenocarcinoma cell line, it was shown that Wnt/β-catenin signaling induces an anergic state in tumor-infiltrating T cells, and the administration of anti-wnt3a antibody holds tumor growth and favors increased numbers of antigen-specific CD8+ T cells expressing both IFN-γ and T-bet [94]. In another elegant study, it was shown that the IFN-γ secretion is decreased in patients with CRC, and this compromised immune response correlated with increased metastasis to the liver and lung. When this observation was adapted into a murine model, the authors detected that such an effect was dependent on the deficiency in ribosomal s6 kinase 2 and that it must phosphorylate T-bet in serine 498 and 502. When T-bet phosphorylation was restored, the IFN-γ mRNA levels were also increased and the metastasis was remarkably reduced [95]. Again, together all these observations strongly suggest a protective or at least beneficial role of the inflammatory transcription factors associated with Th1-biased responses at later stages of CAC or CRC development.

Inflammatory immune cells involved in protection

In addition to the transcription factors and cytokines, the inflammatory-associated immune cells play an important role as protectors of both CAC and CRC, when the tumors have been established. For example, macrophages are cells that have an active role in inflammation and anti-inflammation; they display an enormous plasticity according to the microenvironment where they are activated. Macrophages can be differentiated as an inflammatory or M1 profile, when they are stimulated by lipopolysaccharides through the Toll-like receptor-4 (TLR-4) or by TNF-α through the TNF-α receptor (TNFR) or by IFN-γ through the IFN-γR [96]. Also, macrophages are able to be differentiated as an anti-inflammatory profile or M2, when they are activated by IL-4 or IL-13 through IL-4R [96]. In fact, in tumors of various types of cancer, macrophages are one of the main leukocytes infiltrated and are called tumor-associated macrophages (TAMs) [97]. The participation of TAMs has been associated with both protumor processes and antitumor processes. For example, in patients with CRC, high infiltration of M1 macrophages has been associated with a better prognosis [98]; similarly, TAMs infiltrated in CRC tumor tissue maintain an M1 profile, being able to inhibit proliferation of tumor cells through IFN-γ secretion [98]. Likewise, it has been shown that macrophages differentiated as an M1 profile can inhibit the cell division of different colon cancer cell lines [99]. Interestingly, like macrophages, neutrophils have been associated with protumor processes. However, it was shown that in patients with stage II CRC, high neutrophil infiltration is associated with a better prognosis [100]. In a mice model of CAC, it was shown that specific genetic depletion of neutrophils induced an increased growth, proliferation, and invasiveness of colon tumors; these results correlate with an increased bacterial load, DNA damage, and a higher inflammatory Th17-biased immune response [101]. A correlation study between neutrophils in different stages of CRC from patients surgically resected showed that poor infiltration of neutrophils, defined as CD66+ cells, indicated a worse prognosis during early CRC stages [102]. DCs are the major antigen-presenting cells; their origin, subtypes, and immune mechanisms are heterogeneous [103]. In general, they have been associated with anti-tumor immunity when DCs have an inflammatory phenotype characterized by a high IL-12 secretion and the expression of some activation molecules such as CD80 and CD86 [103]. However, tumor-associated DCs are defective in their antigen-presenting capacity because DCs activation phenotype is affected by the anti-inflammatory microenvironment [104], contributing in this way to the suppression of the local immune response [103]. Accordingly, the number of mature DCs, defined by CD83 and HLA-DR surface markers on 145 tissue samples of patients with CRC, was found in a lower level in the infiltrated immune cells, while immature DCs, defined as CD1a+ and S100+ protein surface markers, were found in higher numbers in the tumor stroma, mainly in advanced tumor stage [105]. It was suggested that a low level of maturation of DCs in the tumor microenvironment is caused by increased amounts of the chemokines CCL2, CXCL1, and CXCL5 and the growth factor VEGF, which together inhibit the upregulation of CD86, CD83, CD54, and HLA-DR surface markers and also the secretion of IL-12p70 [106]. Also, peripheral DCs from patients with CRC deteriorated in both numbers and functions, correlating with advanced stages of the disease [107]. There are few studies showing DCs as a tool for immunotherapy during CRC [107]. Experimental strategies from these studies focused on the isolation of the peripheral blood monocytes to differentiate them in DCs, and these DCs were loaded with tumor-derived antigens and reintroduced to the host or patient [107]. However, the success of this therapy also depends on the immunosuppressive environment established by the tumor [107], so DCs were unable to reverse such unfavorable microenvironment. Maybe, it is necessary to redesign the DCs used to have a better prognosis, and more basic immunological studies are also required to have a better knowledge on how the DCs activation impacts the CRC. Indeed, in a murine model, a major role for WNT/β-catenin pathway in the maturation phenotype of DCs was shown, because the anti-wnt3a antibody recovered the maturation capacity of DCs, which correlated with both lower tumor numbers and also an expansion of tumor antigen-specific CD8+ effector/memory T cells expressing T-bet and IFN-γ [94].

Regulatory T cells (Treg) have an important role in promoting immunosuppressive microenvironments; therefore, this cell type has a key impact on CRC and CAC [108,109]. Treg cells are a specialized subpopulation of CD4+ T cells expressing the transcription factor Foxp3 and the IL-2 receptor CD25, which modulate the immune response, aimed to avoid autoimmunity and exacerbated inflammation [110], but as a consequence, they induce strong immunosuppressive environments [110]. We and other researchers have been performing studies to understand the role of Treg cells during CAC and CRC development. In general, it has been observed that there was an increased number and immunosuppressive phenotype for Treg cells both at systemic level (blood) and at in situ colon tumorigenic microenvironment of CAC mice [108,109]; using a specific or indirect depletion of Treg cells, we and other researchers observed both an increased survival in the mice and a reduction in the number of tumors [108,109,111] Blood samples of CRC patients had an increased systemic and tumor-specific number of Treg cells, which was associated with later disease stages of CRC [110].

It is clear that immunosuppression during later stages of both CRC and CAC is a major event avoiding the correct action for the immune cells associated with inflammatory responses, which included M1 macrophages, DCs, and neutrophils, and understanding the close relationships between them with suppressive cells such as Treg cells, immature DCs, and M2 macrophages, we may probably be able to clarify the mechanisms necessary to define the best immunotherapy against CAC and CRC.

In recent years, an increasing amount of scientific research has been focused on how to manipulate the immune system in order to improve anticancer responses at both preclinical and clinical levels. Maybe, now is the time to take a look on how to modify the inflammatory and anti-inflammatory mechanisms to adequately impact the features of the tumor microenvironment (usually immunosuppressive) to improve clinical responses against CRC.

Immunotherapy and chemotherapy in colon cancer

The first written reference to cancer is in the Edwin Smith Papyrus, which is calculated to have been written in the 17th century BC by Imhotep, an Egyptian doctor. In this papyrus, a wide variety of medical conditions are described, along with their diagnosis, prognosis, and treatments. However, when mentioning cancer treatment, it is written "there is none" [112]. Until today, the available treatments for cancer patients are the following: surgery, of which there are records written by the historian Herodotus in the year 440 a.C. [112]; radiotherapy that began in the year 1895 [113]; chemotherapy that is developed at the beginning of the 20th century [114]; and recently, the immunotherapy in 1997 when Rituximab was approved by the FDA as the first antibody to treat cancer [115]. Currently, decisions on treatment of colon cancer patients are made based on the stage of cancer development. The classification of these stages is through the TNM staging system for malignant tumors, and the ranges are from stage 0 to stage IV. The lower stage is associated with greater survival rates and lower spread and severity of CRC [116]. The treatment of choice for patients with early-stage colon cancer and for some late-stage patients is surgery. However, the surgery effectiveness is greater when it is performed in early stages [117]. Also, chemotherapy is another option for patients who are not candidates for surgery and patients with stage II, III, and IV CRC [116,118]. The use of chemicals as a treatment for cancer had its origins in World War II when people exposed to mustard gas were found to have decreased white blood cell counts. Later, in 1943 Alfred Gilman and Louis Goodmanse demonstrated that nitrogen mustard, a chemical component of mustard gas, was responsible for the therapeutic effects that favored the regression of lymphoid tumors [114]. However, in 1960, one of the main topics of discussion among doctors was whether chemotherapy caused more harm than good [114]. Among the chemotherapy drugs most commonly used to treat colon cancer are oxaliplatin, 5-fluorouracil (5-FU), and leucovorin, which can be used in combination or individually. 5-FU has been available for over 60 years and is a drug with anti-proliferative activity, mainly inhibiting the enzyme thymidylate synthase (TS) [119], thus blocking the formation of thymidine required for DNA synthesis. 5-FU is a much more effective drug in combined therapy since the response rates for 5-FU-based chemotherapy as a first-line treatment for advanced colon cancer are only 10–15% [120]. This is due to the fact that 5-FU administered is rapidly broken down in the body by the enzyme dihydropyrimidine dehydrogenase. Therefore, the effects of 5-FU are low even at high concentrations [121]. Moreover, combined therapy administering 5-FU plus leucovorin as an adjuvant has had better efficacy since leucovorin is able to prevent the rapid degradation, increase the bioavailability, and increase the response rate of 5-FU. However, this combination has not shown a significant impact on survival [122]. In a study of patients with stage II CRC, it was evaluated whether 5-FU in combination with leucovorin (5-FU + LV) was an effective therapy after surgical resection. During a 5-year follow-up period, the average survival was observed to be 80% for patients without treatment and 82% for patients with

5-FU + LV, while the disease-free period was 73% for patients without treatment and 76% for patients with 5-FU + LV, showing a low benefit [123]. Likewise, according to the WHO classification, the toxicity of these drugs in men and women is grade 3/4 neutropenia (41.7% v 5.3% of patients), grade 3/4 diarrhea (11.9% v 5.3%), grade 1/2 alopecia, and grade 3 sensorineural toxicity (18.2% v 0%) [124,125]. However, chemotherapy has been one of the most promising treatments for cancer patients. Without chemotherapy, the survival rate in patients with metastatic colon cancer was 8 months. With 5-FU, this rate increased up to 12 months [126]. Furthermore, emerging adjuvant therapies such as immunotherapy has shown promise treating various types of cancer. Immunotherapy aims to activate or modulate the immune system to respond to tumor-specific antigens [127]. In colon cancer, patient's immunotherapy has been shown to prolong remission periods and increase life expectancy. The main immunotherapeutic approaches are grouped into six categories: (1) monoclonal antibodies (mAbs), (2) checkpoint inhibitors, (3) cytokines, (4) adoptive cell therapy, (5) therapeutic vaccines, and (6) oncolytic viruses [128]. Cetuximab, bevacizumab, and panitumumab are mAbs that block molecules displayed on tumor cells surface, such as epidermal growth factor (EGFR) and vascular endothelial growth factor (VEGF). In monotherapy, cetuximab and panitumunab have shown significantly better response rates and progression-free survival compared to patients treated with chemotherapy [129]. Furthermore, the use of immune checkpoint inhibitors such as ipilimumab and pembrolizumab that block CTLA-4 and PD1 (molecules that inhibit immune cell activation), in a combined immunotherapy, is able to improve specific anti-tumor immune responses compared to any drug alone [130]. Likewise, the efficacy and tolerability of these drugs have been shown to be better than those of conventional treatments [131–133]. On the other hand, cytokines are low-molecular-weight proteins that can control anti-tumor response activities in immune system cells. Currently, only two cytokines IL-2 and IFN-α have been approved by the Food and Drug Administration (FDA) for the treatment of various malignant diseases, including colon cancer [134,135]. Vaccines and adoptive cell therapy are other types of immunotherapies whose development is based on the use of tumor cells to activate natural killer (NK) cells and DCs to produce antigens. The antigens produced are used as vaccines [128,136]. These vaccines target tumor-associated antigens such as carcinoembryonic antigen (CEA) and human chorionic gonadotropin (hCG) beta. However, so far, most of these vaccines have shown no benefit with respect to patient survival [137]. On the other hand, adoptive cell therapy, with activated NK cells reintroduced to patients to improve anti-tumor response, has been considered as one of the most promising immunotherapy treatments [138] because they could generate immune memory [139]. In addition to NK cells, peripheral blood mononuclear cells as well as αβ T cells and γδ T cells are used for adoptive cell therapy. Oncolytic viruses are an innovative therapy and act by tumor dissemination where the virus selectively replicates and kills cancer cells without damaging the normal tissue. Natural immunogenicity toward viral vectors acts as an adjuvant by increasing the inflammatory immune response, which induces anti-tumor immunity [140]. Furthermore, treatment with oncolytic viruses alone or in combination with chemotherapy in patients with colon cancer has been shown to be effective in activating T cells [141,142]. Therefore, chemotherapy has been one of the most promising therapies for cancer patients. However, factors such as toxicity, reduced survival, and in some cases shorter disease-free periods make immunotherapy a promising option because it removes restrictions for the immune system to find and destroy cancer cells, it provides long-term protection due to immune memory, and the toxicity is significantly less. Thus, a combination of these two approaches (immunochemotherapy) may improve both strategies. An approximation to this idea has been tested recently, by blocking STAT6 activity and administrating 5-FU in a mice model of CAC. In the study, it was demonstrated that the blockade of STAT6 activity significantly enhanced the effect of 5-FU by modulating the inflammatory microenvironment that leads to a significant reduction in colon tumorigenesis [143].

Is the microbiota involved in the promotion of inflammation during CAC and CRC?

The colon comprises the final part of the digestive tract. Histologically, it is composed of submucosa, muscular mucosa, lamina propria, and mucosa. The lamina propria is a region harboring immune cells such as macrophages, neutrophils, dendritic cells, and T lymphocytes, while the mucosa is a region in the intestinal lumen and is composed of type II mucins (MUC2), which are glycosylated proteins produced by goblet cells. The mucosa plays a very important role as a protective barrier between immune system cells and microbiota. The microbiota is a group of microorganisms that range from bacteria, viruses, fungi, and protozoa [144]. It is estimated that in the colon, there are about 10 billion bacteria of which 99% are anaerobic and belong to the *Firmicutes* (64%), *Bacteroides* (23%), *Proteobacteria* (8%), and *Actinobacteria* (8%) classes [145]. Colonization of the microbiota in the colon begins immediately after birth, and slowly, its composition begins to vary between stages of life, individuals, and geographic regions. In babies, the intestinal microbiota has great dynamics in its diversity due to the influence of diet or antibiotic treatments, which results in large changes in the relative abundance of different taxonomic groups [146]. In adulthood, the environment may influence the abundance of some species over others, which in some individuals may be stable in the short term (<1 year), while in others may be stable in the long term (>10 years) [147]. An analysis of 13,355 prokaryotic ribosomal RNA gene sequences from multiple sites in the colon mucosa and feces from healthy subjects demonstrated a great variability in the composition of the microbiota between each one of them [145]. Similarly, this variability can also be influenced by geographic region due to the differences in social structures, such as cultural traditions, diet, exposure to pets, livestock, and many other factors that could affect the flow of microbes and microbial genes among the members of a household [148]. Furthermore, microbiota is essential for the maintenance and intestinal epithelium function, homeostasis, and host defense. In fact, a large part of microbiota functions in the intestinal epithelium is through metabolic reactions of carbohydrates, proteins, and vitamins that were not digested and absorbed during digestion. For example, during carbohydrate digestion, some of them are not digested because human enzymes are not capable of breaking down complex carbohydrates. Therefore, they are metabolized by the microbiota to grow and survive, while the resulting metabolites such as short-chain fatty acids, mainly acetate, propionate, and butyrate, are finally used as substrates by humans for processes such as lipogenesis and gluconeogenesis [149] as well as for epithelial barrier maintenance and immune system regulation [150,151]. For example, butyrate decreases the expression of inflammatory cytokines such as IL-6, IL-1β, TNF-α, and TNF-β. Furthermore, it is capable of inhibiting NF-kB activation in immune cells of the lamina propria [152]. Likewise, dietary proteins ingested can be broken down to amino acids in the large intestine, due to the proteolytic activity of bacteria. This leads to the production of metabolites such as short-chain fatty acids, polyamines, hydrogen sulfate, and carbon dioxide [153], which modulate the expression of intercellular junction proteins such as E-cadherin, that is critical for maintaining intestinal epithelial barrier integrity [154]. Microbiota is also capable of synthesizing vitamin K and vitamin B12 that are essential for the host's metabolic reactions. In fact, there is a high prevalence of genes related to biosynthetic pathways of these vitamins in dominant species' edges of colon microbiota [155]. This mutually beneficial interaction between microbiota, intestinal epithelium, and immune system in colon mucosa generates homeostasis, also known as symbiosis. This symbiosis is maintained through detection of microbial signals from intestinal lumen. The two main receptors responsible for this detection are pattern recognition receptors such as isoforms of nucleotide-binding oligomerization domain/caspase recruitment domain (NOD/CARD) in cytoplasm cells, and TLRs on epithelial cells and immune cells from lamina propria. In addition, some molecules such as immunoglobulin A (IgA), defensins, and antimicrobial peptides are secreted by epithelial cells from the mucosa toward the intestinal lumen. These molecules and TLRs and NOD/CARD receptors help to limit the exponential growth of microbiota microorganisms and prevent intense inflammatory responses triggered by the activation of immune cells [156]. However, when there is a sudden change in the composition of resident microbiota in healthy individuals, known as dysbiosis [157], it can be a result of medical interventions due to the use of antibiotics, changes in diet, as well as defects in the immune system; these microbiota changes may also contribute to develop CRC. Genetic studies have suggested an association between polymorphisms in genes for TLR-2 and TLR-4 and the risk of developing colon cancer [158]. Other studies have shown that ethnic groups with a low risk of developing colon cancer can acquire a high risk of developing it when they migrate to countries or regions where there is a high incidence [159]. This association between dysbiosis and the risk of developing colon cancer has extensively been studied, and it has been shown that the microbiota of CRC patients and healthy individuals differ significantly. For example, a significantly higher abundance of *Firmicutes* and *Fusobacteria* in cancerous tissues was found in healthy individuals, while *Proteobacteria* was less abundant in CRC patients [160]. In fecal samples of patients with CD and UC, the following bacterial families types were found: *Faecalibacterium*, *Eubacterium*, *Roseburia*, *Lachnospiraceae*, and *Ruminococcaceae* [161]. On the other hand, the microbiota can also have both a promoter role and a protective role during colon cancer development. For example, *Bacteroides fragilis*, as well as a group of clostridial strains within clusters IV, XIVa, and XVIII, are known to induce Treg cells and anti-inflammatory cytokines, which can protect against experimental colitis [162]. Likewise, *Faecalibacterium prausnitzii* is a clostridial microorganism whose presence in the human gut is correlated with protection from IBD, although the mechanism is not understood [163,164]. In contrast, *Fusobacterium nucleatum* is an oral symbiont, an occasional pathogen that has been identified in, and cultured from, intestinal tumors [165], and a progressive increase in the prevalence of *Fusobacterium nucleatum* has been observed through the CRC stages beginning with highly dysplastic adenomas [166]. In fact, the *Fusobacterium nucleatum* colonization has been associated with a shorter survival in CRC patients [167]. *Enterococcus faecalis* induces colitis in experimentally susceptible animal models through DNA damage and also induces chromosomal instability in epithelial cells [168]. In fact, the use of the intestinal microbiota as a screening biomarker in fecal samples has been suggested at present to detect and forecast CRC. In addition, it is possible to take advantage of the intestinal microbiota for chemotherapy and immunotherapy in the CRC, since it may improve the efficacy and toxicity of immunotherapy and chemotherapy, respectively [169,170].

Conclusion

Inflammation is a physiological response of the body whose main function is to protect the host during infectious and non-infectious processes. In the past, it had been suggested that when the inflammatory response is exacerbated, it can contribute to the development of diseases such as CRC. However, in some recent works that we have mentioned in this chapter, it has been shown that Th1-mediated inflammation plays an important and major role in CRC protection since the number of tumors and the aggressiveness of the disease in CRC increase in the absence of cells, soluble molecules, and transcription factors associated with this type of inflammation, emphasizing that this process occurs once the tumor is established. Interestingly, when these inflammatory molecules have a primary role in chronic tumor development processes, during either the CAC or CRC, there is resistance to the tumor development and the survival rate is increased. Therefore, the suggestion that the generation of an inflammatory response will finalize in CAC or CRC may not be correct. However, if in such an inflammatory response a dominant Th17-biased activity is observed, then the outcome dramatically changes and an accelerated CRC development may be expected. The development of CAC or CRC needs mutations that may be indirectly caused by the acute inflammatory response, which changes the sense of the anti-inflammatory response and generates strong immunosuppression. There is an open immunological field, where it will be necessary to investigate in depth the role of inflammation during the development of chronic degenerative diseases, which will help us, in the near future, to establish preventive therapies that will reduce the incidence and mortality of CRC and CAC. In this review, it became clear that basic biomedical science is developing immunotherapy mechanisms, which most likely in the near future will be applied and directed to patients with certain signs, symptoms, and levels of development of CAC and CRC. Finally, due to the cellular heterogeneity and the plasticity of the phenotype of epithelial, immune, and tumor cells, we should not classify inflammation as the unique cause for the promotion of colon cancer development. But we should take advantage of blocking or favoring this response together with chemotherapeutic agents in order to improve treatments, where the necessity of combined therapies is clear. Then, we must begin to dimension the role of inflammation as a part of a protective immune response during CAC and CRC once the disease has been established.

Acknowledgments

This work is part of the requirements for Itzel Medina-Andrade to obtain her PhD degree at Programa de Doctorado en Ciencias Biómedicas, UNAM. This work was supported by Consejo Nacional de Ciencia y Tecnología: A1-S-37879. Dirección General de Asuntos del Personal Académico, PAPIIT, UNAM: IN226519 for LIT and IA209720 for JEO.

References

[1] Jess T, Frisch M, Simonsen J. Trends in overall and cause-specific mortality among patients with inflammatory bowel disease from 1982 to 2010. Clin Gastroenterol Hepatol 2013;11(1):43–8.
[2] Dulai PS, Sandborn WJ, Gupta S. Colorectal cancer and dysplasia in inflammatory bowel disease: a review of disease epidemiology, pathophysiology, and management. Cancer Prev Res (Phila) 2016;9(12):887–94.
[3] Leon-Cabrera SA, et al. Lack of STAT6 attenuates inflammation and drives protection against early steps of colitis-associated colon cancer. Cancer Immunol Res 2017;5(5):385–96.
[4] Leon-Cabrera S, et al. Deficiency in STAT1 signaling predisposes gut inflammation and prompts colorectal cancer development. Cancers (Basel) 2018;10(9):341.
[5] Chen L, et al. Inflammatory responses and inflammation-associated diseases in organs. Oncotarget 2018;9 (6):7204–18.
[6] Mowat AM, Agace WW. Regional specialization within the intestinal immune system. Nat Rev Immunol 2014;14(10):667–85.
[7] Terzic J, et al. Inflammation and colon cancer. Gastroenterology 2010;138(6):2101–14 e5.
[8] Bhattacharya P, et al. Genetically modified live attenuated *Leishmania donovani* parasites induce innate immunity through classical activation of macrophages that direct the Th1 response in mice. Infect Immun 2015;83 (10):3800–15.
[9] Wang S, et al. Functions of macrophages in the maintenance of intestinal homeostasis. J Immunol Res 2019;2019:1512969.
[10] Reid SD, Penna G, Adorini L. The control of T cell responses by dendritic cell subsets. Curr Opin Immunol 2000;12(1):114–21.
[11] Recchiuti A, Mattoscio D, Isopi E. Roles, actions, and therapeutic potential of specialized pro-resolving lipid mediators for the treatment of inflammation in cystic fibrosis. Front Pharmacol 2019;10:252.
[12] Nussbaum MS. Arginine stimulates wound healing and immune function in elderly human beings. JPEN J Parenter Enteral Nutr 1994;18(2):194.
[13] Straus DS, et al. 15-deoxy-delta 12,14-prostaglandin J2 inhibits multiple steps in the NF-kappa B signaling pathway. Proc Natl Acad Sci U S A 2000;97(9):4844–9.
[14] Bell-Parikh LC, et al. Biosynthesis of 15-deoxy-delta12,14-PGJ2 and the ligation of PPARgamma. J Clin Invest 2003;112(6):945–55.
[15] Colville-Nash PR, et al. Prostaglandin F2alpha produced by inducible cyclooxygenase may contribute to the resolution of inflammation. Inflammopharmacology 2005;12(5–6):473–6 discussion 477-80.
[16] Zhu J, Paul WE. Heterogeneity and plasticity of T helper cells. Cell Res 2010;20(1):4–12.
[17] Cosin-Roger J, et al. The activation of Wnt signaling by a STAT6-dependent macrophage phenotype promotes mucosal repair in murine IBD. Mucosal Immunol 2016;9(4):986–98.
[18] Whyte JL, Smith AA, Helms JA. Wnt signaling and injury repair. Cold Spring Harb Perspect Biol 2012;4(8): a008078.
[19] Ciacci C, Lind SE, Podolsky DK. Transforming growth factor beta regulation of migration in wounded rat intestinal epithelial monolayers. Gastroenterology 1993;105(1):93–101.
[20] Beck PL, et al. Transforming growth factor-beta mediates intestinal healing and susceptibility to injury in vitro and in vivo through epithelial cells. Am J Pathol 2003;162(2):597–608.
[21] Sumagin R, et al. Neutrophil interactions with epithelial-expressed ICAM-1 enhances intestinal mucosal wound healing. Mucosal Immunol 2016;9(5):1151–62.
[22] Spiller KL, et al. The role of macrophage phenotype in vascularization of tissue engineering scaffolds. Biomaterials 2014;35(15):4477–88.
[23] Pull SL, et al. Activated macrophages are an adaptive element of the colonic epithelial progenitor niche necessary for regenerative responses to injury. Proc Natl Acad Sci U S A 2005;102(1):99–104.
[24] Bernasconi E, et al. Granulocyte-macrophage colony-stimulating factor elicits bone marrow-derived cells that promote efficient colonic mucosal healing. Inflamm Bowel Dis 2010;16(3):428–41.
[25] Okeke EB, Uzonna JE. The pivotal role of regulatory T cells in the regulation of innate immune cells. Front Immunol 2019;10:680.
[26] Pickert G, et al. STAT3 links IL-22 signaling in intestinal epithelial cells to mucosal wound healing. J Exp Med 2009;206(7):1465–72.
[27] Grivennikov SI. Inflammation and colorectal cancer: colitis-associated neoplasia. Semin Immunopathol 2013;35 (2):229–44.

[28] Grivennikov SI, Cominelli F. Colitis-associated and sporadic colon cancers: different diseases, different mutations? Gastroenterology 2016;150(4):808–10.
[29] Fearon ER, Vogelstein B. A genetic model for colorectal tumorigenesis. Cell 1990;61(5):759–67.
[30] Itzkowitz SH, Yio X. Inflammation and cancer IV. Colorectal cancer in inflammatory bowel disease: the role of inflammation. Am J Physiol Gastrointest Liver Physiol 2004;287(1):G7–17.
[31] Rigas B. Oncogenes and suppressor genes: their involvement in colon cancer. J Clin Gastroenterol 1990;12 (5):494–9.
[32] Li XL, et al. P53 mutations in colorectal cancer – molecular pathogenesis and pharmacological reactivation. World J Gastroenterol 2015;21(1):84–93.
[33] Aoki K, Taketo MM. Adenomatous polyposis coli (APC): a multi-functional tumor suppressor gene. J Cell Sci 2007;120(Pt 19):3327–35.
[34] Barker N, et al. Crypt stem cells as the cells-of-origin of intestinal cancer. Nature 2009;457(7229):608–11.
[35] Kwong LN, Dove WF. APC and its modifiers in colon cancer. Adv Exp Med Biol 2009;656:85–106.
[36] Long AG, Lundsmith ET, Hamilton KE. Inflammation and colorectal cancer. Curr Colorectal Cancer Rep 2017;13(4):341–51.
[37] Moossavi S, Bishehsari F. Inflammation in sporadic colorectal cancer. Arch Iran Med 2012;15(3):166–70.
[38] Ahn B, Ohshima H. Suppression of intestinal polyposis in Apc(Min/+) mice by inhibiting nitric oxide production. Cancer Res 2001;61(23):8357–60.
[39] Galon J, et al. Type, density, and location of immune cells within human colorectal tumors predict clinical outcome. Science 2006;313(5795):1960–4.
[40] Soh JS, et al. Immunoprofiling of colitis-associated and sporadic colorectal cancer and its clinical significance. Sci Rep 2019;9(1):6833.
[41] Lakatos PL, Lakatos L. Risk for colorectal cancer in ulcerative colitis: changes, causes and management strategies. World J Gastroenterol 2008;14(25):3937–47.
[42] Adami HO, et al. The continuing uncertainty about cancer risk in inflammatory bowel disease. Gut 2016;65 (6):889–93.
[43] Fung BM, Lindor KD, Tabibian JH. Cancer risk in primary sclerosing cholangitis: epidemiology, prevention, and surveillance strategies. World J Gastroenterol 2019;25(6):659–71.
[44] Rhodes JM, Campbell BJ. Inflammation and colorectal cancer: IBD-associated and sporadic cancer compared. Trends Mol Med 2002;8(1):10–6.
[45] Lakatos L, et al. Risk factors for ulcerative colitis-associated colorectal cancer in a Hungarian cohort of patients with ulcerative colitis: results of a population-based study. Inflamm Bowel Dis 2006;12(3):205–11.
[46] Rutter MD, et al. Thirty-year analysis of a colonoscopic surveillance program for neoplasia in ulcerative colitis. Gastroenterology 2006;130(4):1030–8.
[47] Monin L, Gaffen SL. Interleukin 17 family cytokines: signaling mechanisms, biological activities, and therapeutic implications. Cold Spring Harb Perspect Biol 2018;10(4)a028522.
[48] Acosta-Rodriguez EV, et al. Surface phenotype and antigenic specificity of human interleukin 17-producing T helper memory cells. Nat Immunol 2007;8(6):639–46.
[49] Wang K, Karin M. Tumor-elicited inflammation and colorectal cancer. Adv Cancer Res 2015;128:173–96.
[50] Razi S, et al. IL-17 and colorectal cancer: from carcinogenesis to treatment. Cytokine 2019;116:7–12.
[51] Martin M, et al. RORgammat(+) hematopoietic cells are necessary for tumor cell proliferation during colitis-associated tumorigenesis in mice. Eur J Immunol 2015;45(6):1667–79.
[52] Hurtado CG, et al. Roles for interleukin 17 and adaptive immunity in pathogenesis of colorectal cancer. Gastroenterology 2018;155(6):1706–15.
[53] Grivennikov SI, et al. Adenoma-linked barrier defects and microbial products drive IL-23/IL-17-mediated tumour growth. Nature 2012;491(7423):254–8.
[54] Dmitrieva-Posocco O, et al. Cell-type-specific responses to interleukin-1 control microbial invasion and tumor-elicited inflammation in colorectal cancer. Immunity 2019;50(1):166–80. e7.
[55] Tosolini M, et al. Clinical impact of different classes of infiltrating T cytotoxic and helper cells (Th1, th2, treg, th17) in patients with colorectal cancer. Cancer Res 2011;71(4):1263–71.
[56] Cui G, et al. IL-17A in the tumor microenvironment of the human colorectal adenoma-carcinoma sequence. Scand J Gastroenterol 2012;47(11):1304–12.
[57] Chung AS, et al. An interleukin-17-mediated paracrine network promotes tumor resistance to anti-angiogenic therapy. Nat Med 2013;19(9):1114–23.

[58] Omrane I, et al. Significant association between interleukin-17A polymorphism and colorectal cancer. Tumour Biol 2014;35(7):6627–32.
[59] Hanahan D, Weinberg RA. The hallmarks of cancer. Cell 2000;100(1):57–70.
[60] Wiseman H, Halliwell B. Damage to DNA by reactive oxygen and nitrogen species: role in inflammatory disease and progression to cancer. Biochem J 1996;313(Pt 1):17–29.
[61] Wink DA, et al. The multifaceted roles of nitric oxide in cancer. Carcinogenesis 1998;19(5):711–21.
[62] Lelchuk R, et al. Constitutive and inducible nitric oxide synthases in human megakaryoblastic cells. J Pharmacol Exp Ther 1992;262(3):1220–4.
[63] Thomsen LL, et al. Nitric oxide synthase activity in human gynecological cancer. Cancer Res 1994;54(5):1352–4.
[64] Thomsen LL, et al. Nitric oxide synthase activity in human breast cancer. Br J Cancer 1995;72(1):41–4.
[65] Chung P, et al. Overexpression of the human inducible nitric oxide synthase gene enhances radiation-induced apoptosis in colorectal cancer cells via a caspase-dependent mechanism. Nitric Oxide 2003;8(2):119–26.
[66] Adams C, et al. Nitric oxide synthase gene therapy enhances the toxicity of cisplatin in cancer cells. J Gene Med 2009;11(2):160–8.
[67] Seril DN, Liao J, Yang GY. Colorectal carcinoma development in inducible nitric oxide synthase-deficient mice with dextran sulfate sodium-induced ulcerative colitis. Mol Carcinog 2007;46(5):341–53.
[68] Zhang R, et al. Induction of inducible nitric oxide synthase: a protective mechanism in colitis-induced adenocarcinoma. Carcinogenesis 2007;28(5):1122–30.
[69] Oguma K, et al. Activated macrophages promote Wnt signalling through tumour necrosis factor-alpha in gastric tumour cells. EMBO J 2008;27(12):1671–81.
[70] Klampfer L. Cytokines, inflammation and colon cancer. Curr Cancer Drug Targets 2011;11(4):451–64.
[71] Mager LF, et al. Cytokine-induced modulation of colorectal cancer. Front Oncol 2016;6:96.
[72] Luissint AC, Parkos CA, Nusrat A. Inflammation and the intestinal barrier: leukocyte-epithelial cell interactions, cell junction remodeling, and mucosal repair. Gastroenterology 2016;151(4):616–32.
[73] Bradford EM, et al. Epithelial TNF receptor signaling promotes mucosal repair in inflammatory bowel disease. J Immunol 2017;199(5):1886–97.
[74] Wang Y, et al. Protective role of tumor necrosis factor (TNF) receptors in chronic intestinal inflammation: TNFR1 ablation boosts systemic inflammatory response. Lab Invest 2013;93(9):1024–35.
[75] Herbst A, et al. Comprehensive analysis of beta-catenin target genes in colorectal carcinoma cell lines with deregulated Wnt/beta-catenin signaling. BMC Genomics 2014;15:74.
[76] Hale LP, Greer PK. A novel murine model of inflammatory bowel disease and inflammation-associated colon cancer with ulcerative colitis-like features. PLoS ONE 2012;7(7):e41797.
[77] Sakai H, et al. Genetic ablation of Tnfalpha demonstrates no detectable suppressive effect on inflammation-related mouse colon tumorigenesis. Chem Biol Interact 2010;184(3):423–30.
[78] Kaler P, Augenlicht L, Klampfer L. Macrophage-derived IL-1beta stimulates Wnt signaling and growth of colon cancer cells: a crosstalk interrupted by vitamin D3. Oncogene 2009;28(44):3892–902.
[79] Bunt SK, et al. Reduced inflammation in the tumor microenvironment delays the accumulation of myeloid-derived suppressor cells and limits tumor progression. Cancer Res 2007;67(20):10019–26.
[80] Matsuo Y, et al. IL-1alpha secreted by colon cancer cells enhances angiogenesis: the relationship between IL-1alpha release and tumor cells' potential for liver metastasis. J Surg Oncol 2009;99(6):361–7.
[81] North RJ, et al. Interleukin 1-induced, T cell-mediated regression of immunogenic murine tumors. Requirement for an adequate level of already acquired host concomitant immunity. J Exp Med 1988;168(6):2031–43.
[82] Allen IC, et al. The NLRP3 inflammasome functions as a negative regulator of tumorigenesis during colitis-associated cancer. J Exp Med 2010;207(5):1045–56.
[83] Aliberti J, et al. Lipoxin-mediated inhibition of IL-12 production by DCs: a mechanism for regulation of microbial immunity. Nat Immunol 2002;3(1):76–82.
[84] Isaacs A, Lindenmann J. Virus interference. I. The interferon. By A. Isaacs and J. Lindenmann, 1957. J Interferon Res 1987;7(5):429–38.
[85] Bach EA, Aguet M, Schreiber RD. The IFN gamma receptor: a paradigm for cytokine receptor signaling. Annu Rev Immunol 1997;15:563–91.
[86] Boehm U, et al. Cellular responses to interferon-gamma. Annu Rev Immunol 1997;15:749–95.
[87] Wang L, et al. Deficiency of interferon-gamma or its receptor promotes colorectal cancer development. J Interferon Cytokine Res 2015;35(4):273–80.

[88] Wall L, et al. IFN-gamma induces apoptosis in ovarian cancer cells in vivo and in vitro. Clin Cancer Res 2003;9 (7):2487–96.
[89] Gonsky R, et al. Distinct IFNG methylation in a subset of ulcerative colitis patients based on reactivity to microbial antigens. Inflamm Bowel Dis 2011;17(1):171–8.
[90] Zhang X, et al. STAT1 inhibits MiR-181a expression to suppress colorectal cancer cell proliferation through PTEN/Akt. J Cell Biochem 2017;118(10):3435–43.
[91] Koller FL, et al. Epithelial interleukin-4 receptor expression promotes colon tumor growth. Carcinogenesis 2010;31(6):1010–7.
[92] Cao H, et al. IL-13/STAT6 signaling plays a critical role in the epithelial-mesenchymal transition of colorectal cancer cells. Oncotarget 2016;7(38):61183–98.
[93] Ott E, et al. The density of Tbet+ tumor-infiltrating T lymphocytes reflects an effective and druggable preexisting adaptive antitumor immune response in colorectal cancer, irrespective of the microsatellite status. Onco Targets Ther 2019;8(4):e1562834.
[94] Pacella I, et al. Wnt3a neutralization enhances T-cell responses through indirect mechanisms and restrains tumor growth. Cancer Immunol Res 2018;6(8):953–64.
[95] Yao K, et al. RSK2 phosphorylates T-bet to attenuate colon cancer metastasis and growth. Proc Natl Acad Sci U S A 2017;114(48):12791–6.
[96] Snyder RJ, et al. Macrophages: a review of their role in wound healing and their therapeutic use. Wound Repair Regen 2016;24(4):613–29.
[97] Siveen KS, Kuttan G. Role of macrophages in tumour progression. Immunol Lett 2009;123(2):97–102.
[98] Ong SM, et al. Macrophages in human colorectal cancer are pro-inflammatory and prime T cells towards an anti-tumour type-1 inflammatory response. Eur J Immunol 2012;42(1):89–100.
[99] Engstrom A, et al. Conditioned media from macrophages of M1, but not M2 phenotype, inhibit the proliferation of the colon cancer cell lines HT-29 and CACO-2. Int J Oncol 2014;44(2):385–92.
[100] Berry RS, et al. High levels of tumor-associated neutrophils are associated with improved overall survival in patients with stage II colorectal cancer. PLoS ONE 2017;12(12):e0188799.
[101] Triner D, et al. Neutrophils restrict tumor-associated microbiota to reduce growth and invasion of colon tumors in mice. Gastroenterology 2019;156(5):1467–82.
[102] Wikberg ML, et al. Neutrophil infiltration is a favorable prognostic factor in early stages of colon cancer. Hum Pathol 2017;68:193–202.
[103] Veglia F, Gabrilovich DI. Dendritic cells in cancer: the role revisited. Curr Opin Immunol 2017;45:43–51.
[104] Tran Janco JM, et al. Tumor-infiltrating dendritic cells in cancer pathogenesis. J Immunol 2015;194 (7):2985–91.
[105] Gulubova MV, et al. Role of dendritic cells in progression and clinical outcome of colon cancer. Int J Colorectal Dis 2012;27(2):159–69.
[106] Michielsen AJ, et al. Tumour tissue microenvironment can inhibit dendritic cell maturation in colorectal cancer. PLoS ONE 2011;6(11):e27944.
[107] Gessani S, Belardelli F. Immune dysfunctions and immunotherapy in colorectal cancer: the role of dendritic cells. Cancers (Basel) 2019;11(10):1491.
[108] Pastille E, et al. Transient ablation of regulatory T cells improves antitumor immunity in colitis-associated colon cancer. Cancer Res 2014;74(16):4258–69.
[109] Olguin JE, et al. Early and partial reduction in CD4(+)Foxp3(+) regulatory T cells during colitis-associated colon cancer induces CD4(+) and CD8(+) T cell activation inhibiting tumorigenesis. J Cancer 2018;9(2):239–49.
[110] Miyara M, Sakaguchi S. Natural regulatory T cells: mechanisms of suppression. Trends Mol Med 2007;13 (3):108–16.
[111] Akeus P, et al. Regulatory T cells control endothelial chemokine production and migration of T cells into intestinal tumors of APC(min/+) mice. Cancer Immunol Immunother 2018;67(7):1067–77.
[112] Mukherjee S. The emperor of all maladies: a biography of cancer. New York, NY, USA: Scribner; 2010.
[113] Lederman M. The early history of radiotherapy: 1895-1939. Int J Radiat Oncol Biol Phys 1981;7(5):639–48.
[114] DeVita Jr. VT, Chu E. A history of cancer chemotherapy. Cancer Res 2008;68(21):8643–53.
[115] Schilder R. Rituximab immunotherapy. Cancer Biother Radiopharm 1999;14(4):237–40.
[116] Labianca R, et al. Primary colon cancer: ESMO clinical practice guidelines for diagnosis, adjuvant treatment and follow-up. Ann Oncol 2010;21(Suppl 5):v70–7.

[117] Mastalier B, et al. Surgical treatment of colon cancer: colentina surgical clinic experience. J Med Life 2012;5 (3):348–53.
[118] Wolpin BM, Mayer RJ. Systemic treatment of colorectal cancer. Gastroenterology 2008;134(5):1296–310.
[119] Longley DB, Harkin DP, Johnston PG. 5-fluorouracil: mechanisms of action and clinical strategies. Nat Rev Cancer 2003;3(5):330–8.
[120] Johnston PG, Kaye S. Capecitabine: a novel agent for the treatment of solid tumors. Anticancer Drugs 2001;12 (8):639–46.
[121] Tanaka F, et al. The history, mechanism and clinical use of oral 5-fluorouracil derivative chemotherapeutic agents. Curr Pharm Biotechnol 2000;1(2):137–64.
[122] Modulation of fluorouracil by leucovorin in patients with advanced colorectal cancer: evidence in terms of response rate. Advanced colorectal cancer meta-analysis project. J Clin Oncol 1992;10(6):896–903.
[123] Efficacy of adjuvant fluorouracil and folinic acid in B2 colon cancer. International Multicentre Pooled Analysis of B2 Colon Cancer Trials (IMPACT B2) Investigators. J Clin Oncol 1999;17(5):1356–63.
[124] de Gramont A, et al. Leucovorin and fluorouracil with or without oxaliplatin as first-line treatment in advanced colorectal cancer. J Clin Oncol 2000;18(16):2938–47.
[125] Wiela-Hojenska A, et al. Evaluation of the toxicity of anticancer chemotherapy in patients with colon cancer. Adv Clin Exp Med 2015;24(1):103–11.
[126] Schrag D. The price tag on progress – chemotherapy for colorectal cancer. N Engl J Med 2004;351(4):317–9.
[127] Koido S, et al. Immunotherapy for colorectal cancer. World J Gastroenterol 2013;19(46):8531–42.
[128] Abakushina EV, et al. Immunotherapeutic approaches for the treatment of colorectal cancer. Biochemistry (Mosc) 2019;84(7):720–8.
[129] Jean GW, Shah SR. Epidermal growth factor receptor monoclonal antibodies for the treatment of metastatic colorectal cancer. Pharmacotherapy 2008;28(6):742–54.
[130] Drake CG. Combination immunotherapy approaches. Ann Oncol 2012;23(Suppl 8):viii. 41–6.
[131] Brahmer JR, et al. Phase I study of single-agent anti-programmed death-1 (MDX-1106) in refractory solid tumors: safety, clinical activity, pharmacodynamics, and immunologic correlates. J Clin Oncol 2010;28 (19):3167–75.
[132] Yaghoubi N, et al. PD-1/PD-L1 blockade as a novel treatment for colorectal cancer. Biomed Pharmacother 2019;110:312–8.
[133] O'Neil BH, et al. Safety and antitumor activity of the anti-PD-1 antibody pembrolizumab in patients with advanced colorectal carcinoma. PLoS ONE 2017;12(12):e0189848.
[134] Berraondo P, et al. Cytokines in clinical cancer immunotherapy. Br J Cancer 2019;120(1):6–15.
[135] Zhu Y, et al. Efficacy of postoperative adjuvant transfusion of cytokine-induced killer cells combined with chemotherapy in patients with colorectal cancer. Cancer Immunol Immunother 2013;62(10):1629–35.
[136] Sun X, Suo J, Yan J. Immunotherapy in human colorectal cancer: challenges and prospective. World J Gastroenterol 2016;22(28):6362–72.
[137] Bilusic M, et al. Phase I trial of a recombinant yeast-CEA vaccine (GI-6207) in adults with metastatic CEA-expressing carcinoma. Cancer Immunol Immunother 2014;63(3):225–34.
[138] Kim JS, et al. Cell-based immunotherapy for colorectal cancer with cytokine-induced killer cells. Immune Netw 2016;16(2):99–108.
[139] Powell Jr. DJ, et al. Transition of late-stage effector T cells to CD27+ CD28+ tumor-reactive effector memory T cells in humans after adoptive cell transfer therapy. Blood 2005;105(1):241–50.
[140] Chiocca EA, Rabkin SD. Oncolytic viruses and their application to cancer immunotherapy. Cancer Immunol Res 2014;2(4):295–300.
[141] Marshall JL, et al. Phase I study of sequential vaccinations with fowlpox-CEA(6D)-TRICOM alone and sequentially with vaccinia-CEA(6D)-TRICOM, with and without granulocyte-macrophage colony-stimulating factor, in patients with carcinoembryonic antigen-expressing carcinomas. J Clin Oncol 2005;23(4):720–31.
[142] Kaufman HL, et al. Combination chemotherapy and ALVAC-CEA/B7.1 vaccine in patients with metastatic colorectal cancer. Clin Cancer Res 2008;14(15):4843–9.
[143] Mendoza-Rodriguez MG, et al. Use of STAT6 phosphorylation inhibitor and trimethylglycine as new adjuvant therapies for 5-fluorouracil in colitis-associated tumorigenesis. Int J Mol Sci 2020;21(6):2130.
[144] Miele L, et al. Increased intestinal permeability and tight junction alterations in nonalcoholic fatty liver disease. Hepatology 2009;49(6):1877–87.

[145] Eckburg PB, et al. Diversity of the human intestinal microbial flora. Science 2005;308(5728):1635–8.
[146] Koenig JE, et al. Succession of microbial consortia in the developing infant gut microbiome. Proc Natl Acad Sci U S A 2011;108(Suppl 1):4578–85.
[147] Rajilic-Stojanovic M, et al. Long-term monitoring of the human intestinal microbiota composition. Environ Microbiol 2012.
[148] Yatsunenko T, et al. Human gut microbiome viewed across age and geography. Nature 2012;486(7402):222–7.
[149] Velagapudi VR, et al. The gut microbiota modulates host energy and lipid metabolism in mice. J Lipid Res 2010;51(5):1101–12.
[150] Furusawa Y, et al. Commensal microbe-derived butyrate induces the differentiation of colonic regulatory T cells. Nature 2013;504(7480):446–50.
[151] Wang HB, et al. Butyrate enhances intestinal epithelial barrier function via up-regulation of tight junction protein Claudin-1 transcription. Dig Dis Sci 2012;57(12):3126–35.
[152] Segain JP, et al. Butyrate inhibits inflammatory responses through NFkappaB inhibition: implications for Crohn's disease. Gut 2000;47(3):397–403.
[153] Zhao J, et al. Dietary protein and gut microbiota composition and function. Curr Protein Pept Sci 2019;20 (2):145–54.
[154] Liu L, et al. Polyamines regulate E-cadherin transcription through c-Myc modulating intestinal epithelial barrier function. Am J Physiol Cell Physiol 2009;296(4):C801–10.
[155] Das P, Babaei P, Nielsen J. Metagenomic analysis of microbe-mediated vitamin metabolism in the human gut microbiome. BMC Genomics 2019;20(1):208.
[156] Sartor RB. Microbial influences in inflammatory bowel diseases. Gastroenterology 2008;134(2):577–94.
[157] Petersen C, Round JL. Defining dysbiosis and its influence on host immunity and disease. Cell Microbiol 2014;16(7):1024–33.
[158] Boraska Jelavic T, et al. Microsatelite GT polymorphism in the toll-like receptor 2 is associated with colorectal cancer. Clin Genet 2006;70(2):156–60.
[159] Bishehsari F, et al. Epidemiological transition of colorectal cancer in developing countries: environmental factors, molecular pathways, and opportunities for prevention. World J Gastroenterol 2014;20(20):6055–72.
[160] Gao Z, et al. Microbiota disbiosis is associated with colorectal cancer. Front Microbiol 2015;6:20.
[161] Yu LC. Microbiota dysbiosis and barrier dysfunction in inflammatory bowel disease and colorectal cancers: exploring a common ground hypothesis. J Biomed Sci 2018;25(1):79.
[162] Kamada N, Nunez G. Regulation of the immune system by the resident intestinal bacteria. Gastroenterology 2014;146(6):1477–88.
[163] Miquel S, et al. Faecalibacterium prausnitzii and human intestinal health. Curr Opin Microbiol 2013;16 (3):255–61.
[164] Irrazabal T, et al. The multifaceted role of the intestinal microbiota in colon cancer. Mol Cell 2014;54(2):309–20.
[165] Alhinai EA, Walton GE, Commane DM. The role of the gut microbiota in colorectal cancer causation. Int J Mol Sci 2019;20(21):5295.
[166] Amitay EL, et al. Fusobacterium and colorectal cancer: causal factor or passenger? Results from a large colorectal cancer screening study. Carcinogenesis 2017;38(8):781–8.
[167] Liu Y, et al. Progress in characterizing the linkage between Fusobacterium nucleatum and gastrointestinal cancer. J Gastroenterol 2019;54(1):33–41.
[168] Yang Y, et al. Colon macrophages polarized by commensal bacteria cause colitis and cancer through the bystander effect. Transl Oncol 2013;6(5):596–606.
[169] Zitvogel L, et al. The microbiome in cancer immunotherapy: diagnostic tools and therapeutic strategies. Science 2018;359(6382):1366–70.
[170] Wong SH, Yu J. Gut microbiota in colorectal cancer: mechanisms of action and clinical applications. Nat Rev Gastroenterol Hepatol 2019;16(11):690–704.

CHAPTER

8

Environmental pollution as a risk factor to develop colorectal cancer: The role of endocrine-disrupting chemicals in the inflammatory process as a risk factor to develop colorectal cancer

Yair Rodriguez-Santiago[a], *Karen Elizabeth Nava-Castro*[b], *and Jorge Morales-Montor*[a]

[a]Departamento de Inmunología, Instituto de Investigaciones Biomédicas, Universidad Nacional Autónoma de México, Ciudad de México, México [b]Departamento de Mutagénesis y Genotoxicidad Ambientales, Centro de Ciencias de la Atmósfera, Universidad Nacional Autónoma de México, Ciudad de México, México

Abstract

Environmental pollution is nowadays the first risk factor associated with the development of lung cancer, stroke, and heart disease. Within the universe of the complex mix of pollutants that shape the environment, the importance of endocrine-disrupting compounds (EDCs), such as bisphenols, phthalates, polycyclic aromatic hydrocarbons (PAHs), and pesticides, has been highlighted in recent years because of our chronic and ubiquitous exposition, particularly by ingestion of food and water, and their interference in hormone and immune functions. EDCs are associated principally with breast cancer development, but there is a poor understanding of the relationship with other neoplasms such as colorectal cancer (CRC). The objective of this chapter is to address the potential carcinogenic role of EDCs in the development of CRC, focusing on inflammation and neuroendocrine-immune regulation. First, we will describe the interaction between inflammatory bowel disease and CRC in terms of disruption of intestinal permeability, dysbiosis, and immune response induced by EDCs. Moreover, we delve into the neuroendocrine axes involved in the progression of this pathology.

Immunotherapy in Resistant Cancer: From the Lab Bench Work to Its Clinical Perspectives
https://doi.org/10.1016/B978-0-12-822028-3.00007-8

Abbreviations

ACTH	corticotropin
BaP	benzopyrene
BPA	bisphenol A
CAC	colitis-associated cancer
CD	Crohn's disease
CRC	colorectal cancer
CRH	corticotropin releasing hormone
DCs	dendritic cells
DDT	dichlorodiphenyltrichloroethane
DEHP	di(2-ethylhexyl) phthalate
EDC	endocrine-disrupting chemicals
ER	estrogen receptor
E2	estradiol
FSH	follicle stimulating hormone
GnRH	gonadotropin releasing hormone
HPA/HPG	hypothalamic-pituitary-adrenal/gonadal
IBD	inflammatory bowel disease
LH	luteinizing hormone
Ig	immunoglobulin
MΦ	macrophages
Muc 2	mucin 2
NK	natural killer
SCFAs	short-chain fatty acids
Tregs	T regulatory cells
T4	testosterone
UC	ulcerative colitis
PAHs	polycyclic aromatic hydrocarbons
P4	progesterone

Conflict of interest

No potential conflicts of interest were disclosed by the authors.

Introduction

Environmental pollution and disease

Environmental pollution is defined as a mix of chemical, physical, and biological components that are harmful to the ecosystem. World Health Organization (WHO) has listed environmental pollution as the leading cause of morbidity and mortality worldwide, and it estimates around seven million deaths due to exposure to it. Among the main diseases that have a direct relation to pollution are both heart and respiratory stroke [1]. However, the health problems potentially developed are more severe because of the exposure to different types of pollutants by different sources such as water, land, industry and health products, and food. The International Agency for Research on Cancer (IARC) established air pollution as the major risk of developing cancer [2]. Lung cancer is the most studied neoplasm that has been correlated with air pollution, but its association with other cancer types has not been widely considered. For example, colorectal cancer (CRC) is widely associated with infection diseases, but not with air pollution. Besides, air pollution is a multicomponent mix of different compounds; for that reason, the study of each component will provide a better understanding of its effects on human health.

Endocrine-disrupting chemicals and cancer

Endocrine-disrupting chemicals (EDCs) are defined as exogenous chemicals that alter the levels, production, and hormonal communication. The most known compounds are bisphenols, phthalates, polycyclic aromatic hydrocarbons, and pesticides. These chemicals are capable of binding to different affinities to sex steroid receptors. Coupled with that, steroid receptors are expressed in several types of cells. Given the large number of processes that hormones control, it is crucial to study these molecules in a disease context. EDCs are associated with multiple illnesses as reproductive, autoimmune, cardiovascular, and brain disease [3]. Research on cancer development by EDC is principally focused on neoplasms that attack hormone-target organs, principally the breast tissue. A great number of scientific evidence has reported that bisphenol A (BPA) exposure promotes an exacerbated growth of mammary tumors in animal models [4,5]. Globally, the role of EDCs in CRC is less recognized; however, this pathology has been recently likened with the estrogen receptor (ER) dominance [6]. Thus, the effects of these compounds in this type of cancer would be an important health issue since humans are potentially exposed to them by multiple food and water sources (Table 1).

Colorectal cancer and inflammation

CRC is the third most common cancer in the world. It is divided into three types: hereditary, colitis-associated cancer (CAC), and sporadic CRC. Inflammation and CRC have intimately related, in fact, inflammatory bowel disease (IBD), predominantly ulcerative colitis (UC), is the principal risk factor to develop CAC. In fact, the use of antiinflammatory agents reduces the risk of developing this disease [18]. But what is inflammation? And why is it so important in the pathophysiology of this disease?

Inflammation is a biological response of body tissues to harmful stimuli, and one of the main functions of this process is to eliminate the initial cause of cell injury. But when this process becomes chronic, the repair mechanisms are insufficient and the tissues suffer permanent and constant damage, which leads to disease such as cancer. CRC is the perfect precedent between inflammation and the development of the disease. In order to understand in depth the role of inflammation in the carcinogenic process, we first describe the function and interactions of the different elements in the bowel and the alterations of these compounds in IBD and CRC pathologies.

The intestine is made up of five cell layers: mucosa, epithelial, submucosa, muscular, and serosa. The epithelial layer is made up of four types of cells that together form the first defense line against pathogens, injury, or microbiota contact. It also contains Paneth cells that release antimicrobial mediators; epithelial cells that absorb nutrients; microfold cells that transport antigens from the lumen, triggering an immune response; goblet cells that secrete mucus; and

TABLE 1 Exposure to endocrine-disrupting chemical in food and water and their association with disease.

Compound	Exposure source	Dose (ng/g/day) or (μg/lt/day)	Oral reference dose EPA	Disease	Reference
Bisphenols					
Bisphenol A	Meat, fruits and vegetables, bread and cereals, packed food, breast milk	Meat (0.48–10.50) Bread and cereals (0.40–1.73) Fruits and vegetables (0.41–23.5) Fast food (1.10–10.9) Packed food (29.1–44.4) Breast milk (0.5–1.3)	1st dose 2 μg/kg/day 2nd dose 50 μg/kg/day	Human association Fertility, neurology, metabolic and autoimmune disease Animal studies 1st dose. Neurological effects and premature puberty 2nd dose. Reproductive problems and cancer	[7–9]
DEHP	Vegetables, fruits, meats, eggs, fish, fat products	Meat (0.3–0.8) Grain and cereals (1.2–1.9) Fruits and vegetables (0.3–0.9) Fish (0.02–0.05) Eggs (0.01–0.03) Fat products (1.9–7.4)	Dose: 0.02 mg/kg/day	Human associations Gastrointestinal distress Animal studies Reproductive and development Increase liver tumors	[10,11]
Benzopyrene	Meat, fruits and vegetables, grains and cereals, fat products and drinks, chicken and fish	Red meat (1.01–4.86) Fish and chicken (1.01–4.57) Fat products and beverage (1.01–0.18) Cereals and grains (0.03–0.051) Fruits and vegetables (0.01–0.48)	Dose 1st dose (300 ng/kg/day) 2nd dose (400 ng/kg/day)	Human studies Association with cancer and atherosclerosis Animal studies 1st dose Developmental and immunological disease 2nd dose Reproductive disease and cancer	[12–14]
Dichlorodiphenyl trichloroethane (DTT)	Meat, fruits and vegetables, grains and cereals, daily products, fish	Meat (1.4–310) Fruits and vegetables (1.22–22.1) Grain and cereals (0.088–0.15) Daily products (1.2–20)	1st dose 0.34 ng/kg/day 2nd dose 500 ng/kg/day	Human studies Carcinogenic Animal studies 1st dose Carcinogenic disease 2nd dose Hepatic lesions	[15–17]

a complex mixture of glycoproteins, mainly mucins, that avoids direct contact with the microbiome and food antigens [19]. The goblet cells are so important because they represent one of the major populations of differentiated cells in the colonic mucosa. In the disease conditions of patients, for example, UC, the thickness of the mucus barrier decreases together with the loss of the principal compound of the mucus, mucin 2 (Muc 2); both factors are directly associated with the severity of inflammation in both humans and mouse tissues. Furthermore, in CRC patients, the loss of Muc 2 has been classified as a poor prognosis factor that favors the progression and metastasis [20–24].

In physiological conditions, when the mucus layer is disrupted, several microorganisms and other food antigens might come into contact with the epithelial cells, disrupting the structural proteins that maintain the integrity and the paracrine communication of the bowel cells. Related to this, three categories of cell communications have been described: adherent, gap, and tight junctions. In the context of the barrier integrity, the most important are tight junctions, which are subdivided into three subtypes according to the proteins that they contain—occludins, JAM, and claudins. These latter subtypes are the principal tight junctions, whose importance lies in the fact that they are most abundant in the epithelial barrier. They also participate in the polarity of the membrane and in the regulation of mitogenic signaling pathways such as PI3K/AKT, Notch, Wnt, kinases, and ERK [25]. Of note, in the scenario of IBD and CRC, the alteration of claudins is related to the pathogenesis of the disease. Claudin 1 and 2 proteins are overexpressed in both pathological conditions; in fact, their presence has been importantly associated with poor patient survival and greater progression.

In addition, in vivo models based on transgenic APC min/CLd1 mice, where a genetic modification leads to the production of lower levels of claudin 1, the development of a high quantity of polyps and elevated TNF-α levels have been reported. This has been correlated with the participation of claudin 1 in the promotion of the epithelial-mesenchymal transition via ERK and Src activation of pathways. Additionally, claudin 2 has been linked with the increase in the proliferation of CRC cells [26–30].

As we mentioned before, the disruption of epithelial cells characterized by the loss of the mucus and the alteration of tight junctions let the contact of immunology cells with microbiome and food antigens, triggering an exacerbated immune response. If the damage is not repaired and becomes chronic, pathologies such as IBD and CRC can develop. In the following paragraphs, the role of immune cells in response to IBD and CRC will be explored.

In the homeostatic conditions, the cell balance is principally maintained by a controlled response of immune tolerance through the interaction of innate and adaptive immune cells. Macrophages (MΦ), the classical phagocytic cells, are part of the innate immune response; they contribute to the immune tolerance-releasing antiinflammatory factors such as IL-10, promoting the differentiation of T regulatory cells (Tregs) and suppressing the Th1 and Th17 immune responses [31]. The participation of MΦ in IBD patients is crucial in order to develop the inflammation process, since they can contain high activation of the transcription factor NF-kβ and in this manner favors the secretion of proinflammatory factors such as TNF-α, IL-6, and IL-8. On the other hand, similar to other types of cancer, in patients with CRC an alternative phenotype of MΦ has been found—the M2 type that helps in the promotion of angiogenesis due to its ability to secrete VEGF. However, the presence of these cells into the CRC microenvironment has been directly associated with a greater survival, indicating that the role of MΦ is more complex, and it needs more studies in order to be elucidated [32–34]. Regarding the promotion of inflammation by other types of cells that also participate in the innate immune response, the natural killer (NK) cells are present in the gut-associated lymphoid tissue and have been associated with the release of proinflammatory cytokines in response to bacterial pathogenesis. Two phenotypes of them with opposite roles have been reported: the inflammatory NKP46+ cells that produce high levels of IFN-γ participating in the promotion of inflammatory response and NKP46− cells that principally segregate IL-22 implicated in the antiinflammatory actions [35]. NK cells have an important role in another inflammatory process present in the colon, UC, and this pathological condition is characterized by severe inflammation. Of note, the presence of a high amount of NKp46+ cells has been reported in it. In addition, in IBD patients, higher levels of activated NK CD16+ cells have been reported too; nevertheless, in CRC patients, the percentage of NKp46+ cells is reduced, whose tissue infiltration is considered as a good marker of survival. In fact, in mice models, the presence of NK cells seems to be crucial for the protection against the induction of several inflammatory diseases such as colitis. This suggests that these cells participate in the carcinogenic process favoring the inflammation but have an antitumoral activity in the advance stages of CRC patients [36–39].

Neutrophils play an important role in the CRC immune response; they serve as proinflammatory secretory cells that release different biological factors such as IL-8 and IL-17A, among others. However, their role in the inflammation process depends on the phenotype of the neutrophil. In this sense, the CD177+ neutrophil cells release a low amount of proinflammatory cytokines and higher levels of IL-22; conversely, CD177− cells produce IL-6, IL-17A, and IFN-γ [40]. Although CD177+ cells are increased in the inflamed intestine of IBD patients, in CRC patients, the role of the neutrophils is controversial since the presence of tumor-associated neutrophils has been associated with both good and bad prognosis in different studies [41,42]. This possibly lies in their interaction with T lymphocytes and the ratio of these infiltrated populations [40–46].

Dendritic cells (DCs) serve as one of the links between innate and adaptive immune responses. Considered as the main antigen-presenting cells through the interaction with several antigens, damage signals, or proinflammatory cytokines, they promote an inflammatory immune response via secretion of IL-6, IL-23, and IL-17 or a tolerant response-inducing Tregs differentiation [47]. Their importance in intestine homeostasis is reflected in patients with UC and Crohn's disease (CD), which express higher levels of activated DCs, which are localized in the lamina propria, where their presence is correlated to bad prognosis. Additionally, in patients with CRC, the intratumoral infiltration of these immune populations on the immature state is linked with shorter disease-free survival; meanwhile, a low percentage of them in a mature state is present in patients with metastatic CRC [48–52].

On the other hand, the adaptive immune response is orchestrated by T lymphocytes that interact with DCs through the TCR-MHC complex; these cells can be a divide in T helper cells, TCD4+, and cytotoxic $TCD8^+$ cells. When the last population is activated, they produce a lot of proinflammatory factors, principally TNF-α, IFNs, perforins, and granzymes. The activation of TCD8+ lymphocytes is a process governed by the gut microenvironment. In the equilibrium state, $TCD8^+$ progressively expresses the homodimer protein, CD8α, that reduces the sensitivity of the TCR. Contrarily, when there is a localized inflammatory process with the presence of cytokines such as IL-2 and IL-15, the activation into cytotoxic phenotype is promoted [53,54]. Interestingly, in patients with IBD, lower percentages of TCD8+ cells are present, while a high amount of intestinal TCD4+ cells are infiltrated. Therefore, in the tumoral epithelium of patients with CRC, higher amounts of $TCD8^+$ cells are present as compared with the epithelium of healthy individuals. It has been described that the intratumoral ratio of $TCD8^+$/Tregs cells is directly associated with the survival rate in CRC patients with liver metastasis [55–58]. Regarding $TCD4^+$ cell function, they can polarize the immune response into TH1, TH2, TH17, or Tregs response. Polarization toward one phenotype or another is stimuli-dependent: for example, IFN-γ favors a TH1 while IL-13 a TH2 phenotype. In nonpathological conditions, there is a balance of these types of responses; however, when there exists an imbalance among them, the inflammatory process is prompted and the loss of intestinal structure is displayed [59]. The disruption in the balance is evident in IBD patients: for example, the purified lamina propria from CD patients presents an increased number of IFN-γ secreting TCD4+ cells and IL-13 overproduction in UC patients. Interestingly, the infiltration of TH1 cells is strongly correlated with an improved prognosis in CRC patients. On the other hand, IL-17 levels in patients with UC are higher and closely correlate with the severity of the disease; in concordance with that, IL-17 protein is correlated with a poor prognosis in the advanced stage of CRC patients [60–63].

Tregs are also T lymphocytes that exhibit an immunosuppressive activity; they maintain the homeostasis in the gut by the secretion of different cytokines such as TGF and IL-10 cytokine [64]. Of note, CD and UD patients had a lower frequency of Tregs in serum in comparison with healthy controls. Interestingly, the intratumoral infiltration of Tregs is associated with bad prognosis and metastasis, and the excision of the tumor normalizes the Tregs population, which suggests that these cells protect the tumor against the cytotoxic immune response [65–69].

Finally, the last immune cell type is the B lymphocytes, which govern the humoral response. These cells are principally located in the lymphoid gut-associated tissue; they secrete immunoglobulin A (IgA), which is uncharged of the regulation of bacteria population in this tissue. Another important function of B lymphocytes is the antigen-presenting capacity with which they activate $TCD4^+$ and $TCD8^+$ cells. Like other types of immune cells, B cells also turn up in different phenotypes: the effector B1 lymphocytes produce IFN-γ, while effector B2 cells produce IL-4. In addition, another category of these cells, the B regulatory (Bregs) cells secrete IL-10 and the last memory B cells that produce antibodies [70]. In UC pathology, the number of activated B cells is higher in comparison with normal epithelium. This is consistent with the reduction of Bregs. It has been described that in sera from patients with IBD exposed to different antigens, the level of antibodies IgA and IgG are higher than those in the samples coming from healthy individuals. CRC patients show high percentage of intratumoral B cells. In fact, in metastatic and advanced stages of CRC patients, these cell populations protect the tumor against the immune response, similar to the Tregs function [71–74].

Another factor that is closely related to the immune system and inflammation is the microbiome that is shaped by fungi, viruses, parasites, archaea, and bacteria. The last group is the predominant microorganism in the bowel. The microbiome is very heterogeneous among individuals and its diversity depends on the genetic predisposition, diet, and environmental factors. There is no consensus about how it is composed of the microbiome in a "nonpathological condition," although it is generalized that individuals without pathology have predominance (>90%) of two phyla: Bacteroidetes and Firmicutes and the rest are composed of different species of bacteria [75]. In patients with IBD, the diversity of bacteria is diminished and even though the results are very variable, in general, more gram-negative anaerobic bacteria are found. For instance, an increase in the percentage of *E. coli* bacteria has been associated with the recruitment of neutrophils and MΦ in the intestine, which potentiates the inflammation response. Moreover, in patients with CRC, a lot of scientific evidence settles the association between the development of disease and the presence of *Fusobacterium*; this microorganism is capable of activating the E-cadherin/β-catenin pathway and polarize the immune repose toward an immunosuppressor phenotype [76–80]. Therefore, one of the most important mechanisms by which the microbiota participates in the inflammatory process is the secretion of short-chain fatty acids (SCFAs). *Bacteroidetes* principally secrete acetate and butyrate, while *firmicutes* produce butyrate [81]. The SCFAs affect both epithelial and immune cell function. For example, butyrate affects the claudin expression in Caco2 and IPEC-J2 colon cells. The effects of SCFAs also impact on the immune cells since these biological components have immunosuppressive effects. In fact, butyrate inhibits the activation of NF-B in derived-MΦ from patients with UC and induces the differentiation of Tregs in rats, diminishing the severity of the colitis induction [82–84]. Consistent with this, butyrate can inhibit the development of CRC through the alteration of key molecules that are important in angiogenesis, metastasis, or survival. Furthermore, this metabolite can stimulate apoptosis via hyperactivation of Wnt pathway in cancer cells. In line with that, another SCFA, such as acetate, has shown to reduce the inflammatory cell phenotype and triggers the apoptosis process in colon cancer cell lines [85,86].

Immunoendocrine interaction in a context of inflammation

The two most important axes that regulate the inflammatory response are the hypothalamic-pituitary-adrenal/gonadal (HPA/HPG). The regulation of these axes is mediated by different pathways. Briefly, the HPA axis starts with the release of corticotropin-releasing hormone (CRH) in the hypothalamus, which stimulates the secretion of corticotropin (ACTH) in the pituitary gland, which exerts its effect on the suprarenal glands releasing cortisol (humans) or corticosterone (murine models). In the case of the HPG axis, the liberation of the gonadotropin-releasing hormone (GnRH) at the hypothalamus level propitiates the secretion of the follicle-stimulating hormone (FSH) and luteinizing hormone (LH) in the pituitary. This gives rise to the stimulation of the production of sex hormones. Both estrogens and androgens exert a negative feedback control at the hypothalamus level, which prevents the ACTH release [87].

In the following paragraphs, we will describe the inflammation-stress interaction in the context of IBD.

The stress is controlled by the HPA axis. In an excellent review article, Reber describes the effect of stress in the context of inflammation in both humans and animal models. In humans, the results are controversial. For instance, some authors show that perceived life stress has a relationship with the remission of UC and CD. Furthermore, the relatively high success of treating UC with placebo drugs suggests that the emotional state of the individual and her or his perception of stress is an important factor that participates in the development of IBD. On the other hand, other works have not found an association between stressful life events and IBD. These results could be due to the differences in the methodology and the analysis of information, so it is necessary for the employment of animal models to clarify this apparent contradiction. In general, the effect of stressful agents in animals induces a reactivation in colitis that was induced with trinitrobenzene sulfonic acid (TNBS) or dextran sulfate sodium (DSS) chemicals. This reinforces the relationship between stress and inflammation [88]. According to this, the perceived stress is associated with the incidence of CRC. In fact, a recent study demonstrated that the corticosterone treatment in a C57BL/6 mouse strain increased the number of colon tumors and the expression of NF-β and COX-2 proteins [89,90]. Thus, it is clear that stress is an important factor in the development of IBD and CRC, but the molecular mechanisms are not well understood. Just a few experimental approaches have been tried in order to understand the mechanism behind the role of stress in the pathogenesis of colitis-associated CRC. In adrenalectomized C57BL/6 mouse tolerogenic DCs, TCD8+ lymphocytes are elevated in the colon tissue with a concomitantly less damage as compared with control groups; however, in glucocorticoid-treated animals, the damaged tissue considerably improves but does not return to basal conditions. These results suggest that other factors such as catecholamine action would be involved [91].

Sex steroids have been associated with hormone-dependent cancer such as breast and prostate cancer, through binding with hormone receptors. However, their participation with the immune system response or intestinal functions such as permeability, motility, and microbiota balance has not a forthright association. In fact, men have a greater risk of developing IBD and CRC in comparison with women [92]. However, the mechanism is not completely understood, but some studies have shown that the participation of these molecules in the inflammation is so important.

For example, APC/min mouse male model develops more polyps than females, and the number of these lesions in orchiectomized animals is diminished. Congruently, ICR mice present the same dimorphic patron and males have a higher level of proinflammatory factors, particularly myeloperoxidase (MPO) and IL-1β. On the other hand, the estradiol (E2) reconstitution reduced the inflammation in terms of NF-β levels and in the histological morphology in male and female mice [93–95].

Disrupting endocrine chemicals as a risk factor to develop colorectal cancer

Bisphenols

BPA is the most used component in polycarbonate plastics and epoxy resins. Food and water are the principal sources of its exposure. The possible participation of BPA in the progression of IBD and CRC can be seen at different levels. First, BPA is capable of altering the hormonal levels; for example, its concentration in the serum of men is inversely associated with androstenedione and testosterone (T4), which could possibly be mediated by the capacity of BPA to induce the activation of the aromatase enzyme [96,97]. Parallelly, lower plasma levels of T4 have been found in men with CRC, which are inversely associated with the development of cancer risk; in the same way, in women, there is an association between the development of this disease and E2/T4ratio. In relation to the above findings, high levels of E2 in patients with IBD present less disease incidence, all of this suggesting that E2 has a protective role, which is consistent with animal models [98,99]. On the other hand, BPA also has a role in the inflammation context; in this sense, it can affect different cell lineages and their biological functions. For instance, it increases permeability in the intestine cells through the alteration of ZO-1, occluding and claudin-1 junctions; besides, it also reduces the goblet cells and directly inhibits the secretion of Muc 2 protein, which is responsible for the intestinal homeostasis maintenance. This disruption in the membrane epithelium induces the dysbiosis [100,101]. As described before, the dysbiosis is one of the important factors in the development of IBD and CRC. The exposition to BPA favors the accumulation of microbial communities that are associated with the progression of the disease. The panorama is so complex and little understood, but in general, this compound also causes a reduction in the diversity and the accumulation of anaerobic gram-negative bacteria. This can be seen in the diminishing of metabolites that reduce the inflammation in response to it [102–105]. In addition, BPA could participate in the intestinal immune hyperresponse through the disruption of epithelium and dysbiosis, as well as a direct effect on immune cells. However, its effects are not deeply understood and they depend on the dose and time of exposure and the immunological challenge. BPA can modulate the cytokine secretory profile; it has been described that isolated lymphocytes from mice exposed to BPA and stimulated with concanavalin produced an elevation of TH1 (IFN-γ) cytokine and decreased TH2 (IL-4) cytokine production [106]. On the contrary, isolated human peripheral blood mononuclear cells (PBMCs) exposed to BPA produced low levels of IL-10 and IL-12, but no changes in IFN-γ or IL-4 [107]. BPA can also modulate the Tregs function. These immune populations have a dual outcome in CRC, and they are correlated to good prognosis in early stages of IBD and with bad prognosis in the advance stage of CRC (Fig. 1; Table 1) [108].

Phthalates

Phthalates are ester derivatives of plastic products that are mainly used to lead the flexibility of plastic devices. The most used compound of this family is the di(2-ethylhexyl) phthalate (DEHP) with more than two million tons produced annually. Phthalates are principally associated with reproductive, cardiovascular, liver, urological, type-2-diabetes, and hypertension diseases [109]. Phthalates participate in the carcinogenic process by different mechanisms such as the interruption of the hormonal balance and the alteration of the anxiety behavior. As was described previously, the variance of these factors is associated with the development of pathological conditions such as IBD and/or CRC [110–113]. A direct relationship between phthalates and IBD was demonstrated by in vivo studies where the administration of DEHP to APC min/+ mice induced intestinal polyps and modified the histopathological features of epithelial cells, where an increase in β-catenin, IL-8, and SOX2 was also observed. Supporting this finding, its administration in rats induced neoplastic lesions and favored the inflammatory response through the activation of β-catenin, COX-2, and VEGF expression. The effect of DEHP alone or in combination with other phthalates impacts the proliferation of different colon cancer cell lines. In addition, DEHP has been related to the resistance to oxaliplatin and irinotecan, drugs used in the treatment of patients with CRC [114–117]. The phthalates can also modulate the proinflammatory cytokine profile and promote the production of proinflammatory cytokines such as IL-1β, IL-6, and TNF-α. Moreover, these compounds also favor the M2 phenotype of MΦ and regulate the activity

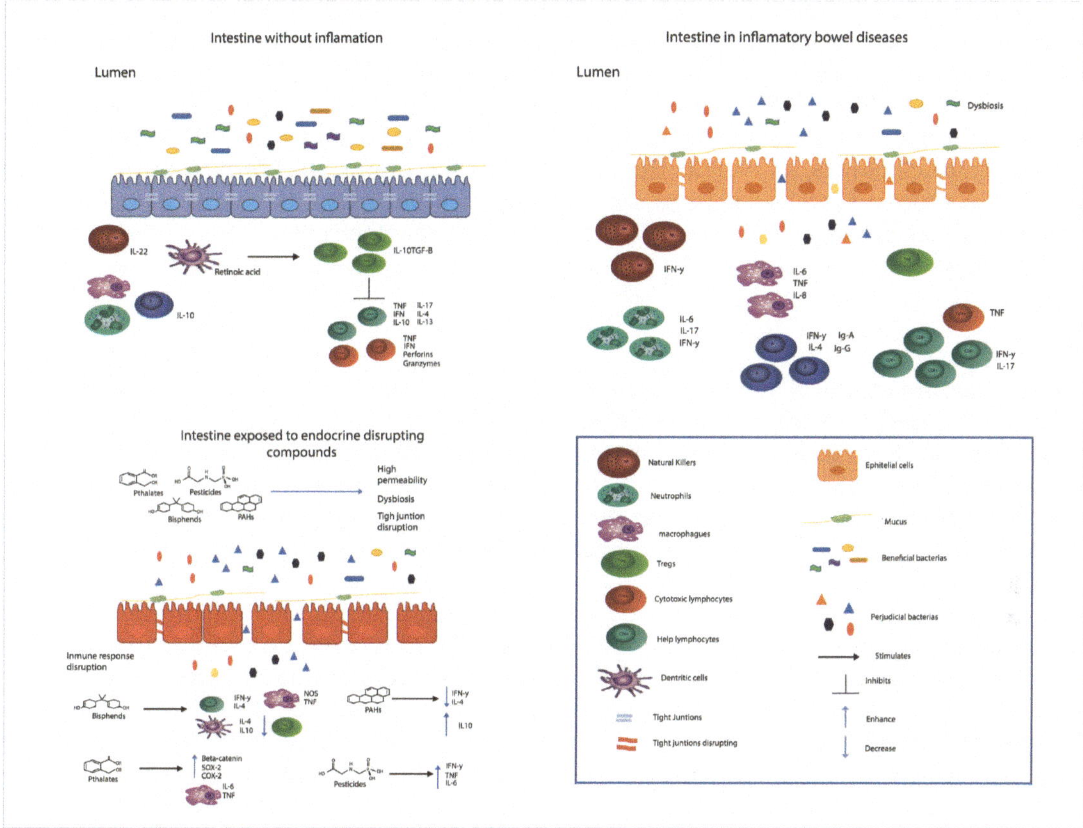

FIG. 1 Simplified model of the immune response in three different conditions in the context of inflammation based on epidemiological, in vitro, and in vivo studies. (A) Intestine in basal condition. (B) Intestine in inflammatory bowel diseases such as ulcerative colitis and Crohn's disease. (C) Intestine exposed to endocrine-disrupting compounds. (D) Description of symbols in (A)–(C).

of neutrophils. Significantly, they can also act in the disruption of cell junctions leading to changes in the microbiota profile; for instance, the exposition of newborns to DEHP has been associated with the reduced levels of *Bifidobacterium longum* and *Rothia* sp. (Fig. 1; Table 1) [118–122].

Polycyclic aromatic hydrocarbons

Polycyclic aromatic hydrocarbons (PAHs) are ubiquitous environmental pollutants that are generated from incomplete combustion when organic materials such as coal, oil, wood, garbage, and tobacco are burned. A direct relationship between the development of CAC and PAHs has been described; specifically, the carcinogenic property of benzopyrene (BaP) is derived from broiled or grilled meat consumption [123]. In this sense, different mice studies have postulated the carcinogenic potential of BaP through the development of polyps as a final event. One of the mechanisms described in this process is that BaP metabolites have the capacity to form DNA adducts evoking mutations in the epithelial cells [124–131]. Additionally, this compound has also been involved in the modulation of both sex and glucocorticoid hormones [132,133]. It is important to mention that BaP can also modulate the immune response; for instance, mouse MΦ exposed to it presented a decrement in the production of inflammatory cytokines, as well as the inhibition of the monocyte's differentiation toward MΦ. This compound can also suppress the activity of NF-β. They could participate in the reduction of tissue damage, which in early phases of the pathological condition is the cause of a poor immune response breaking tissue homeostasis (Fig. 1; Table 1) [134–136].

Pesticides

Pesticides are chemical substances used in the control of plagues. Their classification depends on the target organism (insecticides, fungicides, herbicides, etc.). Although there are many types of these substances, they share a chemical structure that is capable of mimicking the actions of endogen hormones. Epidemiological studies have found an association between pesticide exposure and CRC risk. Supporting this fact, the exposure of dichlorodiphenyltrichloroethane (DDT) in colon cancer cells has been demonstrated to inhibit the apoptosis activation regulated by the PI3K/AKT and Wnt/β-catenin pathways [137]. Additionally, mice exposed to monocrotophos showed an outstanding disruption of the intestinal epithelium. Moreover, different studies have demonstrated that this chemical compound causes hyperplasia of goblet cells with a concomitant tissue infiltration of inflammatory cells and the disruption of cell permeability mediated by the reduction of Muc 2 protein. The exposure to the chemical fungicide such as imazalil has also been associated with the induction of dysbiosis and intestinal inflammation [138–140]. Besides, an exposure to pesticides can also have an impact on the regulation of cortisol levels. Interestingly, this molecule is the main factor that controls the immune hyperresponse, which could lead to more aggressive tissue damage. The employment of the synthetic organochloride insecticide, methoxychlor, in mice models affects the hormonal levels E2, T4, and progesterone (P4), and the enzymes involved in sex steroidogenesis [141–143]. Pesticides can also cause apoptosis of the neutrophils, NK cells, and B lymphocytes, making the body vulnerable to infections or other prejudicial agents. Additionally, the catalytic metabolism of these chemicals prompts the production of reactive oxygen, which are also considered as the key mediators in the inflammatory response (Fig. 1; Table 1) [144–146].

Conclusion

Based principally on in vivo studies with APC min/+ mice models and in vitro assays, EDCs have been proved to play an important role in the development of CRC through different mechanisms such as modification of the inflammatory profile, endocrine homeostasis, and epithelial junctions. Because there is no direct link between the EDC actions and the severity and therapy resistance of the CRC disease, we considered that it is important to encourage future research and development in this area, taking into account the multiple effects of these compounds discussed in this chapter on this disease.

Acknowledgments

This study was supported by grants from Programa de Apoyo a Proyectos de Investigación e Innovación Tecnológica (PAPIIT), Dirección General de Asuntos del Personal Académico (DGAPA), Universidad Nacional Autónoma de México (UNAM), grant/award number IN-209719, and from Fronteras en la Ciencia, Consejo Nacional de Ciencia y Tecnología (CONACYT), grant No FC 2016 2125, both to Jorge Morales-Montor. Yair Rodríguez-Santiago is a PhD student at Programa de Doctorado en Ciencias Biomedicas, UNAM and has a PhD fellowship from CONACYT (964693).

References

[1] World Health Organization. Air pollution infographics.
[2] Weiderpass E. Air pollution as a major risk factor for cancer. Lyon, France: International Agency for Research on Cancer, World Health Organization; 2016.
[3] Schug TT, et al. Endocrine disrupting chemicals and disease susceptibility. J Steroid Biochem Mol Biol 2011;127 (3–5):204–15.
[4] Snedeker SM. Pesticides and breast cancer risk: a review of DDT, DDE, and dieldrin. Environ Health Perspect 2001;109(Suppl 1):35–47.
[5] Weber Lozada K, Keri RA. Bisphenol A increases mammary cancer risk in two distinct mouse models of breast cancer. Biol Reprod 2011;85(3):490–7.
[6] Caiazza F, et al. Estrogen receptors and their implications in colorectal carcinogenesis. Front Oncol 2015;5:19.
[7] Almeida S, et al. Bisphenol A: food exposure and impact on human health. Compr Rev Food Sci Food Saf 2018;17(6):1503–17.
[8] Vilarinho F, et al. Bisphenol A in food as a result of its migration from food packaging. Trends Food Sci Technol 2019;91:33–65.
[9] Mendonca K, et al. Bisphenol A concentrations in maternal breast milk and infant urine. Int Arch Occup Environ Health 2014;87(1):13–20.
[10] Serrano SE, et al. Phthalates and diet: a review of the food monitoring and epidemiology data. Environ Health 2014;13(1):43.
[11] Di(2-etilhexil) ftalato (DEHP), Available from: https://www.atsdr.cdc.gov/es/phs/es_phs9.pdf; 2002.
[12] Kazerouni N, et al. Analysis of 200 food items for benzo[a]pyrene and estimation of its intake in an epidemiologic study. Food Chem Toxicol 2001;39(5):423–36.
[13] Toxicological review of benzo[a]pyrene, Available from: https://cfpub.epa.gov/ncea/iris/iris_documents/documents/subst/0136_summary.pdf; 2017.
[14] Ekere NR, et al. Levels and risk assessment of polycyclic aromatic hydrocarbons in water and fish of Rivers Niger and Benue confluence Lokoja, Nigeria. J Environ Health Sci Eng 2019;17(1):383–92.
[15] p,p′-Dichlorodiphenyltrichloroethane (DDT), Available from: https://cfpub.epa.gov/ncea/iris2/chemicalLanding.cfm?substance_nmbr=147; 1987.
[16] Haque R, et al. Intake of DDT and its metabolites through food items among reproductive age women in Bangladesh. Chemosphere 2017;189:744–51.
[17] Fang Y, et al. Human health risk assessment of pesticide residues in market-sold vegetables and fish in a northern metropolis of China. Environ Sci Pollut Res Int 2015;22(8):6135–43.
[18] Stidham RW, Higgins PDR. Colorectal cancer in inflammatory bowel disease. Clin Colon Rectal Surg 2018;31 (3):168–78.
[19] Buckley A, Turner JR. Cell biology of tight junction barrier regulation and mucosal disease. Cold Spring Harb Perspect Biol 2018;10(1);a029314.
[20] Betge J, et al. MUC1, MUC2, MUC5AC, and MUC6 in colorectal cancer: expression profiles and clinical significance. Virchows Arch 2016;469(3):255–65.
[21] Swidsinski A, et al. Comparative study of the intestinal mucus barrier in normal and inflamed colon. Gut 2007;56(3):343–50.
[22] Dharmani P, Leung P, Chadee K. Tumor necrosis factor-alpha and Muc2 mucin play major roles in disease onset and progression in dextran sodium sulphate-induced colitis. PLoS One 2011;6(9);e25058.

[23] Bresalier RS, et al. Mucin production by human colonic carcinoma cells correlates with their metastatic potential in animal models of colon cancer metastasis. J Clin Invest 1991;87(3):1037–45.
[24] Kang H, et al. Loss of E-cadherin and MUC2 expressions correlated with poor survival in patients with stages II and III colorectal carcinoma. Ann Surg Oncol 2011;18(3):711–9.
[25] Bhat AA, et al. Tight junction proteins and signaling pathways in cancer and inflammation: a functional crosstalk. Front Physiol 2018;9:1942.
[26] Kinugasa T, et al. Increased claudin-1 protein expression contributes to tumorigenesis in ulcerative colitis-associated colorectal cancer. Anticancer Res 2010;30(8):3181–6.
[27] Oshima T, Miwa H, Joh T. Changes in the expression of claudins in active ulcerative colitis. J Gastroenterol Hepatol 2008;23(Suppl 2):S146–50.
[28] Zeissig S, et al. Changes in expression and distribution of claudin 2, 5 and 8 lead to discontinuous tight junctions and barrier dysfunction in active Crohn's disease. Gut 2007;56(1):61–72.
[29] Bhat AA, et al. Claudin-1 promotes TNF-alpha-induced epithelial-mesenchymal transition and migration in colorectal adenocarcinoma cells. Exp Cell Res 2016;349(1):119–27.
[30] Dhawan P, et al. Claudin-2 expression increases tumorigenicity of colon cancer cells: role of epidermal growth factor receptor activation. Oncogene 2011;30(29):3234–47.
[31] Wang S, et al. Functions of macrophages in the maintenance of intestinal homeostasis. J Immunol Res 2019;2019:1512969.
[32] Schenk M, et al. TREM-1–expressing intestinal macrophages crucially amplify chronic inflammation in experimental colitis and inflammatory bowel diseases. J Clin Invest 2007;117(10):3097–106.
[33] Rogler G, et al. Nuclear factor kappaB is activated in macrophages and epithelial cells of inflamed intestinal mucosa. Gastroenterology 1998;115(2):357–69.
[34] Khorana AA, et al. Vascular endothelial growth factor, CD68, and epidermal growth factor receptor expression and survival in patients with stage II and stage III colon carcinoma: a role for the host response in prognosis. Cancer 2003;97(4):960–8.
[35] Poggi A, et al. Human gut-associated natural killer cells in health and disease. Front Immunol 2019;10:961.
[36] Yadav PK, Chen C, Liu Z. Potential role of NK cells in the pathogenesis of inflammatory bowel disease. J Biomed Biotechnol 2011;2011:348530.
[37] Steel AW, et al. Increased proportion of CD16(+) NK cells in the colonic lamina propria of inflammatory bowel disease patients, but not after azathioprine treatment. Aliment Pharmacol Ther 2011;33(1):115–26.
[38] Halama N, et al. Natural killer cells are scarce in colorectal carcinoma tissue despite high levels of chemokines and cytokines. Clin Cancer Res 2011;17(4):678–89.
[39] Hall LJ, et al. Natural killer cells protect mice from DSS-induced colitis by regulating neutrophil function via the NKG2A receptor. Mucosal Immunol 2013;6(5):1016–26.
[40] Zhou GX, Liu ZJ. Potential roles of neutrophils in regulating intestinal mucosal inflammation of inflammatory bowel disease. J Dig Dis 2017;18(9):495–503.
[41] Wera O, Lancellotti P, Oury C. The dual role of neutrophils in inflammatory bowel diseases. J Clin Med 2016;5(12):118.
[42] Berry RS, et al. High levels of tumor-associated neutrophils are associated with improved overall survival in patients with stage II colorectal cancer. PLoS One 2017;12(12)e0188799.
[43] Rao HL, et al. Increased intratumoral neutrophil in colorectal carcinomas correlates closely with malignant phenotype and predicts patients' adverse prognosis. PLoS One 2012;7(1)e30806.
[44] Dell'Aquila E, et al. Prognostic and predictive role of neutrophil/lymphocytes ratio in metastatic colorectal cancer: a retrospective analysis of the TRIBE study by GONO. Ann Oncol 2018;29(4):924–30.
[45] Halazun KJ, et al. Elevated preoperative neutrophil to lymphocyte ratio predicts survival following hepatic resection for colorectal liver metastases. Eur J Surg Oncol 2008;34(1):55–60.
[46] Rashtak S, et al. Peripheral neutrophil to lymphocyte ratio improves prognostication in colon cancer. Clin Colorectal Cancer 2017;16(2):115–23 e3.
[47] Stagg AJ. Intestinal dendritic cells in health and gut inflammation. Front Immunol 2018;9:2883.
[48] Hart AL, et al. Characteristics of intestinal dendritic cells in inflammatory bowel diseases. Gastroenterology 2005;129(1):50–65.
[49] Ikeda Y, et al. Characterization of antigen-presenting dendritic cells in the peripheral blood and colonic mucosa of patients with ulcerative colitis. Eur J Gastroenterol Hepatol 2001;13(7):841–50.

[50] te Velde AA, et al. Increased expression of DC-SIGN+IL-12+IL-18+ and CD83+IL-12-IL-18- dendritic cell populations in the colonic mucosa of patients with Crohn's disease. Eur J Immunol 2003;33(1):143–51.
[51] Sandel MH, et al. Prognostic value of tumor-infiltrating dendritic cells in colorectal cancer: role of maturation status and intratumoral localization. Clin Cancer Res 2005;11(7):2576–82.
[52] Kusume A, et al. Suppression of dendritic cells by HMGB1 is associated with lymph node metastasis of human colon cancer. Pathobiology 2009;76(4):155–62.
[53] Kurd N, Robey EA. Unconventional intraepithelial gut T cells: the TCR says it all. Immunity 2014;41 (2):167–8.
[54] van Wijk F, Cheroutre H. Mucosal T cells in gut homeostasis and inflammation. Expert Rev Clin Immunol 2010;6(4):559–66.
[55] Smids C, et al. Intestinal T cell profiling in inflammatory bowel disease: linking T cell subsets to disease activity and disease course. J Crohns Colitis 2018;12(4):465–75.
[56] Roosenboom B, et al. Intestinal CD103+CD4+ and CD103+CD8+ T-cell subsets in the gut of inflammatory bowel disease patients at diagnosis and during follow-up. Inflamm Bowel Dis 2019;25(9):1497–509.
[57] Reissfelder C, et al. Tumor-specific cytotoxic T lymphocyte activity determines colorectal cancer patient prognosis. J Clin Invest 2015;125(2):739–51.
[58] Sideras K, et al. Prognostic value of intra-tumoral CD8(+)/FoxP3(+) lymphocyte ratio in patients with resected colorectal cancer liver metastasis. J Surg Oncol 2018;118(1):68–76.
[59] Brucklacher-Waldert V, et al. Cellular plasticity of CD4+ T cells in the intestine. Front Immunol 2014;5:488.
[60] Tosolini M, et al. Clinical impact of different classes of infiltrating T cytotoxic and helper cells (Th1, th2, treg, th17) in patients with colorectal cancer. Cancer Res 2011;71(4):1263–71.
[61] Fuss IJ, et al. Disparate CD4+ lamina propria (LP) lymphokine secretion profiles in inflammatory bowel disease. Crohn's disease LP cells manifest increased secretion of IFN-gamma, whereas ulcerative colitis LP cells manifest increased secretion of IL-5. J Immunol 1996;157(3):1261–70.
[62] Heller F, et al. Interleukin-13 is the key effector Th2 cytokine in ulcerative colitis that affects epithelial tight junctions, apoptosis, and cell restitution. Gastroenterology 2005;129(2):550–64.
[63] Ling A, et al. The infiltration, and prognostic importance, of Th1 lymphocytes vary in molecular subgroups of colorectal cancer. J Pathol Clin Res 2016;2(1):21–31.
[64] Harrison OJ, Powrie FM. Regulatory T cells and immune tolerance in the intestine. Cold Spring Harb Perspect Biol 2013;5(7)a018341.
[65] Khalili A, et al. CD4+CD25+CD127low FoxP3+ regulatory T cells in Crohn's disease. Rom J Intern Med 2018;56 (3):158–66.
[66] Mohammadnia-Afrouzi M, et al. Decrease of CD4(+) CD25(+) CD127(low) FoxP3(+) regulatory T cells with impaired suppressive function in untreated ulcerative colitis patients. Autoimmunity 2015;48(8):556–61.
[67] Betts G, et al. Suppression of tumour-specific CD4(+) T cells by regulatory T cells is associated with progression of human colorectal cancer. Gut 2012;61(8):1163–71.
[68] Michel S, et al. High density of FOXP3-positive T cells infiltrating colorectal cancers with microsatellite instability. Br J Cancer 2008;99(11):1867–73.
[69] Wang Q, et al. Intratumoral regulatory T cells are associated with suppression of colorectal carcinoma metastasis after resection through overcoming IL-17 producing T cells. Cell Immunol 2014;287(2):100–5.
[70] Ratajczak W, et al. Immunological memory cells. Cent Eur J Immunol 2018;43(2):194–203.
[71] Yacyshyn BR. Activated CD19+ B cell lamina propria lymphocytes in ulcerative colitis. Immunol Cell Biol 1993;71(Pt 4):265–74.
[72] Wang X, et al. Ulcerative colitis is characterized by a decrease in regulatory B cells. J Crohns Colitis 2016;10 (10):1212–23.
[73] Hevia A, et al. Association of levels of antibodies from patients with inflammatory bowel disease with extracellular proteins of food and probiotic bacteria. Biomed Res Int 2014;2014:351204.
[74] Shimabukuro-Vornhagen A, et al. Characterization of tumor-associated B-cell subsets in patients with colorectal cancer. Oncotarget 2014;5(13):4651–64.
[75] Dieterich W, Schink M, Zopf Y. Microbiota in the gastrointestinal tract. Med Sci (Basel) 2018;6(4):116.
[76] Tamboli CP, et al. Dysbiosis in inflammatory bowel disease. Gut 2004;53(1):1–4.
[77] Mirsepasi-Lauridsen HC, et al. Disease-specific enteric microbiome dysbiosis in inflammatory bowel disease. Front Med (Lausanne) 2018;5:304.

[78] Shang FM, Liu HL. Fusobacterium nucleatum and colorectal cancer: a review. World J Gastrointest Oncol 2018;10(3):71–81.
[79] Komiya Y, et al. Patients with colorectal cancer have identical strains of *Fusobacterium nucleatum* in their colorectal cancer and oral cavity. Gut 2019;68(7):1335–7.
[80] Kosumi K, et al. Dysbiosis of the gut microbiota and colorectal cancer: the key target of molecular pathological epidemiology. J Lab Precis Med 2018;3:76.
[81] Luhrs H, et al. Butyrate inhibits NF-kappaB activation in lamina propria macrophages of patients with ulcerative colitis. Scand J Gastroenterol 2002;37(4):458–66.
[82] Zhang M, et al. Butyrate inhibits interleukin-17 and generates Tregs to ameliorate colorectal colitis in rats. BMC Gastroenterol 2016;16(1):84.
[83] Yan H, Ajuwon KM. Butyrate modifies intestinal barrier function in IPEC-J2 cells through a selective upregulation of tight junction proteins and activation of the Akt signaling pathway. PLoS One 2017;12(6) e0179586.
[84] Valenzano MC, et al. Remodeling of tight junctions and enhancement of barrier integrity of the CACO-2 intestinal epithelial cell layer by micronutrients. PLoS One 2015;10(7)e0133926.
[85] Wu X, et al. Effects of the intestinal microbial metabolite butyrate on the development of colorectal cancer. J Cancer 2018;9(14):2510–7.
[86] Marques C, et al. Acetate-induced apoptosis in colorectal carcinoma cells involves lysosomal membrane permeabilization and cathepsin D release. Cell Death Dis 2013;4:e507.
[87] Oyola MG, Handa RJ. Hypothalamic-pituitary-adrenal and hypothalamic-pituitary-gonadal axes: sex differences in regulation of stress responsivity. Stress 2017;20(5):476–94.
[88] Reber SO. Stress and animal models of inflammatory bowel disease–an update on the role of the hypothalamo-pituitary-adrenal axis. Psychoneuroendocrinology 2012;37(1):1–19.
[89] Kikuchi N, et al. Perceived stress and colorectal cancer incidence: the Japan collaborative cohort study. Sci Rep 2017;7:40363.
[90] Li B, et al. Glucocorticoids promote the development of azoxymethane and dextran sulfate sodium-induced colorectal carcinoma in mice. BMC Cancer 2019;19(1):94.
[91] de Souza PR, et al. Adrenal-derived hormones differentially modulate intestinal immunity in experimental colitis. Mediators Inflamm 2016;2016:4936370.
[92] Mulak A, Tache Y, Larauche M. Sex hormones in the modulation of irritable bowel syndrome. World J Gastroenterol 2014;20(10):2433–48.
[93] Amos-Landgraf JM, et al. Sex disparity in colonic adenomagenesis involves promotion by male hormones, not protection by female hormones. Proc Natl Acad Sci U S A 2014;111(46):16514–9.
[94] Lee SM, et al. The effect of sex on the Azoxymethane/dextran sulfate sodium-treated mice model of colon cancer. J Cancer Prev 2016;21(4):271–8.
[95] Son HJ, et al. Effect of estradiol in an azoxymethane/dextran sulfate sodium-treated mouse model of colorectal cancer: implication for sex difference in colorectal cancer development. Cancer Res Treat 2019;51(2):632–48.
[96] Zhou Q, et al. Serum bisphenol-A concentration and sex hormone levels in men. Fertil Steril 2013;100(2):478–82.
[97] Kim JY, et al. Bisphenol A-induced aromatase activation is mediated by cyclooxygenase-2 up-regulation in rat testicular Leydig cells. Toxicol Lett 2010;193(2):200–8.
[98] Lin JH, et al. Association between sex hormones and colorectal cancer risk in men and women. Clin Gastroenterol Hepatol 2013;11(4):419–24 e1.
[99] Shah SC, et al. Sex-based differences in incidence of inflammatory bowel diseases-pooled analysis of population-based studies from western countries. Gastroenterology 2018;155(4):1079–89 e3.
[100] Zhao Z, et al. Bisphenol A inhibits mucin 2 secretion in intestinal goblet cells through mitochondrial dysfunction and oxidative stress. Biomed Pharmacother 2019;111:901–8.
[101] Feng L, et al. Bisphenol A increases intestinal permeability through disrupting intestinal barrier function in mice. Environ Pollut 2019;254(Pt A):112960.
[102] Braniste V, et al. Impact of oral bisphenol A at reference doses on intestinal barrier function and sex differences after perinatal exposure in rats. Proc Natl Acad Sci U S A 2010;107(1):448–53.
[103] Javurek AB, et al. Effects of exposure to bisphenol A and ethinyl estradiol on the gut microbiota of parents and their offspring in a rodent model. Gut Microbes 2016;7(6):471–85.
[104] Lai KP, et al. Bisphenol A alters gut microbiome: comparative metagenomics analysis. Environ Pollut 2016;218:923–30.

[105] DeLuca JA, et al. Bisphenol-A alters microbiota metabolites derived from aromatic amino acids and worsens disease activity during colitis. Exp Biol Med (Maywood) 2018;243(10):864–75.
[106] Youn JY, et al. Evaluation of the immune response following exposure of mice to bisphenol A: induction of Th1 cytokine and prolactin by BPA exposure in the mouse spleen cells. Arch Pharm Res 2002;25(6):946–53.
[107] Camarca A, et al. Human peripheral blood mononuclear cell function and dendritic cell differentiation are affected by bisphenol-A exposure. PLoS One 2016;11(8);e0161122.
[108] Yan H, Takamoto M, Sugane K. Exposure to bisphenol A prenatally or in adulthood promotes T(H)2 cytokine production associated with reduction of CD4CD25 regulatory T cells. Environ Health Perspect 2008;116 (4):514–9.
[109] Singh S, Li SS. Phthalates: toxicogenomics and inferred human diseases. Genomics 2011;97(3):148–57.
[110] Wang DC, et al. Exercise prevents the increased anxiety-like behavior in lactational di-(2-ethylhexyl) phthalate-exposed female rats in late adolescence by improving the regulation of hypothalamus-pituitary-adrenal axis. Horm Behav 2014;66(4):674–84.
[111] Rattan S, et al. Prenatal exposure to di(2-ethylhexyl) phthalate disrupts ovarian function in a transgenerational manner in female mice. Biol Reprod 2018;98(1):130–45.
[112] Araki A, et al. Prenatal di(2-ethylhexyl) phthalate exposure and disruption of adrenal androgens and glucocorticoids levels in cord blood: the Hokkaido study. Sci Total Environ 2017;581-582:297–304.
[113] Davis BJ, Maronpot RR, Heindel JJ. Di-(2-ethylhexyl) phthalate suppresses estradiol and ovulation in cycling rats. Toxicol Appl Pharmacol 1994;128(2):216–23.
[114] Chen HP, et al. Effects of di(2-ethylhexyl)phthalate exposure on 1,2-dimethyhydrazine-induced colon tumor promotion in rats. Food Chem Toxicol 2017;103:157–67.
[115] Chen HP, et al. Phthalate exposure promotes chemotherapeutic drug resistance in colon cancer cells. Oncotarget 2018;9(17):13167–80.
[116] Moussa L, et al. A low dose of fermented soy germ alleviates gut barrier injury, hyperalgesia and faecal protease activity in a rat model of inflammatory bowel disease. PLoS One 2012;7(11);e49547.
[117] Yurdakok Dikmen B, et al. In vitro effects of phthalate mixtures on colorectal adenocarcinoma cell lines. J Environ Pathol Toxicol Oncol 2015;34(2):115–23.
[118] Vetrano AM, et al. Inflammatory effects of phthalates in neonatal neutrophils. Pediatr Res 2010;68(2):134–9.
[119] Lee JW, et al. Di-(2-ethylhexyl) phthalate enhances melanoma tumor growth via differential effect on M1-and M2-polarized macrophages in mouse model. Environ Pollut 2018;233:833–43.
[120] Nishioka J, et al. Di-(2-ethylhexyl) phthalate induces production of inflammatory molecules in human macrophages. Inflamm Res 2012;61(1):69–78.
[121] Hansen JF, et al. Influence of phthalates on in vitro innate and adaptive immune responses. PLoS One 2015; 10(6);e0131168.
[122] Yang YN, et al. Phthalate exposure alters gut microbiota composition and IgM vaccine response in human newborns. Food Chem Toxicol 2019;132:110700.
[123] Sinha R, et al. Dietary benzo[a]pyrene intake and risk of colorectal adenoma. Cancer Epidemiol Biomarkers Prev 2005;14(8):2030–4.
[124] Alexandrov K, et al. Evidence of anti-benzo[a]pyrene diolepoxide-DNA adduct formation in human colon mucosa. Carcinogenesis 1996;17(9):2081–3.
[125] Tabatabaei SM, et al. Dietary benzo[a]pyrene intake from meat and the risk of colorectal cancer. Cancer Epidemiol Biomarkers Prev 2010;19(12):3182–4.
[126] Mayhew JW, et al. Increased oxidation of a chemical carcinogen, benzo(a)pyrene, by colon tissue biopsy specimens from patients with ulcerative colitis. Gastroenterology 1983;85(2):328–34.
[127] Hakura A, et al. Rapid induction of colonic adenocarcinoma in mice exposed to benzo[a]pyrene and dextran sulfate sodium. Food Chem Toxicol 2011;49(11):2997–3001.
[128] Niestroy J, et al. Single and concerted effects of benzo[a]pyrene and flavonoids on the AhR and Nrf2-pathway in the human colon carcinoma cell line Caco-2. Toxicol In Vitro 2011;25(3):671–83.
[129] Ghazali AR, et al. Effects of pterostilbene on activities and protein expression of cytochrome P450 1A1 (CYP1A1) and glutathione S-transferase (GST) in benzo[a]pyrene-induced HT-29 colorectal cancer cell line. J Sains Kesihatan Malaysia Isu Khas 2018;5:27–33.
[130] Myers JN, Rekhadevi PV, Ramesh A. Comparative evaluation of different cell lysis and extraction methods for studying benzo(a)pyrene metabolism in HT-29 colon cancer cell cultures. Cell Physiol Biochem 2011;28 (2):209–18.

[131] Wu JC, et al. Polymethoxyflavones prevent benzo[a]pyrene/dextran sodium sulfate-induced colorectal carcinogenesis through modulating xenobiotic metabolism and ameliorate autophagic defect in ICR mice. Int J Cancer 2018;142(8):1689–701.

[132] Tintos A, et al. Beta-naphthoflavone and benzo(a)pyrene treatment affect liver intermediary metabolism and plasma cortisol levels in rainbow trout *Oncorhynchus mykiss*. Ecotoxicol Environ Saf 2008;69(2):180–6.

[133] Kang SC, Lee BM. Effect of estrogen receptor (ER) on benzo[a]pyrene-DNA adduct formation in human breast cancer cells. J Toxicol Environ Health A 2005;68(21):1833–40.

[134] van Grevenynghe J, et al. Polycyclic aromatic hydrocarbons inhibit differentiation of human monocytes into macrophages. J Immunol 2003;170(5):2374–81.

[135] Volkov MS, Kobliakov VA. Activation of transcription factor NF-kappaB by carcinogenic polycyclic aromatic hydricarbons. Tsitologiia 2011;53(5):418–22.

[136] Fueldner C, et al. Benzo(a)pyrene attenuates the pattern-recognition-receptor induced proinflammatory phenotype of murine macrophages by inducing IL-10 expression in an aryl hydrocarbon receptor-dependent manner. Toxicology 2018;409:80–90.

[137] Song L, et al. p,p′-Dichlorodiphenyltrichloroethane inhibits the apoptosis of colorectal adenocarcinoma DLD1 cells through PI3K/AKT and Hedgehog/Gli1 signaling pathways. Toxicol Res (Camb) 2015;4:1214–24.

[138] Jin C, et al. Oral imazalil exposure induces gut microbiota dysbiosis and colonic inflammation in mice. Chemosphere 2016;160:349–58.

[139] Jin C, et al. Imazalil exposure induces gut microbiota dysbiosis and hepatic metabolism disorder in zebrafish. Comp Biochem Physiol C Toxicol Pharmacol 2017;202:85–93.

[140] Vismaya, Rajini PS. Oral exposure to the organophosphorus insecticide, monocrotophos induces intestinal dysfunction in rats. Food Chem Toxicol 2014;71:236–43.

[141] Basavarajappa MS, et al. Methoxychlor reduces estradiol levels by altering steroidogenesis and metabolism in mouse antral follicles in vitro. Toxicol Appl Pharmacol 2011;253(3):161–9.

[142] Tian H, et al. Monocrotophos pesticide affects synthesis and conversion of sex steroids through multiple targets in male goldfish (*Carassius auratus*). Sci Rep 2017;7(1):2306.

[143] Zhang X, et al. Impairment of the cortisol stress response mediated by the hypothalamus-pituitary-interrenal (HPI) axis in zebrafish (*Danio rerio*) exposed to monocrotophos pesticide. Comp Biochem Physiol C Toxicol Pharmacol 2015;176–177:10–6.

[144] Lee GH, Choi KC. Adverse effects of pesticides on the functions of immune system. Comp Biochem Physiol C Toxicol Pharmacol 2020;235:108789.

[145] Mokarizadeh A, et al. A comprehensive review of pesticides and the immune dysregulation: mechanisms, evidence and consequences. Toxicol Mech Methods 2015;25(4):258–78.

[146] Cortés-Iza SC, Rodríguez AI. Oxidative stress and pesticide disease: a challenge for toxicology. Rev Fac Med 2018;66(2):261–7.

CHAPTER

9

Targeting the STAT6 signaling pathway as a therapy against colon cancer

Ana Catalina Rivera Rugeles[a], *Yael Delgado-Ramirez*[a], *Luis I. Terrazas*[b,c], *and Sonia Leon-Cabrera*[a,d]

[a]Unidad de Investigación Biomédica, Universidad Nacional Autónoma de México, Tlalnepantla, México [b]Laboratorio Nacional en Salud: Diagnóstico Molecular y Efecto Ambiental en Enfermedades Crónico-degenerativas, Facultad de Estudios Superiores Iztacala, Universidad Nacional Autónoma de México, Tlalnepantla, México [c]Unidad de Biomedicina, Facultad de Estudios Superiores Iztacala, Universidad Nacional Autónoma de México, Tlalnepantla, México [d]Escuela de Medicina, Facultad de Estudios Superiores Iztacala, Universidad Nacional Autónoma de México, Tlalnepantla, Mexico

Abstract

Ongoing evidence suggests that signal transducer and activator of transcription 6 (STAT6) plays a pivotal role in colorectal cancer (CRC) development. Compelling evidence from both man and experimental models shows that STAT6 not only contributes in mediating immune response but is also involved in the pathology associated with disease by altering the epithelial barrier function, promoting the proliferation of intestinal epithelial cells, and regulating the expression of pro-survival and pro-metastatic proteins. These studies have led to the approach of targeting STAT6 as an effective treatment strategy in alleviating CRC. Thus, silencing or hindering STAT6 signaling may strengthen the chemotherapy response with positive effects in the current treatments. Therefore, this chapter describes the role of STAT6 in colorectal cancer biology and their potential as a new therapeutic target for the prevention and treatment of this disease.

Abbreviations

5-FU	5-fluorouracil
11βHSD2	11β-hydroxysteroid dehydrogenase type II
AOM	azoxymethane
AP1	atherosclerotic plaque-specific peptide-1
ARG1	arginase-1

Immunotherapy in Resistant Cancer: From the Lab Bench Work to Its Clinical Perspectives
https://doi.org/10.1016/B978-0-12-822028-3.00017-0

BAX	Bcl-2-associated X protein
CAC	colitis-associated carcinogenesis
CAV1	caveolin-1
CCL	chemokine (C—C motif) ligand
CD	Crohn's disease
CIC	cancer initiating cell
COX2	cyclooxygenase-2
CR-CSC	colorectal cancer stem cells
CRC	colorectal cancer
CXCR	CXC chemokine receptors
DSS	dextran sulfate sodium
EMT	epithelial-to-mesenchymal transition
FDA	Food and Drug Administration
IBD	inflammatory bowel disease
IEC	intestinal epithelial cell
IFN	interferon
IG	immunoglobulin
IL	interleukin
ILC	innate lymphoid cell
IL-R	interleukin receptor
JAK	Janus kinase
LPL	lamina propria leukocyte
MDM2	mouse double minute 2 homolog protein
MDSC	myeloid-derived suppressor cell
MLCK1	myosin light-chain kinase 1
MRC1	mannose receptor 1
NADPH	nicotinamide adenine dinucleotide phosphate
NOX1	NADPH oxidase 1
OS	overall survival
OXA	oxazolone
PI3K	phosphoinositide 3-kinase
P-STAT	phosphorylated STAT
ROS	reactive oxygen species
SAHA	suberoylanilide hydroxamic acid
siRNA	small interfering RNA
STAT	signal transducer and activator of transcription
TAM	tumor-associated macrophage
TMPRSS4	transmembrane protease, serine 4
TNBS	2,4,6-trinitrobenzenesulfonic acid
TNFα	tumor necrosis factor α
TYK2	tyrosine kinase 2
UC	ulcerative colitis
WT	wild type
ZEB1	zinc finger E-box-binding homeobox 1

Conflict of interest

No potential conflicts of interest were disclosed by the authors.

Introduction

Colorectal cancer (CRC) is one of the most lethal neoplastic diseases worldwide [1]. Roughly 1.8 million cases were diagnosed globally in 2018 [1], making CRC the third most commonly detected cancer and the second most common cause of cancer-associated mortality rate with 881.00 deaths per year [2]. Besides the improvements made in diagnostic, treatment, and increased screening, 40%–50% of all CRC patients present with metastasis [3]. Additionally, a 5-year relative survival rate for patients with stage IIIC and IV is approximately 53% and 11%, respectively [3]. Actual CRC management involves different approaches based on personalized treatment strategies. Surgical resection, adjuvant chemotherapy with cytotoxic agents, radiotherapy, monoclonal antibodies, recombinant fusion proteins, and oral multikinase inhibitors are the main treatment options [4]. Additionally, most CRC patients with distant metastasis typically in the liver or lungs are not apt for usual interventions and exhibit a poor 5-year survival rate [5]. Conversely, the prognosis in these cases remains reduced and the treatment is mainly palliative. From a therapeutic perspective, finding new biomarkers of CRC and defining the molecular mechanisms underlying metastatic progression is vital to decreasing morbidity and mortality during this disease.

STAT6 is a signal transducer and activator of transcription which can be activated by phosphorylation in the presence of interleukin (IL)-4, IL-13, and some growth factors [6]. Previous studies have shown that STAT6 is mutated in various types of cancers like colon, lung, breast, pancreas, solitary fibrous tumor, and endometrial adenocarcinoma [7]. Additionally, high levels of STAT6 protein have been reported in different types of cancers like cervical, prostate, breast, and CRC [8]. Furthermore, tumoral phospho-STAT6 (p-STAT6) levels have been commonly detected in the colon of patients with clinically detectable inflammatory bowel diseases (IBD), and tumoral p-STAT6 positively correlates to the clinical stage and poor prognosis of human CRC [9,10].

Experimental studies have demonstrated that STAT6 is involved in the early steps of colitis-associated carcinogenesis (CAC) and that it modulates inflammatory responses, as well as controls cell recruitment and proliferation [11,12]. T cells, macrophages, and natural killer T cells exhibit increased STAT6 phosphorylation during colitis development [13]. Also, STAT6 promotes an immunosuppressive microenvironment by the expansion of myeloid-derived suppressor cells (MDSCs) and M2-type tumor-associated macrophages during cancer, implying regulation of antitumor T-cell response [12,14].

In non-immune cells, STAT6 signaling pathway favors the expression of anti-apoptotic proteins [15] and promotes the proliferation of polyp epithelial cells in the colon [9]. Persistent activation of STAT6 modifies the expression of proteins involved in epithelial barrier permeability and interrupts tight junction integrity, resulting in recurrent exposure of luminal microbiota favoring inflammation and CAC development [9,13].

Experimental evidences have also shown that blocking IL-4 and IL-13, as well as their receptors, prevents STAT6 activation and has different consequences on colon tumor development [16–18]. Additionally, IL-4/IL-13/STAT6 signaling enhances the expression of epithelial-to-mesenchymal transition (EMT)-promoting factors in colon cancer cell lines, resulting in a decrease in adhesion, improving migration, invasion, and metastatic colonization [19].

Therefore, STAT6 may be considered a key regulatory molecule with potential as a therapeutic target for the treatment of colon cancer. Recent studies have shown that interrupting the expression of STAT6 in CRC cell lines by small interfering RNAs (siRNAs) decreases cell proliferation and migration and induces apoptosis [20]. Similar observations have also been made in lung cancer [21]. Also, it has been shown that, when STAT6 inhibitor is used as an adjuvant during CAC, the tumor response to chemotherapeutic drugs is enhanced [22] and inflammatory markers and tumor development are decreased [11].

Therefore, this chapter describes the role of STAT6 in colorectal cancer biology and their potential as a new therapeutic target for the prevention and treatment of this disease.

STAT6 signaling pathway

STAT6 serves important roles in signal transduction throughout the cytoplasm and is a transcription factor in the nucleus. STAT6 activation occurs by IL-4 or IL-13 binding to their receptors. Currently, three types of receptors have been described. The IL-4 type I receptor is composed of the IL-4 receptor α (IL-4Rα) chain and the common γ chain (γc). The IL-4Rα chain has a high affinity for IL-4, while the γc is a component of receptors that bind IL-2 and IL-9. IL-4 binds to the IL-4Rα chain, resulting in dimerization with the γc chain and receptor activation [23]. Additionally, IL-13 can bind to the IL-13 receptor α variant 2 (IL-13Rα2) chain; this receptor is a monomer that binds IL-13 with higher affinity than IL-13Rα1 and is expressed in T cells, B cells, and endothelial cells, among others, where it has been demonstrated to promote tumorigenesis [24]. IL-13Rα2 is considered as a decoy receptor, as it is able to bind IL-13 and prevent it from binding to IL-13Rα1 [24].

After stimulation of the IL-4Rα receptor, JAK1-3 and tyrosine kinase 2 (Tyk2) are phosphorylated in the cytoplasmic tails of the receptor. Once active, JAKs phosphorylate the tyrosine residues Y575, Y603, and Y631 on the receptor, generating docking sites for STAT6. When monomers of STAT6 become phosphorylated on tyrosine Y641, the C-terminal of the SH2 domain usually forms homodimers [25] that are translocated to the nucleus, where they can bind DNA and activate or repress target genes.

Dysregulation of STAT6 signaling in IBD and CRC

IL-4/IL-13/STAT6 pathway has been shown to be a key component for the mucosa's response to injury and the progression of IBD and CRC (Table 1). STAT6 was initially described as part of a pathway involved in T helper-type (Th) 2 cell activation and differentiation. However, it has been established that STAT6 is present in immune and non-immune cells and plays diverse biological activities [48]. STAT6 promotes CRC due to its ability to enhance gut permeability and microbiota translocation, which leads to an increase in inflammatory cell recruitment at the intestine. Additionally, STAT6 promotes proliferation of intestinal epithelial cells, regulates the expression of anti-apoptotic and pro-metastatic proteins, and promotes the expansion of immunosuppressive mediators in the intestine preventing an active cytotoxic response capable of containing tumor growth (Fig. 1).

STAT6 signaling and epithelial barrier dysfunction

IBD is a group of chronic, relapsing, and systemic diseases that mainly affect the gastrointestinal tract. This pathology includes two clinical entities, Crohn's disease (CD) and ulcerative colitis (UC), which differ in some aspects but share the epithelial barrier disruption and the exacerbated response of the immune system to gut microbiota. Among the main causes of the IBD development, a breakdown of intestinal homeostasis by uncontrolled interactions between immune cells, intestinal epithelial cells, and gut microbiota has been proposed. During IBD, the chronic colonic inflammatory conditions constitute an important risk factor to CAC development [49].

TABLE 1 Activation of IL-4/IL-13/STAT6 signaling pathway during IBD and colorectal cancer.

		Immune cells			
Pathology	Cell lines	Monocytes, granulocytes, and macrophages	Lymphocytes and neutrophils	Epithelial colonic cells	Reference
Inflammatory bowel disease	NK human cell lines and CD161 (+) LPTs from UC patients are cytotoxic to HT-29 colonic cells; this is augmented with IL-13 application		LPT cells bearing NK markers from UC patients produce high amounts of IL-13		[26]
				IL-13Rα1 and IL-4Rα are expressed in mucosal biopsies of UC patients Claudin-2 expression is considerably higher in UC patients compared to normal patients	[27]
				IL-13-mediated STAT6 phosphorylation is upregulated in pediatric patients with UC	[28]
			IL-4 induces CD4+ Th2 cells to produce IL-13, which in turn induces IgE production by B cells, in a UC mouse model		[29]
		In a mouse colitis model, IL-13 knockout mice had fewer macrophage infiltrates in the colonic tissue compared to WT mice		In a mouse colitis model, IL-13 administration increased the severity of colitis and diminished the production of serotonin by enterochromaffin cells, which is important for epithelial cell secretion	[30]
		STAT6 phosphorylation is increased in	STAT6 phosphorylation is increased in T cells and natural killer cells in a	In a mouse colitis model, IL-13 drives STAT6 phosphorylation which increases claudin-2 activity, thus altering the epithelial barrier	[13]

Continued

TABLE 1 Activation of IL-4/IL-13/STAT6 signaling pathway during IBD and colorectal cancer—cont'd

		Immune cells			
Pathology	**Cell lines**	**Monocytes, granulocytes, and macrophages**	**Lymphocytes and neutrophils**	**Epithelial colonic cells**	**Reference**
		macrophages in a UC mouse model	UC mouse model IL-4, IL-5, IL-13, and IFN-γ secretion by mesenteric lymph node cells is decreased in STAT6 knockout mice with UC		
				Treatment with anrukinzumab, an anti-interleukin 13 monoclonal antibody, does not affect wound healing in UC patients compared to a placebo treatment	[18]
		STAT6 mediates wound healing through Wnt signaling and M2 macrophage activity in an IBD mouse model			[31]
	In colonic T84 cells, the phosphorylation of STAT6 caused by treatment with IL-4 and IL-13 cytokines downregulates matriptase, a protein involved in epithelial barrier dysfunction				[32]
	MLCK1 expression, which increases epithelial permeability, is induced by IL-13 and decreased by STAT6 knockout, in the Caco-2 colonic cell line		STAT6 knockout mice show a reduction in pro-inflammatory cytokines and neutrophil infiltration in the colon compared to WT mice, in a colitis murine model ILC2 are the major source of IL-13 in the colon of colitis-induced mice	STAT6 induces the transcription of MLCK1, which in turn relates to tight junction dysfunction and microbiota infiltration in a colitis mouse model	[9]

Colon cancer	IL-4 and IL-13 inhibit cell-cell adhesion through E-cadherin downregulation in the colo-205 CRC cell line	Tumor-infiltrating lymphocytes isolated from CRC tissue samples express IL-4 and IL-13		[33]
	Application of IL-13 enhances the apoptotic rate of HT-29 CRC cells through caspase activation. It also dysregulates the pore-forming tight junction protein claudin-2, which is important to the intestinal barrier integrity			[34]
	IL-4-induced STAT6 activity induces the expression of anti-apoptotic and pro-metastatic genes, while its absence is correlated to the expression of pro-apoptotic and anti-metastatic genes, in CRC cell lines			[15]
	Autocrine IL-4 production is important to block apoptosis in primary colon cancer cells in culture through the expression of anti-apoptotic proteins such as PED, cFLIP, Bcl-xL, and Bcl-2			[35]
			Patients with high STAT6 expression have lower survival rates. STAT6 expression is correlated with lymph node metastasis and distant metastasis	[10]
	Abnormal IL-4/STAT6 signaling increases the levels of survivin and its nuclear localization in primary CR-CSC in culture			[36]
			Patients with CRC had an elevated expression of Th2-related genes (IL-4, IL-6, IL-17, and STAT6); however, it was not correlated with a poor prognosis	[37]
	Metastatic CRC cell lines secrete more IL-4 and IL-13 and have a higher expression of IL-13Rα2, in comparison with non-metastatic CRC cell lines Metastatic CRC cell lines also exhibit a higher activation of PI3K, Akt, and SRC in response to IL-13 signaling		Late-stage CRC patients have a higher IL-13Rα2 expression than early-stage CRC patients	[38]
			Patients with CRC had a high IL-4, IL-13, IL-4R, and IL-13R	[39]

Continued

TABLE 1 Activation of IL-4/IL-13/STAT6 signaling pathway during IBD and colorectal cancer—cont'd

Pathology	Cell lines	Immune cells: Monocytes, granulocytes, and macrophages	Immune cells: Lymphocytes and neutrophils	Epithelial colonic cells	Reference
				expression. IL-13 high expression was correlated with a better OS	
	$STAT6^{high}$ HT-29 CRC cell line displays an expression profile favoring Th2-type cytokines, while $STAT6^{low}$ Caco-2 cell line leans towards a Th1-Th17 expression profile				[40]
				pSTAT6 expression in epithelial and immune cells is lower in CAC patients, compared to those with inflammatory disease, contrary to the expression of pSTAT3, which is higher in CAC patients	[41]
	CICs isolated from CRC patients express IL-4, which mediates antitumor immunity; its blockage increased a Th1-type CD8+ T cell response				[42]
	IL-13/STAT6 signaling enhances the expression of EMT-promoting factor ZEB1 in the HT29 and SW480 cell CRC cell lines, which is reversed with STAT6 knockdown			IL-13Rα1 and ZEB1 mRNA levels are correlated in patients with CRC	[19]
	IL-13/ IL-13Rα2 signaling promotes the expression and activity of 11βHSD2, Akt, and COX2 in the CT26 and SW480 CRC cell lines IL-13/IL-13Rα2 signaling also confers metastatic abilities in the same cell lines				[43]
		Circulating inflammatory monocytes and granulocytes were decreased in STAT6 knockout mice compared to WT in a CAC model STAT6-specific inhibitor AS1517499 decreased these same cells in CAC-induced WT mice		STAT6 knockout mice exhibited reduced tumorigenicity and decreased COX2 and β-catenin protein levels (associated with proliferation) in the colon epithelia in a CAC model mRNA IL-17A and TNF-α levels were also reduced AS1517499 application decreased the tumor load in CAC-induced WY mice	[11]

	IL-4, IL-13/JAK1/STAT6 signaling increases NOX1 expression, which is correlated with ROS production, DNA damage, and proliferation in HT-29 CRC cell line			CRC tissue samples showed an increase in NOX1 and IL-4Rα2 compared to normal tissue	[44]
			STAT6 expression is correlated with high levels of MDSC in the spleen and lamina propria, and a decreased CD8-mediated cytotoxic response towards tumor cells in the APCmin/$^{+}$ mouse model	The inhibition of STAT6 in ApcMin/$^{+}$ mice reduces polyp cell proliferation	[12]
	IL-4 induces an E2F1/SP3/STAT6 dependent expression of the EMT-promoting factors ZEB1 and ZEB2 and increases the invasiveness in HCT116 CRC cell lines			SP3, E2F1, and STAT6 are correlated in CRC patients	[45]
				STAT6 deficiency increases the tumor load and proliferation of epithelial cells in a CAC mouse model STAT6 deficiency leads to chromatin compaction with direct repercussions in the apoptosis of epithelial cells and mucosal damage	[46]
		Impaired STAT6 acetylation in Trim24 knockout mice leads to a debilitated TAM antitumor response and increased tumor growth, after MC-38 adenocarcinoma cell injection			[47]

Continued

TABLE 1 Activation of IL-4/IL-13/STAT6 signaling pathway during IBD and colorectal cancer—cont'd

Pathology	Cell lines	Immune cells: Monocytes, granulocytes, and macrophages	Immune cells: Lymphocytes and neutrophils	Epithelial colonic cells	Reference
	STAT6 knockdown by siRNA in the human HT-29 adenocarcinoma cell line reduces proliferation and induces apoptosis				[20]
	Trimethylglycine, an anti-inflammatory compound, decreases STAT6 phosphorylation after treatment with IL-4 in the HCT-116 CRC cell line			The combination of the CRC drug 5-fluorouracil with trimethylglycine and the STAT6-inhibitor AS1517499 reduces tumor growth in a CAC mouse model The combination of these drugs also reduces STAT6 phosphorylation and EMT markers (SNAI1 and β- catenin) and increases the levels of apoptosis and cell adhesion markers in the same model	[22]

Abbreviations: IBD, *inflammatory bowel disease;* IFN, *interferon;* IL, *interleukin;* ILC2, *type 2 innate lymphoid cells;* LPT cells, *lamina propria T cells;* NK, *natural killer cells;* pSTAT, *phosphorylated signal transducer and activator of transcription;* STAT, *signal transducer and activator of transcription;* UC, *ulcerative colitis;* CAC, *colitis-associated colon cancer;* CIC, *cancer-initiating cell;* CRC, *colorectal cancer;* CR-CSC, *colorectal cancer stem cells;* EMT, *epithelial-mesenchymal transition;* OS, *overall survival;* ROS, *reactive oxygen species;* pSTAT, *phosphorylated signal transducer and activator of transcription;* TAM, *tumor-associated macrophage;* WT, *wild type.*

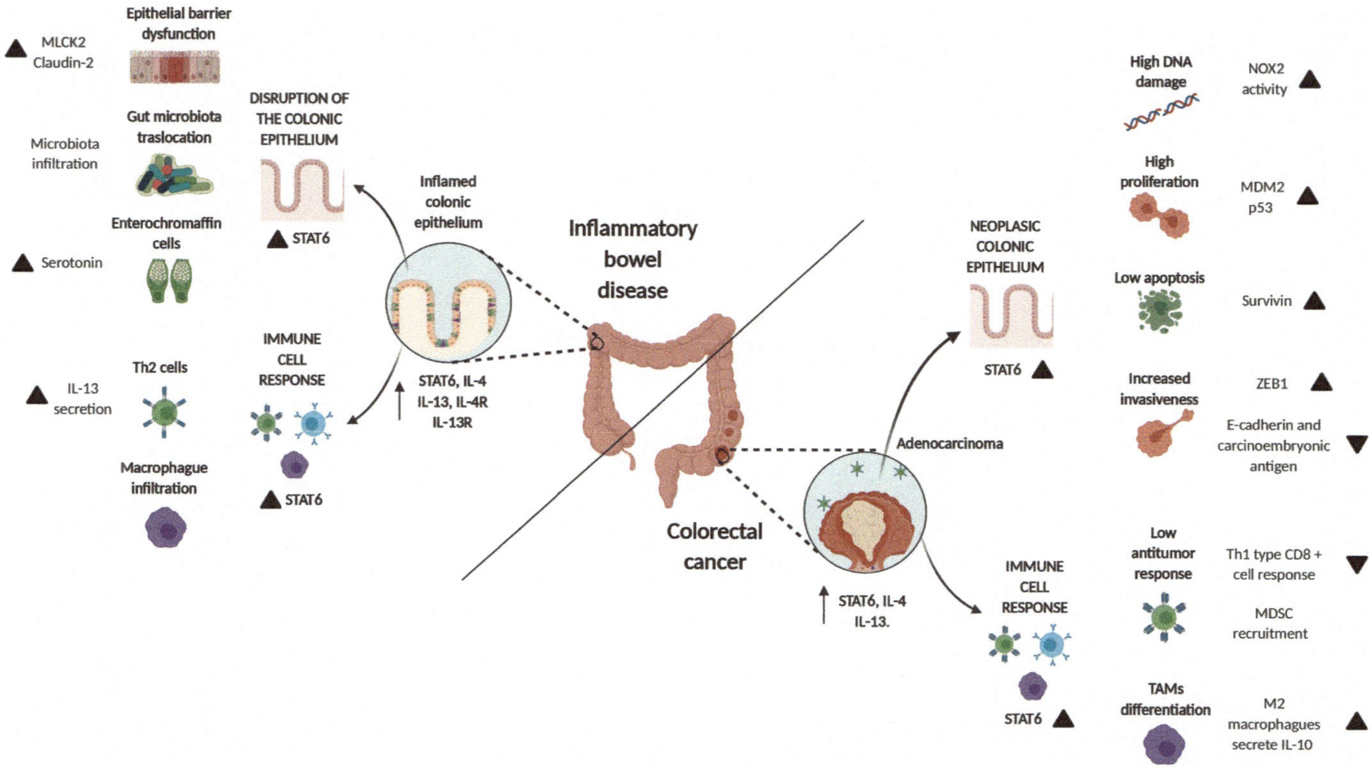

FIG. 1 Activation of STAT6 in immune cells and non-immune cells contributes to the pathology associated with IBD and CRC. STAT6 plays a role in the development of inflammatory bowel diseases (IBD) and colorectal cancer (CRC). In IBD *(left)*, the expression of proteins of the IL-4/IL-13/STAT6 axis is reported to be high in epithelial and immune cells. In the colonic epithelium, a high STAT6 expression is related to epithelial barrier dysfunction and microbiota infiltration; macrophage infiltration is also observed, which contributes to the tissue's inflammation. In colorectal cancer *(right)*, there is also a high expression of proteins of the IL-4/IL-13/STAT6 axis. A high STAT6 expression is related to neoplastic characteristics such as high DNA damage, high proliferation, low apoptosis, and high invasiveness; it also has effects in the immune response, such as the diminishment of Th2-type antitumor response.

Clinical and experimental studies have demonstrated that IL-4/IL-13/STAT6 pathway is constitutively activated during UC, CD, and in colon cancer tissues [39]. In colonic tissues of pediatric subjects with UC or CD, immunohistochemistry demonstrated that nuclear pSTAT6 was significantly upregulated in the colonic epithelium [28]. Also, the upregulation of STAT6 in neoplastic tissues of patients with CRC has been associated with lower survival rates and with lymph node metastasis [10]. However, one study reported that patients with CRC had high IL-4, IL-13, IL-4R, and IL-13R levels of protein expression in tumors compared with normal tissues and this increase was co-related with a decrease in metastasis and IL-13 high expression was associated with a better overall survival rate [39].

Additionally, IL-13 has been identified as an important cytokine in UC that damages epithelial barrier function by affecting tight junctions and inducing epithelial apoptosis [34]. Lamina propria lymphocytes from UC patients produced increased levels of IL-13 [26]. IL-13Ralpha1 and IL-4Ralpha receptors are importantly expressed in colonic epithelial cells of patients with UC [27].

Epithelial barrier integrity is essential to keep away bacteria able to produce genotoxic molecules capable to damage DNA. Ongoing evidence has determined that the gut-dwelling bacterium *Escherichia coli* can accelerate colon tumor formation through a genotoxin called colibactin [50]. Therefore, a disruption in epithelial barrier could favor the exposure of luminal bacteria causing an excessive innate inflammatory response and DNA damage, leading to CRC development. In a model of dextran sulfate sodium (DSS)-induced colitis, p-STAT6 levels were importantly increased in intestinal epithelial cells (IECs) and lamina propria leukocytes (LPL) during the progress of the disease and these levels were in accordance with epithelial disfunction and bacterial translocation [9]. In the same model, during STAT6 deficiency, colitis was significantly abrogated, gene expression of the tight junction permeability protein MLCK1 was decreased, and a better-preserved epithelial integrity was observed [9]. In an oxazolone (OXA)-induced colitis, IL-13 drives STAT6 phosphorylation which increases the pore-forming tight junction protein claudin-2 activity, increasing epithelial barrier permeability, and compromising gastrointestinal barrier function [13]. Furthermore, when colon cell lines HT-29 and T84 were treated with an STAT6 inhibitor or transfected with STAT6 small interfering RNA (siRNA), IL-13-induced claudin-2 expression and barrier disruption were prevented [13]. Recently, Lin et al. [9] demonstrated that during experimental colitis, IL-13 was derived from gut-resident type 2 innate lymphoid cells (ILC2) in response to intestinal microbiota and the increase in IL-13 was the leading cause of epithelial STAT6 activation. When IL-13/STAT6 signaling was activated in epithelial cells, increasing activity of MLCK1 could mediate the dysregulated epithelial tight junction, increasing permeability, bacterial translocation, mucosal inflammation, and colitis development [9].

STAT6 has also been implicated in the loss of the barrier-protective protease pathway at the intestine [32]. Patients with active UC and CD, and mice with DSS-induced colitis, presented a decrease in the expression of matriptase and prostasin, two membrane-anchored serine proteases important for the epithelial barrier development and for homeostasis [32]. The use of an STAT6 inhibitor suberoylanilide hydroxamic acid (SAHA) restored the expression levels of matriptase and prostasin and the epithelial barrier function in experimental colitis [32]. Additionally, in the same model when IL-13 was administrated to mice, enterochromaffin cells which express IL-13R showed high serotonin production and an increase in intestinal damage was observed [30]. These evidences point IL-13/STAT6 signaling pathway as an important contributor to alterations of barrier permeability and gastrointestinal barrier disfunction with consequences in tumor development.

STAT6 axis is required for apoptosis resistance, proliferation, and epithelial-mesenchymal transition of colorectal cancer cells

STAT6 like other members of the signal transducer and activator of transcription (STAT) family of proteins acts as a signaling molecule and a transcription factor in the nucleus. A correlation between IL-4/IL-13/STAT6 signaling and apoptosis and metastasis during CRC has been reported [15,38].

IL-4-induced activation of STAT6 has been implicated in the expression of anti-apoptotic proteins and the growth of epithelial colon cancer cells in vivo and in vitro [15,40]. CRC cell lines with STAT6 (high) phenotype expressed more mRNA of anti-apoptotic and pro-metastatic genes survivin, MDM2, and TMPRSS4, to the contrary, STAT6 (null) cells expressed more mRNA of pro-apoptotic and anti-metastatic genes BAX, CAV1, and P53 [15]. Highly apoptosis-resistant stem-like cells are responsible for colon carcinogenesis. The autocrine production of IL-4 by colorectal cancer stem cells is able to increase STAT6 activation and promote the expression of survivin allowing them to evade cell death [36]. To the contrary, the treatment of these cells with an IL-4 neutralizing antibody or with an STAT6 inhibitor induces nuclear survivin accumulation decreasing apoptosis resistance [36]. Additionally, the exogenous administration of IL-4 to epithelial cancer cells from colon, breast, and lung carcinomas, in addition to increasing the expression of anti-apoptotic proteins favors growth and made them resistant to chemotherapy [35]. NADPH oxidase 1 (NOX1) catalyzes the production of reactive oxygen species (ROS) during CRC contributing to DNA damage and neoplastic transformation [44]. Human colon cancer cells that express type II IL-4 receptor increased the expression of NOX1 via the activation of IL-4/STAT6 signaling pathway and cell proliferation enhancement [44]. Therefore, IL-4 via STAT6 signaling acts as an autocrine survival factor in epithelial cells.

Epithelial-to-mesenchymal transition (EMT) is a process where epithelial cells enhanced migratory capacity and invasiveness by transforming into aggressive phenotypes [51]. In CRC, EMT is associated with an invasive or metastatic phenotype characterized by the loss of cell-cell contact and polarity in epithelial cells [51]. The role of IL-4 and IL-13 in colon cancer cell-cell adhesion has been previously reported [33]. Both cytokines are considered like negative regulators in the expression of E-cadherin and carcinoembryonic antigen, key components of adherent junctions, and in the maintaining of epithelial cell adhesion and decreasing invasiveness [33]. E2F transcription factor 1 (E2F1), a critical transcription factor for CRC development, increases the susceptibility of colon cancer cells to IL-4 by inducing the expression of STAT6 [45]. When an increase in total STAT6 protein occurs in CRC cells, the IL-4 administration induces STAT6 phosphorylation increasing the expression of several EMT drivers [45]. Despite the fact that uncontrolled STAT6 activation seems to be detrimental for cancer development, the role of STAT6 in intestinal homeostasis has not been completely elucidated. Oliveira et al. [46] reported that STAT6 could also act like a protecting pathway in the intestinal epithelium from apoptosis and severe tissue damage by modifications in chromatin condensation in intestinal epithelial cells [46].

Additionally, STAT6 phosphorylation is induced by IL-13. Exposure to IL-13 enhances the expression of the EMT-promoting factor zinc finger *E*-box binding homeobox 1

(ZEB1) in the colon cancer HT29 and SW480 cell lines. When STAT6 is blocked or knocked down, the IL-13-induced EMT and ZEB1 induction in CRC cells is reversed [40]. Additionally, a positive association between IL-13Rα1 and ZEB1 at the mRNA level has been observed in human CRC samples, demonstrating that the IL-13/STAT6 signaling pathway serves a critical role in promoting EMT and the aggressiveness of CRC [19]. 11β-hydroxysteroid dehydrogenase type II (11βHSD2) is a key enzyme induced in an IL-13Rα2-dependent manner that promotes the expression of Akt and COX2; upon inhibiting 11βHSD2, liver metastasis is decreased, suggesting that IL-13 may regulate malignancy via 11βHSD2 during CRC [43]. In addition, highly metastatic CRC cells express high levels of IL-13Rα2 [38]. IL-13Rα2 silencing results in a decrease in adhesion, migration, invasion, and metastatic colonization. In a previous study, the upregulation of IL13Rα2 expression in 66% of tumor samples from patients with colon cancer was associated with late stages of progression (metastasis in lymph nodes or liver) and a poor outcome in patients with CRC. Highly metastatic CRC cells exhibit activation of PI3K, Akt, and SRC proto-oncogene non-receptor tyrosine kinase in response to IL-13, supporting the role of the IL-13/IL-13R/STAT6 signaling pathway in CRC cell invasion and metastasis [38]. Therefore, STAT6 is a key regulatory molecule with tumor proliferating functions that may aid the identification of molecular targets for the treatment of colon cancer.

Implications of STAT6 signaling in immune response during IBD and CRC development

STAT6 signaling pathway is involved in the development and function of the immune system and plays an important role in maintaining immune tolerance and tumor surveillance [52]. It has been fully described that STAT6 mediates Th2 immune response following IL-4 and IL-13 activation and several target genes of STAT6 have been identified and characterized [53]. During the clearance of intestinal parasites and during allergic disease, STAT6 mediates Th2 cells and eosinophil recruitment within the sites of allergic inflammation and is involved in immunoglobulin (Ig) class switching to produce IgE [52].

It is also long noted that STAT6 serves an important role in the regulation of tumor immunity. In the tumor microenvironment, macrophages are the most abundant myeloid cell populations, play pivotal roles in tumor growth, and are associated with tumor stage and metastasis [54]. By the secretion of pro-inflammatory cytokines, these cells could stimulate an inflammatory response, to the contrary by releasing high levels of anti-inflammatory cytokines can suppress immune responses. Classic M1 macrophages have increased expression of cytokines IL-1β and IL-6 promoting tumoricidal responses, whereas M2 macrophages could drive tumor progression by secreting IL-10 [55]. Furthermore, the activation of STAT6 is critical for macrophage function and required for M2-polarized tumor-associated macrophages development from CD11b + cells [56]. In a murine model of primary lung carcinogenesis, STAT6-deficient (STAT6−/−) mice presented few and less aggressive tumors which was correlated with a decreased in mobilization and differentiation of CD11b + cells [57]. In this model, the ablation of STAT6 caused a deficiency in the formation of CD11b + cells resulting in a decrease of M2 and an increase in M1 macrophages contributing to tumor suppression. M2 myeloid cell polarization mediated by STAT6 promotes IL-4 secretion by M2 and tumor cells increasing cancer cell proliferation [57]. In an orthotopic 4T1 mammary carcinoma mouse model, the use of the STAT6-specific inhibitor AS1517499 was shown to attenuate tumor growth and liver metastasis [58]. This result was associated with a reduction in tumor-associated macrophages displaying the M2 phenotype with the decrease of Arg-1 (arginase-1) and Mrc-1 (mannose receptor 1) expressions and arginase activity [58]. Similarly, during CAC, STAT6 −/− mice presented a decrease in colonic tumor load and in the number of circulating inflammatory monocytes (CD11b + Ly6C^{hi}) and granulocytes (CD11b + Ly6G^{+}) compared with wild-type animals [11]. Moreover, the absence of STAT6 also impaired the expression levels of inflammatory cytokines IL-17A and TNFα, the chemokines CCL9, CCL25, and the chemokine receptor CXCR2 during CAC development [11]. CCL9, CCL25, and CXCR2 have been reported to play an important role in the recruitment of inflammatory cells [59,60]. In different cancer models, STAT6 appears to promote the expansion of myeloid-derived suppressor cells (MDSCs) [61]. These cells possess strong immunosuppressive activities and are importantly expand during cancer [62,63]. Cancer tissues with high infiltration of MDSCs are associated with poor patient prognosis and resistance to therapies [64]. In the APC$^{min/+}$ mouse model, which recapitulates the disease observed in patients with familial APC, ApcMin/+STAT6 −/− mice developed few polyps with reduced proliferation at the small intestine [12]. Additionally, PD-1 expression on CD4+ cells was reduced and a strong CD8-mediated cytotoxic response was observed implying the regulation of antitumor T-cell response by MDSCs via STAT6 [12]. In fact, blocking IL-4 in cancer-initiating cells obtained from patients with CRC increased Th1-type CD8^{+} T cells responses in vitro [42] corroborating that IL-4/STAT6 signaling promotes tumor growth by inducing immunosuppressive microenvironment.

The activation and function of STAT6 in the course of colitis at CAC is dynamic. During colitis, myeloid cells are fundamental to maintain mucosal homeostasis and M2 macrophages are in charge of mediating mucosal repair. STAT6 −/− mice that received 2,4,6-trinitrobenzenesulfonic acid (TNBS) developed colitis, exhibited delayed wound healing, and decreased mucosal expression levels of Wnt ligands, which are important responses to epithelial injury [31]. The administration of in vitro M2-polarized macrophages to STAT6 −/− mice with colitis promoted mucosal repair through the activation of the Wnt signaling pathway demonstrating that at least in this scenario, STAT6-M2 polarization constitutes an essential element for mucosal repair [31].

Given the reduction of inflammatory monocytic and granulocytic cells [11,57] and the development of potent antitumor immunity during STAT6 deficiency in cancer [14], STAT6 signaling pathway activation is critical for the recruitment of immune cells and is centrally involved in tumor initiation and progression. The interference of STAT6 function in colitis or in tumor cells may present a novel strategy for the treatment of CRC (Fig. 1; Table 1).

Targeting STAT6 signaling to alter tumor progression

The earlier mentioned evidence suggests that STAT6 has the potential as a biomarker and molecular target in patients with CRC. Silencing or hindering STAT6 signaling during STAT6 high expression may strengthen the chemotherapy response with positive effects in the current treatments. Additionally, the STAT6 activation by phosphorylation largely depends on IL-4/IL-13 bind to their receptors and JAK recruitment. Therefore, the strategies focus on impaired the signaling though the entire pathway has the potential to be an effective treatment of CRC (Table 2). Despite this evidence, the most information has been obtained from cell

TABLE 2 Targeting IL-4/IL-13/STAT6 axis for colorectal cancer therapy.

Strategy	Clinical implication	Reference
IL-4 neutralizing antibodies in combination with chemotherapy	Decreased growth and viability of the Caco cell line and reduced expression levels of the CRC stem cell marker CD133 were observed	[65]
Anti-IL-4Rα antibody in combination with chemotherapy	The antitumor efficacy of these standard chemotherapeutic drugs was strongly increased through selective sensitization of CD133(+) cells to apoptosis	[66]
IL-4 neutralizing antibodies in combination with 5-FU or oxaliplatin	The efficacy of chemotherapy was enhanced in both mature cancer cells and cancer stem-like cells	[67]
AP1 was conjugated with liposomal doxorubicin	Significant inhibition of tumor growth with an increased and selective cytotoxic effect on CRC cells	[68]
IL-13Rα2 D1, a peptide able to block IL-13 binding to IL-13Rα2	Inhibition in tumor growth, invasion, and proliferation in a metastatic mouse model	[69]
Small-molecule Jak1/2 inhibitor AZD1480	Diminished tumor cell proliferation and growth and increased apoptosis during CAC	[70]
In a phase I study, AZD1480 was administered as monotherapy to patients with solid tumors	pSTAT3 inhibition in granulocytes; however, unusual toxicity profile and overall lack of clinical activity led to discontinuation of the study	[71]
Ruxolitinib was tested in a phase 2 study in combination with regorafenib in CRC patients	There was no significant difference in the overall survival rate or progression-free survival rate	[72]
Pacritinib, an oral inhibitor of JAKs and other kinases, was administrated to patients with metastatic CRC	Not conclusive results as the trial was discontinued prior to completion	[73]
Tofacitinib	Enhances cancer metastasis via depletion of NK cells in an experimental lung metastasis mouse model of colon cancer	[74]
STAT6-specific inhibitor AS1517499	Decreased colonic tumor load and circulating inflammatory monocytes and granulocytes	[11]
STAT6 inhibitor AS1517499 in combination with trimethylglycine and 5-FU	A reduction in tumor load was observed coinciding with reduced STAT6-phosphorylation and decrease in the expression of inflammatory cytokines	[22]
Propofol, an agent that interferes with the IL-13/STAT6 signaling pathway	Increased the expression levels of miR-135b and miR-361, STAT6-targeting miRNAs were observed. Decreased cell proliferation and migration in the CRC SW480 and RKO cell lines	[75]
STAT6-specific siRNA sequences decreased mRNA STAT6 expression in human cancer cell line HT29	The treatment inhibited HT29 cell proliferation and induced late apoptosis	[20]
Upregulation of lncRNA- RP11-468E2.5 in human CRC HCT116 and SW480 cells	Suppressed cell proliferation and promote cell apoptosis by targeting STAT5 and STAT6	[76]

Abbreviations: CRC, *colorectal cancer;* IL, *interleukin;* IL-R, *interleukin receptor;* miR, *microRNA;* siRNA, *small interfering RNA;* lncRNA, *long non-coding RNA;* 5-FU, *5-fluorouracil;* AP1, *atherosclerotic plaque-specific peptide-1.*

lines and experimental models. Clinical trials have been implemented in the treatment of UC with conflicting results. Thus, a better understanding of STAT6 signaling and potential pharmaceutical targets for each component could provide a valuable approach for treatment strategies.

Cytokine receptor targeting

IL-4 and IL-13 bind to their receptors and activate the JAK/STAT6 signaling pathway. Hence, the use of molecules that can block binding or impair the signaling of these cytokines to prevent STAT6 activation may be clinically relevant.

In the CRC cell line Caco, the use of IL-4 neutralizing antibodies or an inhibitory form of IL-4 as adjuvant therapies enhanced the activity of chemotherapy in controlling colon cancer development through a decreased expression levels of the CRC stem cell marker CD133 [65]. The treatment of stem-like cells derived from xenografted tumors of colon carcinoma with conventional drugs, such as 5-fluorouracil (5-FU) or oxaliplatin, was improved with the use of IL-4 inhibitors [67]. Nude mice injected with colon cancer spheroids from tumorigenic CD133(+) cells, where treated with an IL-4Rα antagonist or anti-IL-4 neutralizing antibody to inhibit IL-4 and IL-13 responses in combination with 5-FU and oxaliplatin. The antitumor efficacy of these standard chemotherapeutic drugs was strongly increased through selective sensitization of CD133(+) cells to apoptosis [66]. Therefore, IL-4 inhibitors may be considered as potential adjuvant therapies to improve chemotherapy effectivity during CAC development.

Taking into consideration that colorectal tumors have high interleukin-4 receptor alpha (IL-4Ralpha) expression, the use of specific targets to its receptor is a promising therapeutic strategy. The atherosclerotic plaque-specific peptide-1 (AP1) is a peptide characterized by the ability to bind IL-4R [68]. As a therapeutic strategy, AP1 was conjugated with liposomal doxorubicin, a potent chemotherapeutic agent. The administration of AP1-conjugated liposomal doxorubicin to mice produced a significant inhibition of tumor growth with an increased and selective cytotoxic effect on CRC cells [68].

Considering the link between IL-13 and UC and CAC development, various studies have been implemented to target IL-13 or its receptors as a treatment strategy. IL-13Rα2 is overexpressed in metastatic CRC [69]. The administration of the synthetic IL-13Rα2 D1, a peptide able to block IL-13 binding to IL-13Rα2, resulted in inhibited migration, invasion, and proliferation in metastatic colorectal cancer cells. In accordance, IL-13-mediated STAT6 activation was repressed and better survival rates were observed in a metastatic CRC mouse model [69].

JAK inhibition as a therapeutic strategy during IBD and CRC

Many treatment approaches against UC target the JAK signaling pathway with promising results. This is based on the fact that numerous inflammatory cytokines signal through JAKs and are involved in signaling pathways that can stimulate tumor growth. Nevertheless, STAT6 is phosphorylated following JAK1/2 activation and forms homodimers that translocate to the nucleus to start the transcription of multiple genes. So, the inhibition of JAK will result in a decrease in STAT6 activity.

JAK inhibitors have been therapeutically targeting to treat rheumatoid arthritis, psoriasis, and IBD and have already shown that they are safe for human use [77,78]. However, it is difficult to discern and control the effect of JAK inhibition, particularly in cancer development, where the tumor and stromal cells are actively interacting through cytokines. Targeting JAK signaling directly as a treatment for CRC presents several challenges, as normal cells in the intestine use this pathway to maintain the homeostasis and keep the intestine healthy. Blocking JAK may prevent stromal and tumor communication and polyp growth, but it could also cause damage to the healthy tissues. Consequently, the use of JAK inhibitors for the treatment of colon cancer is not widespread. However, it is important to take into consideration the result of the inhibition of JAK in CRC progression like an alternative for the treatment.

The administration of small-molecule Jak1/2 inhibitor AZD1480 diminished tumor cell proliferation and growth and increased apoptosis in inflammation-associated colon cancer mouse model [70]. This result was associated with the impairment of STAT3-driven proliferation; however, the colonic epithelium of AZD1480-treated mice contained less pJak1 and pJak2, but the pStat6 levels were not analyzed [70]. In a phase I study, AZD1480 was administered as monotherapy to patients with solid tumors [71]. The results showed a greater pSTAT3 inhibition in granulocytes; however, unusual toxicity profile and overall lack of clinical activity led to discontinuation of the study [71].

Ruxolitinib, an oral selective inhibitor of JAK1/2, is approved by the Food and Drug Administration (FDA) for use in myelofibrosis [79]. Ruxolitinib was tested in a phase 2 study in combination with regorafenib, an oral multi-targeted kinase inhibitor, in patients with advanced and metastatic adenocarcinoma of the colon or rectum [72]. However, there was no significant difference in the overall survival rate or progression-free survival rate between the regorafenib + placebo versus regorafenib + ruxolitinib groups [72]. Nevertheless, the treatment was administered in the advanced stages of tumor development, and the trial was terminated early per sponsor decision; therefore, further investigation is required, particularly during the early stages of CRC.

Pacritinib, an oral inhibitor of JAKs and other kinases, was administrated to patients with metastatic CRC; however, the study did not produce conclusive results as the trial was discontinued prior to completion [73]. Additionally, the use of tofacitinib, a JAK1/3 inhibitor widely used for the treatment of rheumatoid arthritis, was found to be associated with an increase in lung metastasis accompanied with a decrease in natural killer cell number in a mouse model of colon cancer [74]. Therefore, the development of selective tissue-specific JAK inhibitors is necessary to overcome side effects.

STAT6 inhibitors for CRC

The potential benefits in decreasing the STAT6 signaling in CRC remain an interesting approach in treating disease and may help the actual treatments. Ongoing evidences in cancer cell lines and experimental models have proved that silencing or blocking STAT6 signaling stops cancer progression both in vitro and in vivo [20,22].

Our own studies revealed that the use of the STAT6-specific inhibitor AS1517499 reduced tumor growth and signs of the disease in a mouse model of CAC [11,22]. In vivo, STAT6 inhibition diminished colonic tumor load and colon tissue damage reduced the epithelial cell proliferation and the pro-inflammatory cytokine expression, corresponding to decreased STAT6 phosphorylation in the intestine [11]. When AS1517499 and trimethylglycine were used as adjuvants together with 5-FU in the in vivo AOM/DSS CAC model, induced a significant reduction in tumor growth and EMT markers improving the response to chemotherapy [22]. Similarly, in primary epithelial cells from patients with prostate cancer, exposure to AS1517499 decreased IL-4-induced colony formation [80], pointing STAT6 as a valuable target for adjuvant therapy against cancer.

Tumor-associated macrophages (TAMs) are the most abundant immune population in the tumor microenvironment [81]. These cells resemble M2-polarized macrophages and secrete several growth factors promoting tissue remodeling, angiogenesis, and metastasis [81]. IL-4 is a critical signal for M2 activation, therefore, targeting IL-4/STAT6 axis is an attractive treatment for inhibition of TAMs to suppress tumor development. In breast cancer cells, the use of AS1517499 decreased tumor growth and early liver metastasis in accordance with a reduction in M2 macrophage markers Ym-1 and Mrc-1, and metastatic niche markers Mmp-2 and CD-34 [58]. This result was associated with a decreased capacity of macrophages to differentiate into M2 phenotype via STAT6. In the in vivo AOM/DSS CAC model, the inhibition of STAT6 phosphorylation by AS1517499 and the use of trimethylglycine and 5-FU induced a decrease in CD11b + Ly6Chi monocytic cells and CD11b + Ly6C − Ly6G+ granulocytic cells, both populations associated with suppression of antitumor immunity [22]. The inhibition of STAT6 phosphorylation may favor the response to chemotherapy by modulating immune cells that support tumor growth.

The treatment of the CRC SW480 and RKO cell lines with a common intravenous anesthetic agent propofol was shown to increase the expression levels of microRNA (miRNA/miR)-135b and miR-361, which are STAT6-targeting miRNAs [75]. The administration of propofol induced a decrease in cell proliferation and migration in colon cancer cell lines, suggesting that propofol interferes with the IL-13/STAT6 signaling pathway [75]. However, the in vivo effect of propofol has not been addressed yet.

Another strategy to impaired STAT6 expression is the use of small interfering RNAs (siRNA) sequences. The human colon adenocarcinoma cell line, HT-29, was transfected with STAT6-specific siRNA sequences resulting in the inhibition of cell proliferation and inducing an increase in apoptotic events [20]. STAT5 and STAT6 are target genes of long non-coding RNA (lncRNA)- RP11-468E2.5. CRC tissues showed lower expression of RP11-468E2.5 but higher expression of STAT5 and STAT6 [76]. Experimental studies suggested that the upregulation of RP11-468E2.5 in human CRC HCT116 and SW480 cells could suppress cell proliferation and promote cell apoptosis by targeting STAT5 and STAT6 [76].

Even though siRNAs are a promising tool, many efforts are necessary to use them in clinical trials like efficient delivery methods with target tissue specificity. Healey et al. reported a successful preclinical model to evaluate the therapeutic potential of candidate siRNA targeting STAT6 in lung epithelial cells [21]. In the study, sequence selection and modification strategies for siRNA candidates targeting STAT6 were used for intranasal application attenuating allergic airway inflammation by suppressing STAT6 expression. This study demonstrated that STAT6 may be targeted in specific tissues; however, further studies evaluating the possibility of suppressing STAT6 signaling in the colonic epithelium are required.

Conclusion

STAT6 has been implicated in colorectal cancer development, progression, metastasis, survival, and resistance to treatment. Additionally, the nuclear expression of pSTAT6 in tumors from CRC patients is associated with lower survival rates pointing that this molecule participates in CRC aggressiveness. Experimental evidences in cell lines and mouse models suggest that STAT6 regulates mechanisms that promote epithelial barrier disruption and the exacerbated response of the immune system to the gut microbiota, promoting colitis development. In consequence, STAT6 seems to play an important role in the progression of early tumorigenesis during colitis-associated cancer. STAT6 has an oncogenic role in the intestinal tumor progression by the suppression of antitumor immunity and preventing an active cytotoxic process. In non-immune cells, STAT6 activation drives epithelial cell proliferation and increases the expression of several EMT drivers enhancing migratory capacity and invasiveness of tumor cells. Experimental observations by silencing or inhibiting STAT6 have resulted in a decrease in cancer-associated processes. Similarly, the administration of STAT6 inhibitors strengthens the chemotherapy response, suggesting that STAT6 may serve as a potential target in the treatment of colon cancer. In the future, preclinical and clinical studies investigating specific STAT6 inhibitors with a targeted organ distribution are required to evaluate the potential of STAT6 as a target for colon cancer therapy.

Acknowledgments

The present study was supported by grants from (PAPPIT, UNAM) (Grant No. IA204218) and CONACYT (Grant No. A1-S-23944). ACRR is a student from Programa de Maestría en Ciencias Biológicas and receiver of a fellowship (1012872) from CONACYT.

YDR is a doctoral student from Programa de Doctorado en Ciencias Biomédicas, Universidad Nacional Autónoma de México (UNAM) and receiver of a fellowship (606590) from CONACYT.

References

[1] Ferlay J, Colombet M, Soerjomataram I, Mathers C, Parkin DM, Piñeros M, Znaor A, Bray F. Estimating the global cancer incidence and mortality in 2018: GLOBOCAN sources and methods. Int J Cancer 2019;144(8):1941–53.

[2] Bray F, Ferlay J, Soerjomataram I, Siegel RL, Torre LA, Jemal A. Global cancer statistics 2018: GLOBOCAN estimates of incidence and mortality worldwide for 36 cancers in 185 countries. CA Cancer J Clin 2018;68 (6):394–424.

[3] American Cancer Society. Cancer facts & figures.

[4] Mármol I, Sánchez-de-Diego C, Pradilla Dieste A, Cerrada E, Rodriguez Yoldi MJ. Colorectal carcinoma: a general overview and future perspectives in colorectal cancer. Int J Mol Sci 2017;18(1):197.

[5] Manfredi S, Lepage C, Hatem C, Coatmeur O, Faivre J, Bouvier AM. Epidemiology and management of liver metastases from colorectal cancer. Ann Surg 2006;244(2):254–9.

[6] Yu H, Pardoll D, Jove R. STATs in cancer inflammation and immunity: a leading role for STAT3. Nat Rev Cancer 2009;9(11):798–809.

[7] Consortium APG. AACR project GENIE: powering precision medicine through an international Consortium. Cancer Discov 2017;7(8):818–31.

[8] Uhlen M, Zhang C, Lee S, Sjostedt E, Fagerberg L, Bidkhori G, Benfeitas R, Arif M, Liu Z, Edfors F, Sanli K, von Feilitzen K, Oksvold P, Lundberg E, Hober S, Nilsson P, Mattsson J, Schwenk JM, Brunnstrom H, Glimelius B, Sjoblom T, Edqvist PH, Djureinovic D, Micke P, Lindskog C, Mardinoglu A, Ponten F. A pathology atlas of the human cancer transcriptome. Science 2017;357(6352):eaan2507. https://doi.org/10.1126/science.aan2507.

[9] Lin Y, Li B, Yang X, Liu T, Shi T, Deng B, Zhang Y, Jia L, Jiang Z, He R. Non-hematopoietic STAT6 induces epithelial tight junction dysfunction and promotes intestinal inflammation and tumorigenesis. Mucosal Immunol 2019;12(6):1304–15.

[10] Wang CG, Ye YJ, Yuan J, Liu FF, Zhang H, Wang S. EZH2 and STAT6 expression profiles are correlated with colorectal cancer stage and prognosis. World J Gastroenterol 2010;16(19):2421–7.

[11] Leon-Cabrera SA, Molina-Guzman E, Delgado-Ramirez YG, Vazquez-Sandoval A, Ledesma-Soto Y, Perez-Plasencia CG, Chirino YI, Delgado-Buenrostro NL, Rodriguez-Sosa M, Vaca-Paniagua F, Avila-Moreno F, Gutierrez-Cirlos EB, Arias-Romero LE, Terrazas LI. Lack of STAT6 attenuates inflammation and drives protection against early steps of colitis-associated colon cancer. Cancer Immunol Res 2017;5(5):385–96.

[12] Jayakumar A, Bothwell ALM. Stat6 promotes intestinal tumorigenesis in a mouse model of adenomatous polyposis by expansion of MDSCs and inhibition of cytotoxic CD8 response. Neoplasia 2017;19(8):595–605.

[13] Rosen MJ, Chaturvedi R, Washington MK, Kuhnhein LA, Moore PD, Coggeshall SS, McDonough EM, Weitkamp JH, Singh AB, Coburn LA, Williams CS, Yan F, Van Kaer L, Peebles Jr. RS, Wilson KT. STAT6 deficiency ameliorates severity of oxazolone colitis by decreasing expression of claudin-2 and Th2-inducing cytokines. J Immunol 2013;190(4):1849–58.

[14] Ostrand-Rosenberg S, Grusby MJ, Clements VK. Cutting edge: STAT6-deficient mice have enhanced tumor immunity to primary and metastatic mammary carcinoma. J Immunol 2000;165(11):6015–9.

[15] Li BH, Yang XZ, Li PD, Yuan Q, Liu XH, Yuan J, Zhang WJ. IL-4/Stat6 activities correlate with apoptosis and metastasis in colon cancer cells. Biochem Biophys Res Commun 2008;369(2):554–60.

[16] Hoving JC. Targeting IL-13 as a host-directed therapy against ulcerative colitis. Front Cell Infect Microbiol 2018;8:395.

[17] Camelo A, Barlow JL, Drynan LF, Neill DR, Ballantyne SJ, Wong SH, Pannell R, Gao W, Wrigley K, Sprenkle J, McKenzie AN. Blocking IL-25 signalling protects against gut inflammation in a type-2 model of colitis by suppressing nuocyte and NKT derived IL-13. J Gastroenterol 2012;47(11):1198–211.

[18] Reinisch W, Panes J, Khurana S, Toth G, Hua F, Comer GM, Hinz M, Page K, O'Toole M, Moorehead TM, Zhu H, Sun Y, Cataldi F. Anrukinzumab, an anti-interleukin 13 monoclonal antibody, in active UC: efficacy and safety from a phase IIa randomised multicentre study. Gut 2015;64(6):894–900.

[19] Cao H, Zhang J, Liu H, Wan L, Zhang H, Huang Q, Xu E, Lai M. IL-13/STAT6 signaling plays a critical role in the epithelial-mesenchymal transition of colorectal cancer cells. Oncotarget 2016;7(38):61183–98.

[20] Salguero-Aranda C, Sancho-Mensat D, Canals-Lorente B, Sultan S, Reginald A, Chapman L. STAT6 knockdown using multiple siRNA sequences inhibits proliferation and induces apoptosis of human colorectal and breast cancer cell lines. PLoS ONE 2019;14(5):e0207558.

[21] Healey GD, Lockridge JA, Zinnen S, Hopkin JM, Richards I, Walker W. Development of pre-clinical models for evaluating the therapeutic potential of candidate siRNA targeting STAT6. PLoS One 2014;9(2):e90338.

[22] Mendoza-Rodriguez MG, Sanchez-Barrera CA, Callejas BE, Garcia-Castillo V, Beristain-Terrazas DL, Delgado-Buenrostro NL, Chirino YI, Leon-Cabrera SA, Rodriguez-Sosa M, Gutierrez-Cirlos EB, Perez-Plasencia C, Vaca-Paniagua F, Meraz-Rios MA, Terrazas LI. Use of STAT6 phosphorylation inhibitor and trimethylglycine as new adjuvant therapies for 5-fluorouracil in colitis-associated tumorigenesis. Int J Mol Sci 2020;21(6):2130.

[23] Mueller TD, Zhang JL, Sebald W, Duschl A. Structure, binding, and antagonists in the IL-4/IL-13 receptor system. Biochim Biophys Acta 2002;1592(3):237–50.

[24] Sengupta S, Thaci B, Crawford AC, Sampath P. Interleukin-13 receptor alpha 2-targeted glioblastoma immunotherapy. Biomed Res Int 2014;2014:952128.

[25] Mikita T, Campbell D, Wu P, Williamson K, Schindler U. Requirements for interleukin-4-induced gene expression and functional characterization of Stat6. Mol Cell Biol 1996;16(10):5811–20.

[26] Fuss IJ, Heller F, Boirivant M, Leon F, Yoshida M, Fichtner-Feigl S, Yang Z, Exley M, Kitani A, Blumberg RS, Mannon P, Strober W. Nonclassical CD1d-restricted NK T cells that produce IL-13 characterize an atypical Th2 response in ulcerative colitis. J Clin Invest 2004;113(10):1490–7.

[27] Heller F, Florian P, Bojarski C, Richter J, Christ M, Hillenbrand B, Mankertz J, Gitter AH, Burgel N, Fromm M, Zeitz M, Fuss I, Strober W, Schulzke JD. Interleukin-13 is the key effector Th2 cytokine in ulcerative colitis that affects epithelial tight junctions, apoptosis, and cell restitution. Gastroenterology 2005;129(2):550–64.

[28] Rosen MJ, Frey MR, Washington MK, Chaturvedi R, Kuhnhein LA, Matta P, Revetta FL, Wilson KT, Polk DB. STAT6 activation in ulcerative colitis: a new target for prevention of IL-13-induced colon epithelial cell dysfunction. Inflamm Bowel Dis 2011;17(11):2224–34.
[29] Hoving JC, Kirstein F, Nieuwenhuizen NE, Fick LCE, Hobeika E, Reth M, Brombacher F. B cells that produce immunoglobulin E mediate colitis in BALB/c mice. Gastroenterology 2012;142(1):96–108.
[30] Shajib MS, Wang HQ, Kim JJ, Sunjic I, Ghia JE, Denou E, Collins M, Denburg JA, Khan WI. Interleukin 13 and serotonin: linking the immune and endocrine systems in murine models of intestinal inflammation. PLoS ONE 2013;8(8):e72774.
[31] Cosin-Roger J, Ortiz-Masia D, Calatayud S, Hernandez C, Esplugues JV, Barrachina MD. The activation of Wnt signaling by a STAT6-dependent macrophage phenotype promotes mucosal repair in murine IBD. Mucosal Immunol 2016;9(4):986–98.
[32] Buzza MS, Johnson TA, Conway GD, Martin EW, Mukhopadhyay S, Shea-Donohue T, Antalis TM. Inflammatory cytokines down-regulate the barrier-protective prostasin-matriptase proteolytic cascade early in experimental colitis. J Biol Chem 2017;292(26):10801–12.
[33] Kanai T, Watanabe M, Hayashi A, Nakazawa A, Yajima T, Okazawa A, Yamazaki M, Ishii H, Hibi T. Regulatory effect of interleukin-4 and interleukin-13 on colon cancer cell adhesion. Br J Cancer 2000;82(10):1717–23.
[34] Heller F, Fromm A, Gitter AH, Mankertz J, Schulzke JD. Epithelial apoptosis is a prominent feature of the epithelial barrier disturbance in intestinal inflammation: effect of pro-inflammatory interleukin-13 on epithelial cell function. Mucosal Immunol 2008;1(Suppl 1):S58–61.
[35] Todaro M, Lombardo Y, Francipane MG, Alea MP, Cammareri P, Iovino F, Di Stefano AB, Di Bernardo C, Agrusa A, Condorelli G, Walczak H, Stassi G. Apoptosis resistance in epithelial tumors is mediated by tumor-cell-derived interleukin-4. Cell Death Differ 2008;15(4):762–72.
[36] Di Stefano AB, Iovino F, Lombardo Y, Eterno V, Hoger T, Dieli F, Stassi G, Todaro M. Survivin is regulated by interleukin-4 in colon cancer stem cells. J Cell Physiol 2010;225(2):555–61.
[37] Tosolini M, Kirilovsky A, Mlecnik B, Fredriksen T, Mauger S, Bindea G, Berger A, Bruneval P, Fridman W-H, Pagès F, Galon J. Clinical impact of different classes of infiltrating T cytotoxic and helper cells (Th1, Th2, Treg, Th17) in patients with colorectal cancer. Cancer Res 2012;71(4):1263–71.
[38] Barderas R, Bartolome RA, Fernandez-Acenero MJ, Torres S, Casal JI. High expression of IL-13 receptor alpha2 in colorectal cancer is associated with invasion, liver metastasis, and poor prognosis. Cancer Res 2012;72 (11):2780–90.
[39] Formentini A, Braun P, Fricke H, Link KH, Henne-Bruns D, Kornmann M. Expression of interleukin-4 and interleukin-13 and their receptors in colorectal cancer. Int J Colorectal Dis 2012;27(10):1369–76.
[40] Li BH, Xu SB, Li F, Zou XG, Saimaiti A, Simayi D, Wang YH, Zhang Y, Yuan J, Zhang WJ. Stat6 activity-related Th2 cytokine profile and tumor growth advantage of human colorectal cancer cells in vitro and in vivo. Cell Signal 2012;24(3):718–25.
[41] Wick EC, LeBlanc RE, Ortega G, Robinson C, Platz E, Pardoll DM, Iacobuzio-Donahue C, Sears CL. A shift from pStat6 to pStat3 predominance is associated with inflammatory bowel disease-associated dysplasia. Inflamm Bowel Dis 2012;18(7):1267–74.
[42] Volonte A, Di Tomaso T, Spinelli M, Todaro M, Sanvito F, Albarello L, Bissolati M, Ghirardelli L, Orsenigo E, Ferrone S, Doglioni C, Stassi G, Dellabona P, Staudacher C, Parmiani G, Maccalli C. Cancer-initiating cells from colorectal cancer patients escape from T cell-mediated immunosurveillance in vitro through membrane-bound IL-4. J Immunol 2014;192(1):523–32.
[43] Jiang L, Cheng Q, Zhang B, Zhang M. IL-13 induces the expression of 11betaHSD2 in IL-13Ralpha2 dependent manner and promotes the malignancy of colorectal cancer. Am J Transl Res 2016;8(2):1064–72.
[44] Liu H, Antony S, Roy K, Juhasz A, Wu Y, Lu J, Meitzler JL, Jiang G, Polley E, Doroshow JH. Interleukin-4 and interleukin-13 increase NADPH oxidase 1-related proliferation of human colon cancer cells. Oncotarget 2017;8 (24):38113–35.
[45] Chen J, Gong C, Mao H, Li Z, Fang Z, Chen Q, Lin M, Jiang X, Hu Y, Wang W, Zhang X, Chen X, Li H. E2F1/SP3/STAT6 axis is required for IL-4-induced epithelial-mesenchymal transition of colorectal cancer cells. Int J Oncol 2018;53(2):567–78.
[46] De Oliveira T, Ramakrishnan M, Diamanti MA, Ziegler PK, Brombacher F, Greten FR. Loss of Stat6 affects chromatin condensation in intestinal epithelial cells causing diverse outcome in murine models of inflammation-associated and sporadic colon carcinogenesis. Oncogene 2019;38(11):1787–801.

[47] Yu T, Gan S, Zhu Q, Dai D, Li N, Wang H, Chen X, Hou D, Wang Y, Pan Q, Xu J, Zhang X, Liu J, Pei S, Peng C, Wu P, Romano S, Mao C, Huang M, et al. Modulation of M2 macrophage polarization by the crosstalk between Stat6 and Trim24. Nat Commun 2019;10(1):4353.
[48] Goswami R, Kaplan MH. STAT transcription factors in T cell control of health and disease. Int Rev Cell Mol Biol 2017;331:123–80.
[49] Jess T, Rungoe C, Peyrin-Biroulet L. Risk of colorectal cancer in patients with ulcerative colitis: a meta-analysis of population-based cohort studies. Clin Gastroenterol Hepatol 2012;10(6):639–45.
[50] Pleguezuelos-Manzano C, Puschhof J, Huber AR, van Hoeck A, Wood HM, Nomburg J, Gurjao C, Manders F, Dalmasso G, Stege PB, Paganelli FL, Geurts MH, Beumer J, Mizutani T, Miao Y, van der Linden R, van der Elst S, Genomics England Research Consortium, Garcia KC, Top J, Willems RJL, Giannakis M, Bonnet R, Quirke P, Meyerson M, Cuppen E, van Boxtel R, Clevers H. Mutational signature in colorectal cancer caused by genotoxic pks(+) *E. coli*. Nature 2020;580(7802):269–73.
[51] Vu T, Datta PK. Regulation of EMT in colorectal cancer: a culprit in metastasis. Cancers (Basel) 2017;9(12):171.
[52] Hebenstreit D, Wirnsberger G, Horejs-Hoeck J, Duschl A. Signaling mechanisms, interaction partners, and target genes of STAT6. Cytokine Growth Factor Rev 2006;17(3):173–88.
[53] Bao K, Reinhardt RL. The differential expression of IL-4 and IL-13 and its impact on type-2 immunity. Cytokine 2015;75(1):25–37.
[54] Yahaya MAF, Lila MAM, Ismail S, Zainol M, Afizan N. Tumour-associated macrophages (TAMs) in colon cancer and how to reeducate them. J Immunol Res 2019;2019:2368249.
[55] Huang S, Dong D, Zhang Y, Chen Z, Geng J, Zhao Y. NEAT1 regulates Th2 cell development by targeting STAT6 for degradation. Cell Cycle 2019;18(3):312–9.
[56] Jablonski KA, Amici SA, Webb LM, Ruiz-Rosado Jde D, Popovich PG, Partida-Sanchez S, Guerau-de-Arellano M. Novel markers to delineate murine M1 and M2 macrophages. PLoS One 2015;10(12):e0145342.
[57] Fu C, Jiang L, Hao S, Liu Z, Ding S, Zhang W, Yang X, Li S. Activation of the IL-4/STAT6 signaling pathway promotes lung cancer progression by increasing M2 myeloid cells. Front Immunol 2019;10:2638.
[58] Binnemars-Postma K, Bansal R, Storm G, Prakash J. Targeting the Stat6 pathway in tumor-associated macrophages reduces tumor growth and metastatic niche formation in breast cancer. FASEB J 2018;32(2):969–78.
[59] Mukaida N, Sasaki S, Baba T. Chemokines in cancer development and progression and their potential as targeting molecules for cancer treatment. Mediators Inflamm 2014;2014:170381.
[60] Kitamura T, Fujishita T, Loetscher P, Revesz L, Hashida H, Kizaka-Kondoh S, Aoki M, Taketo MM. Inactivation of chemokine (C-C motif) receptor 1 (CCR1) suppresses colon cancer liver metastasis by blocking accumulation of immature myeloid cells in a mouse model. Proc Natl Acad Sci U S A 2010;107(29):13063–8.
[61] Ostrand-Rosenberg S, Sinha P, Clements V, Dissanayake SI, Miller S, Davis C, Danna E. Signal transducer and activator of transcription 6 (Stat6) and CD1: inhibitors of immunosurveillance against primary tumors and metastatic disease. Cancer Immunol Immunother 2004;53(2):86–91.
[62] Gabrilovich DI, Ostrand-Rosenberg S, Bronte V. Coordinated regulation of myeloid cells by tumours. Nat Rev Immunol 2012;12(4):253–68.
[63] Noman MZ, Desantis G, Janji B, Hasmim M, Karray S, Dessen P, Bronte V, Chouaib S. PD-L1 is a novel direct target of HIF-1alpha, and its blockade under hypoxia enhanced MDSC-mediated T cell activation. J Exp Med 2014;211(5):781–90.
[64] Ma P, Beatty PL, McKolanis J, Brand R, Schoen RE, Finn OJ. Circulating myeloid derived suppressor cells (MDSC) that accumulate in premalignancy share phenotypic and functional characteristics with MDSC in cancer. Front Immunol 2019;10:1401.
[65] Gharib AF, Shalaby SM, Raafat N, Fawzy WMS, Abdel Hakim NH. Assessment of neutralizing interleukin-4 effect on CD133 gene expression in colon cancer cell line. Cytokine 2017;97:66–72.
[66] Todaro M, Alea MP, Di Stefano AB, Cammareri P, Vermeulen L, Iovino F, Tripodo C, Russo A, Gulotta G, Medema JP, Stassi G. Colon cancer stem cells dictate tumor growth and resist cell death by production of interleukin-4. Cell Stem Cell 2007;1(4):389–402.
[67] Todaro M, Perez Alea M, Scopelliti A, Medema JP, Stassi G. IL-4-mediated drug resistance in colon cancer stem cells. Cell Cycle 2008;7(3):309–13.
[68] Yang CY, Liu HW, Tsai YC, Tseng JY, Liang SC, Chen CY, Lian WN, Wei MC, Lu M, Lu RH, Lin CH, Jiang JK. Interleukin-4 receptor-targeted liposomal doxorubicin as a model for enhancing cellular uptake and antitumor efficacy in murine colorectal cancer. Cancer Biol Ther 2015;16(11):1641–50.

[69] Bartolome RA, Jaen M, Casal JI. An IL13Ralpha2 peptide exhibits therapeutic activity against metastatic colorectal cancer. Br J Cancer 2018;119(8):940–9.
[70] Stuart E, Buchert M, Putoczki T, Thiem S, Farid R, Elzer J, Huszar D, Waring PM, Phesse TJ, Ernst M. Therapeutic inhibition of Jak activity inhibits progression of gastrointestinal tumors in mice. Mol Cancer Ther 2014;13(2):468–74.
[71] Plimack ER, Lorusso PM, McCoon P, Tang W, Krebs AD, Curt G, Eckhardt SG. AZD1480: a phase I study of a novel JAK2 inhibitor in solid tumors. Oncologist 2013;18(7):819–20.
[72] Fogelman D, Cubillo A, Garcia-Alfonso P, Miron MLL, Nemunaitis J, Flora D, Borg C, Mineur L, Vieitez JM, Cohn A, Saylors G, Assad A, Switzky J, Zhou L, Bendell J. Randomized, double-blind, phase two study of ruxolitinib plus regorafenib in patients with relapsed/refractory metastatic colorectal cancer. Cancer Med 2018;7(11):5382–93.
[73] Regenbogen T, Chen L, Trinkaus K, Wang-Gillam A, Tan BR, Amin M, Pedersen KS, Park H, Suresh R, Lim KH, Ratchford E, Brown A, Lockhart AC. Pacritinib to inhibit JAK/STAT signaling in refractory metastatic colon and rectal cancer. J Gastrointest Oncol 2017;8(6):985–9.
[74] Shimaoka H, Takeno S, Maki K, Sasaki T, Hasegawa S, Yamashita Y. A cytokine signal inhibitor for rheumatoid arthritis enhances cancer metastasis via depletion of NK cells in an experimental lung metastasis mouse model of colon cancer. Oncol Lett 2017;14(3):3019–27.
[75] Xu K, Tao W, Su Z. Propofol prevents IL-13-induced epithelial-mesenchymal transition in human colorectal cancer cells. Cell Biol Int 2018;42(8):985–93.
[76] Jiang L, Zhao XH, Mao YL, Wang JF, Zheng HJ, You QS. Long non-coding RNA RP11-468E2.5 curtails colorectal cancer cell proliferation and stimulates apoptosis via the JAK/STAT signaling pathway by targeting STAT5 and STAT6. J Exp Clin Cancer Res 2019;38(1):465.
[77] O'Shea JJ, Kontzias A, Yamaoka K, Tanaka Y, Laurence A. Janus kinase inhibitors in autoimmune diseases. Ann Rheum Dis 2013;72(Suppl 2):ii. 111–5.
[78] Regenbogen T, Chen L, Trinkaus K, Wang-Gillam A, Tan BR, Amin M, Pedersen KS, Park H, Suresh R, Lim KH, Ratchford E, Brown A, Lockhart AC. Pacritinib to inhibit JAK/STAT signaling in refractory metastatic colon and rectal cancer. J Gastrointest Oncol 2017;8(6):985–9.
[79] Pardanani A, Tefferi A. How I treat myelofibrosis after failure of JAK inhibitors. Blood 2018;132(5):492–500.
[80] Nappo G, Handle F, Santer FR, McNeill RV, Seed RI, Collins AT, Morrone G, Culig Z, Maitland NJ, Erb HHH. The immunosuppressive cytokine interleukin-4 increases the clonogenic potential of prostate stem-like cells by activation of STAT6 signalling. Oncogenesis 2017;6(5):e342.
[81] Pinto ML, Rios E, Duraes C, Ribeiro R, Machado JC, Mantovani A, Barbosa MA, Carneiro F, Oliveira MJ. The two faces of tumor-associated macrophages and their clinical significance in colorectal cancer. Front Immunol 2019;10:1875.

CHAPTER

10

Macrophage migration inhibitory factor (MIF): Its role in the genesis and progression of colorectal cancer

Imelda Juárez-Avelar, Tonathiu Rodríguez, Ana P. García-García, and Miriam Rodríguez-Sosa

Unidad de Biomedicina, Facultad de Estudios Superiores Iztacala, Universidad Nacional Autónoma de México, Tlalnepantla, México

Abstract

Macrophage migration inhibitory factor (MIF) is a pleiotropic protein with cytokine and chemokine properties that regulate a diverse range of physiological functions related to innate immunity, inflammation, and glucocorticoid-mediated immunosuppression and is highly expressed by cancer cells, via which it affects angiogenesis, tumor growth, and metastasis. The role of MIF in colorectal cancer (CRC) is underscored by data showing that its overexpression in the chronic stages of CRC is associated with clinical severity. The specific functions of MIF are now being defined in CRC, and MIF-targeted biologic therapeutics are in early-stage trials. In this chapter, we summarize current knowledge about the role of MIF in cancer with special emphasis on CRC.

Abbreviations

APC	antigen-presenting cell
Bcl	B-cell lymphoma
CD	cluster of differentiation
COX	cyclooxygenase-2
CRC	colorectal cancer
CTGF	connective tissue growth factor
CXCR	chemokine receptor
DC	dendritic cells
ERK	extracellular signal-regulated kinase
FGF	fibroblast growth factor
HIF	hypoxia-inducible factor
HLA	human leukocyte antigen

Immunotherapy in Resistant Cancer:
From the Lab Bench Work to Its Clinical Perspectives
https://doi.org/10.1016/B978-0-12-822028-3.00012-1

IBD inflammatory bowel disease
ICAM intercellular adhesion molecule
IFNγ interferon gamma
IL interleukin
JNK c-Jun N-terminal kinase
LOX lysyl oxidase
LPS lipopolysaccharide
MMP matrix metalloproteinases
mo monocytes
Mφ macrophage
PDGF platelet-derived growth factor
PGE prostaglandin E
TGF transforming growth factor
TLR toll-like receptor
TNF tumor necrosis factor
VCAM vascular cell adhesion molecule
VEGF vascular endothelial growth factor

Conflict of interest

No potential conflicts of interest were disclosed by the authors.

Introduction

Cancer is a major public health problem worldwide; in 2018, it was estimated that there were 18 million new cases of cancer and 9.6 million cancer-related deaths. In addition, it is estimated that the current number of people living for at least five years after diagnosis is 43.8 million [1]. Due to changes in population growth and aging, the WHO estimates that by 2030, the number of new cases of cancer will exceed 20 million annually [2]. Of the different types of cancers, lung cancer, breast cancer, and colorectal cancer (CRC) rank first, second, and third in incidence, respectively. CRC, although it ranks third in incidence, ranks second in mortality in the world (Table 1).

Cytokines are responsible for the regulation of cellular function and have been implicated in the development of different types of cancers, especially in tissues with chronic inflammation. Some of these cytokines can directly affect tumor, endothelial, and stromal cells, which can affect cancer development and progression [4].

It is important to note that some of these molecules have the dual function of acting as cytokines and chemokines, that is, in addition to activating some cellular mechanisms, they can also mediate the recruitment of cells of the immune system. Depending on the tissue context, these molecules can favor the development of an exacerbated inflammatory microenvironment, or the recruitment of immune effector cells that help to eliminate tumor cells. These characteristics have awakened interest in the study of proinflammatory cytokines as determining factors in cancer biology. However, the study of chemokines that can attract cells that control the development of cancer is also of great interest.

TABLE 1 Number of cases of the 10 most common cancers in the world.

Cancer type	Incidence	Mortality	Prevalence
Lung	2,093,876	1,761,007	2,129,964
Brest	2,088,849	626,679	6,875,009
Colorectal	1,849,518	880,792	4,789,635
Prostate	1,276,106	358,989	3,724,658
Stomach	1,033,701	782,685	1,589,752
Liver	841,080	781,631	675,210
Esophagus	572,034	508,585	547,104
Uterine cervical	569,847	311,365	1,474,265
Thyroid	567,233	41,071	1,997,846
Bladder	549,393	199,922	1648 482

Data on incidence, mortality, and prevalence were collected in 2018, and the data are for both sexes and all ages [3].

In this chapter, we will focus specifically on describing the knowledge that has been accumulated about the MIF molecule, which has characteristics of both proinflammatory cytokines and chemokines involved in the recruitment of T and Mφ cells. We subdivide the chapter into a brief description of the structure, receptors, and general functions of MIF and focus on recent knowledge of its role in cancer genesis and progression, with special emphasis on the influence that MIF has on CRC.

Macrophage migration inhibitory factor (MIF)

MIF was one of the first functional cytokines identified in the mid-1960s, when it was recognized as a soluble factor of activated T lymphocytes that inhibited random migration of macrophages in vitro during experiments to characterize the mechanisms involved in the delayed-type hypersensitivity reaction [5,6]. However, its biochemical characterization was not performed until its cloning in 1989 by J. David et al. [7]. Thereafter, as a result of investigations of glucocorticoid action regulators, a human MIF homolog in mice was described [8]. MIF protein, neutralizing monoclonal antibodies, and MIF knockout (*Mif−/−*) mice were produced in a short time, and several in vitro and in vivo studies emerged that established that MIF is a proinflammatory pleiotropic mediator, with an important role in the innate as well as adaptive immune response, and that its expression is positively regulated during situations such as inflammation, infection, and stress [9,10].

MIF is a cytokine preserved in evolution [11] and is a soluble factor secreted by a wide variety of cells of the immune system, such as T and B lymphocytes [12], monocytes (mo), macrophages (Mφ) [13], dendritic cells (DC), neutrophils, eosinophils, basophils, and mast cells [14]. It is also secreted by nonimmune cells, including epithelial cells [15], endothelial cells [16], fibroblasts, insulin-secreting pancreatic β cells, pituitary cells [17–19], neurons, and stem cell progenitors [20–22], in areas that come into direct contact with the environment, such as the skin, lung, gastrointestinal, and genitourinary tracts, as well as in the eye, ear, liver, testis, prostate, bones, kidneys, joints, and endocrine system.

MIF expression is induced in response to antigenic challenges, such as lipopolysaccharide (LPS) and gram-positive exotoxins, or in response to cytokines such as tumor necrosis factor (TNF)-α and interferon (IFN)-γ, as well as physiological stress [23,24].

Biological structure of MIF

MIF is a relatively small 12.5 kDa protein made up of three monomers. Each monomer consists of 114 amino acids, and the MIF gene is located on chromosome 22. At the amino acid level, mouse and human MIF are 90% identical [14]. Its α/β structure is composed of two antiparallel α-helices packed against a four-chain β-sheet. The monomer has two additional β strands that interact with the β-sheets of the adjacent subunits to form the interface between the monomers. The three β-sheets are arranged to form a barrel containing a solvent-accessible channel that runs through the center of the protein along the axis formed by the three β-sheets (Fig. 1) [25].

MIF in the immune response

MIF is an activator of mo, Mφ, and DC [13,26,27]; Mφ in contact with MIF increases their phagocytic and lytic capacity, which increases their ability to eliminate intracellular pathogens [28] such as *Leishmania major*, *Trypanosoma cruzi*, *Toxoplasma gondii*, and extracellular such as *Taenia crassiceps* [29–32]. This function of activating Mφ and DC has been corroborated in *Mif−/−* mice, which have Mφs and CD with significantly diminished capacity to eliminate these pathogens, which makes *Mif−/−* mice highly susceptible to infections with these agents [33,34].

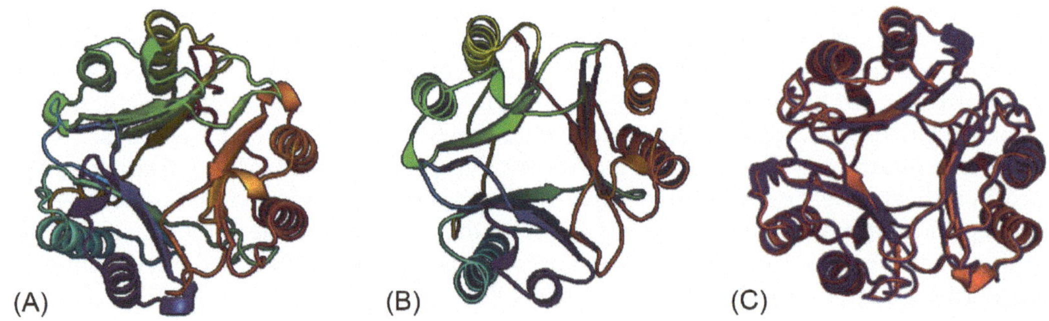

FIG. 1 Comparison of human and mouse MIF protein structures. Trimer structures of (A) human MIF and (B) mouse MIF. In the homotrimer, each of the monomers is shown in a different color. (C) Alignment of human with mouse MIF; mouse MIF is shown in *red* and human MIF is shown in *blue*. The model of the homotrimer secondary structure of the protein in ".pdb" format was obtained using MODELLER software (https://salilab.org/modeller). The alignment was obtained using PyMOL software (https://pymol.org/2).

The use of *Mif* −/− mice also revealed new MIF activities, such as upregulation of expression of the toll-like receptor (TLR4), a gram-negative pathogen recognition receptor [34]. Furthermore, the use of MIF antagonist antibodies proved that MIF neutralization inhibits delayed-type hypersensitivity and T cell recruitment [12,35].

However, the ability of MIF to activate cells and its proinflammatory actions that make it an essential element of antimicrobial defense contribute significantly to various pathologies associated with excessive inflammation and autoimmunity, such as septic shock [36], arthritis [37], diabetes type 1 [27], type 2 diabetes and obesity [38,39], asthma [40], atherosclerosis, acute respiratory distress syndrome, and inflammatory autoimmune diseases [41–43].

MIF receptors

MIF binds to the cell membrane receptor CD74, also known as histocompatibility antigen (HLA) α chain class II [44]. Phosphorylation of CD74 through MIF binding leads to signaling mediated by activation of the CD44 coreceptor [45]. Thus, the CD74/CD44 complex initiates a signaling cascade that favors cell survival [46], since it contributes to cell proliferation by facilitating entry into the S-phase of the cell cycle and favors the elevation of cyclin E, followed by a cascade that increases the expression of survival genes such as Bcl-XL and Bcl-2 [47].

Furthermore, when MIF binds to CD74/CD44, it stimulates ICAM-1 expression by endothelial cells, as well as ICAM-1 and VCAM-1 expressions by monocytes [48]. This recognition also activates the extracellular signal-regulated kinase (ERK)-1/2 MAPK pathway [49] and consequently the production of prostaglandin E (PGE)-2, cell proliferation [46] and the production of inflammatory cytokines such as TNF-α, interleukin (IL)-1β, IL-6, IL-12, IL-8, IL-17, and IFN-γ, as well as nitric oxide (NO), and cyclooxygenase (COX)-2 [42,50]. MIF also increases the expression of cytokine receptors (IL-1R and TNFR) [51] and inhibits the anti-inflammatory effects of glucocorticoids [52]; in this way, MIF has proinflammatory function [53].

On the other hand, MIF is an unconventional ligand of C-X-C motif chemokine receptor (CXCR) 2, 4, and 7. Thus, it participates in the recruitment of inflammatory and atherogenic cells. This ability of MIF to function as a chemokine has been recognized since its discovery and its ability to recruit Mφ, T cells, B cells, and fibroblasts were subsequently verified [54–56]. Furthermore, there is evidence that MIF binding to CXCR7 contributes to the activation of the ERK1/2 signaling pathway. The binding of MIF with these noncanonical receptors can contribute to cell recruitment in inflammatory processes [57].

Role of MIF in cancer genetics, development, and malignancy

MIF is a promoter of inflammation, proliferation, and tumor development in CRC

In this section, we will focus on providing experimental evidence of the biological mechanisms that MIF regulates to contribute to inflammation, angiogenesis, metastasis, tumorigenesis, and carcinogenesis. We will place special emphasis on its influence on the development and malignancy of CRC.

As already mentioned in the previous sections, MIF possesses protumoral functional characteristics that make it a central molecule in cancer. The first evidence for this role of MIF was derived from the observation of the presence of high levels of MIF in different tumors, such as squamous cell carcinoma [58], CRC [59], lung cancer [60], breast cancer [61], prostate cancer [62], nasopharyngeal cancer [63], glioblastomas [64], and cervical adenocarcinoma [65]. It is important to note that MIF is not only increased in tumor tissue but also significantly increased in the serum and plasma of cancer patients (Table 2).

MIF has been found to be largely localized to the cytoplasm of tumor cells and appears to be overexpressed in response to the release of growth factors, such as transforming growth factor (TGF-β), fibroblast growth factor (FGF-β), and platelet-derived growth factor (PDGF), which play a fundamental role in the regulation of cell proliferation and transformation [66].

In this way, the presence of MIF joins to other factors to promote tumor growth and cell proliferation via various mechanisms, including the activation of the MAPK/PI3K/AKT pathway and subsequent inflammation, inhibition of p53-dependent apoptosis, induction of vascular endothelial growth factor (VEGF) production, as well as the inhibition of the antitumor immune response [67,68].

TABLE 2 MIF expression in the serum and plasma of patients with different types of cancer.

	MIF levels in cancer				
Cancer type	**Women (ng/mL)**	**Men (ng/mL)**	**Both (ng/mL)**	**Healthy (ng/mL)**	**Reference**
Hepatocellular carcinoma	ND	ND	35.3–90	15	[66]
Oral carcinoma of squamous cell	ND	ND	58.6	ND	[67]
Head and neck squamous cell carcinoma	ND	ND	0.2179	0.075	[68]
Ovarian epithelial cancer	6.6			2.62	[69–71]
	6.8	–	–	0.4	
	2.6			0.69	
Colorectal cancer			20	3	[72]
Prostate cancer	ND	5.87 ± 4	ND	2.19 ± 2.6	[73,74]
		10.8		2.1	
Gastric cancer	ND	682.7		22.04	[75,76]
			1.06	0.285	
Squamous cell carcinoma of the esophagus	ND	ND	100	2.5	[77]
Chronic obstructive pulmonary disease (COPD) and lung cancer	ND	ND	1.54	0.950	[78]
Breast cancer	91.1	ND	ND	8.3	[79,80]
	1.51			0.38	

The quantification of the MIF protein in all patients was carried out by the ELISA method. ND, *not determined.*

The proinflammatory characteristics of MIF and its association with cancer [69–72] have motivated the study of its influence on the genesis of CRC because it is recognized that the risk of developing CRC is substantially higher in patients with chronic inflammatory bowel diseases. The inflammation promotes the proliferation and survival of malignant cells, stimulates angiogenesis and metastasis, and modifies the antitumor immune response and the response to hormones [73].

MIF, as a proinflammatory cytokine is overexpressed in ulcerative colitis [74,75], as well as in the inflammation of the colon that precedes the development of CRC [76–78]. Epidemiological studies in humans show the direct relationship between *Helicobacter* gastritis and gastric cancer [79,80], Barrett's esophagus and esophageal cancer [81], and chronic pancreatitis and pancreatic cancer [82,83]. Furthermore, 18% of patients suffering from inflammatory bowel disease (IBD), including ulcerative colitis and Crohn's disease, develop CRC [73,84]. These observations have been supported experimentally with *Mif*$-^{-}/-$ mice, which did not develop induced ulcerative colitis, and with the observation that anti-MIF therapy reduced the severity of the pathology [85,86]. Furthermore, mice that overexpressed MIF showed an increase in the severity of colitis associated with increased myeloperoxidase activity and reduced corticosterone levels [87,88].

Importantly, the high levels of MIF present in the pathologies that precede CRC are also present during the development of CRC. In humans, there is a significant increase in MIF expression during the development of colonic metastatic polyps and adenomas [89,90]. Similarly, the analysis of adenomatous intestinal polyps from $APC^{Min/+}$ mice demonstrated an association of the development of colonic tumorigenesis with MIF expression [91].

MIF receptors also play an important role in tumor development, for example, expression of the CD74 receptor in gastric cells [78,92], colon epithelial cells and colon carcinoma cells (CT-26) [93,94] has been reported. Moreover, in CRC tumor cell cultures, transformed cells have been found to be the major source of MIF and overexpress the CXCR4 chemokine receptor [95,96], which has led to the suggestion that the MIF/CXCR4 complex can promote tumor progression, with roles in tumor growth and degree of malignancy, angiogenesis, and metastasis [97,98], in addition to promoting the activation of the AKT survival pathway [54].

In addition, it has shown that the CXCR7 receptor is expressed in endothelial and cancer cells [99]; therefore, the MIF/CXCR7 complex could also play a crucial role in tumor growth and metastasis [100]. A schematic summary of the functions of MIF as a promoter of cancer genesis, development, and malignancy is shown in Fig. 2.

MIF as a chemokine involved in carrying effector immune cells against CRC

The immune response is decisive in the development and metastasis of CRC. Lymphocytic cells ($CD8^+$ T cells, $CD4^+$ T cells, and NK cells) and Mφs are able to identify and eliminate the tumor in early stages [101]. However, tumor cells that manage to escape from immune effector cells proliferate and spread to develop CRC. Therefore, the recruitment of immune effector cells is of utmost importance, especially in the early stages, to contain tumor development and metastasis. This recruitment is mediated by proteins called chemokines.

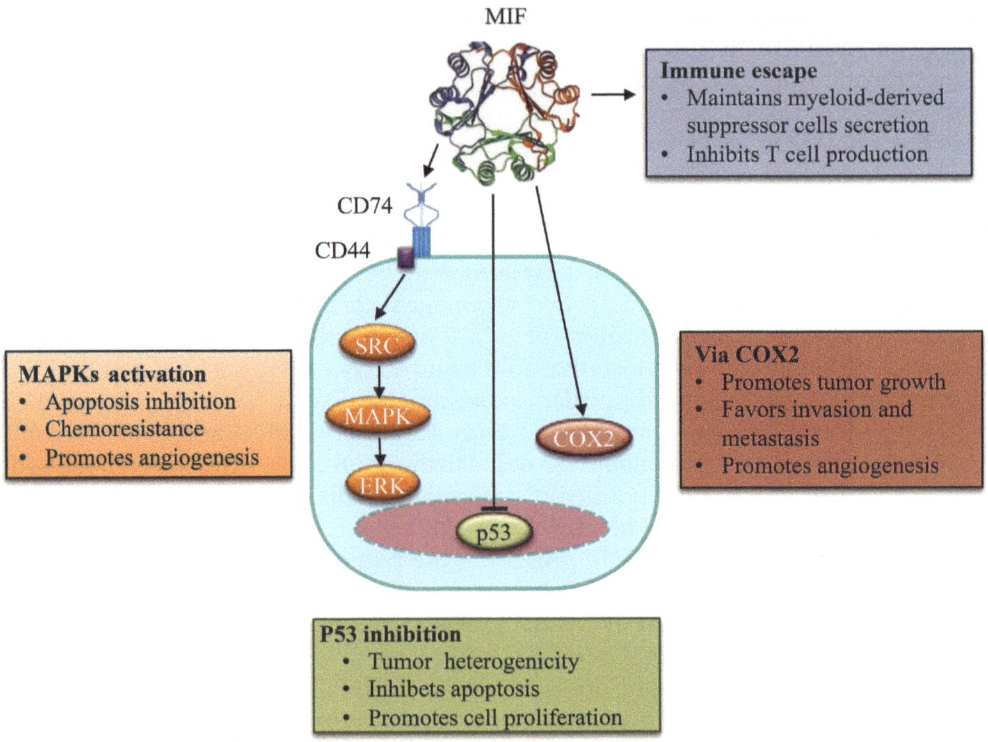

FIG. 2 Mechanisms that regulate cancer genesis, development, and malignancy promotion by MIF. The binding of MIF to its canonical receptor CD74/CD44 promotes the activation of the MAPK/PI3K/AKT pathway; consequently, the production of inflammatory mediators promotes the inhibition of apoptosis and favors chemoresistance and angiogenesis. Inhibition of p53 favors tumor heterogeneity, inhibits apoptosis, and promotes cell proliferation. The activation of the COX-2 pathway promotes tumor growth, angiogenesis, and metastasis.

In humans with CRC, the importance of cells infiltrating the tumor for predicting survival and time to relapse after colectomy is recognized [102]. The presence of Mφs and $CD8^+$ T cells in the tumor increases the survival of patients with CRC (78 months), while those with more IL-17-producing cells have an early recurrence (18 months) [103]. However, at the time this report was published, there were no human studies, relating increased levels of MIF, derived from tumor cells, with infiltrating immune cells in CRC.

In murine CRC models, there is evidence that MIF favors the recruitment of immune cells to the tumor site. *Mif−/−* mice implanted with the CRC-derived CT-26 cell line developed smaller tumors than wild-type mice. This reduction was associated with increased recruitment of $CD8^+$ T cells by MIF secreted by the implanted tumor cells. That is, even when the mouse was devoid of MIF, the implanted CT-26 cells could produce high amounts of MIF and attract $CD8^+$ T cells, thereby establishing a regulatory role of MIF in T-lymphocyte trafficking [104]. This is in line with that observed in CRC patients; those with higher concentrations of MIF in connective tissue have a survival of more than 5 years, significantly higher than that of the patients with decreased levels of MIF [105].

Moreover, we found that *Mif−/−* mice develop a greater number of tumors and have lower percentages of intestinal epithelium tumor-associated Mφs than wild-type mice in a chemically induced CRC murine model. These results suggested that MIF plays a role in controlling the initial development of CRC by attracting Mφs to the tumor, which is a condition that favors the initial antitumor response [106]. It is worth noting that the findings in well-established CRC cases (in which it has been suggested that MIF is a promoter of malignancy and metastasis) and the findings in cases of spontaneous CRC development induced in *Mif−/−* mice (in which MIF has been suggested to participate in the control of the development of CRC) regarding the role of MIF appear contradictory. However, the discrepancy in results might be rectified if these observations were confirmed and extended in different cell types, which could indicate that the role of the MIF differs according to the temporal progression of tumorigenesis. In other words, MIF could suppress the development of CRC at the beginning of its development, while it participates in malignancy and metastasis if CRC is already established.

MIF is a promoter of angiogenesis

For the growth of the tumor, the formation of new blood vessels is essential, and this growth is controlled by angiogenic factors that are secreted by the tumor cells. Among these factors is VEGF, which is very important for increased microvascular density and is directly related to the increase in metastasis in different types of cancer, including CRC [107,108]. Some evidence suggests that MIF is capable of inducing angiogenesis by promoting the differentiation of endothelial cells into blood vessels [109,110]. This is supported by the observation of an important correlation between increased MIF mRNA and VEGF expression in CRC metastasis, as well as the observation that suppression of MIF with anti-MIF antibodies reduces angiogenesis [111]. In murine models of CRC generated by implantation of cell lines, in already established cancers, the inhibition of MIF reduced the number and size of tumors, as well as the aggressiveness and invasiveness of tumor cells, angiogenesis, and the expression of metalloproteinase (MMP) 9, via a Rho-dependent pathway [112,113].

Taken together, these observations support the idea that MIF, along with growth factors released by tumor cells, tumor-oriented blood vessels, and tumor-associated Mφ are involved in the growth of tumor cells through the induction of angiogenesis.

Role of MIF in metastasis

Angiogenesis is an important condition for tumor metastasis, so it is not surprising that MIF overexpression is closely correlated with the aggressiveness of cancer and its metastatic potential [114]. There are reports of the overexpression of MIF in the metastasis of different cancers, such as prostate cancer [115], breast cancer [116], hepatocellular cancer [117], and lung adenocarcinoma [65].

High levels of MIF have been observed in the serum and epithelial cells of tumor tissue from patients, murine models, and CRC cell lines, suggesting that MIF can promote malignancy and metastasis [95,97,118–120]. MIF overexpression has been associated with elevated levels of collagenase I and with the transformation of tumor-associated fibroblasts into epithelial cells, conditions that favor the invasion of tumor cells into other tissues [121].

Furthermore, some members of the metalloproteinase family such as MMP2 and MMP9 have been reported to selectively degrade-type IV collagen, an important component of the extracellular matrix, which is significantly associated with tumor cell invasion and metastasis [122,123]. In vitro experiments have shown that MIF promotes MMP secretion, which favors cell invasion [124]. High MIF expression associated with MMP2 and MMP9 expressions has been observed in CRC, which may favor genesis [125], development [126], progression, and metastasis of tumor [127].

Hypoxia

Hypoxia (low oxygen concentration) is an important feature of the tumor microenvironment that promotes invasion, metastasis, and resistance to therapy [128]. Various studies have shown that in normal cells, MIF expression is regulated by hypoxia. For example, smooth muscle vascular and endothelial cells release MIF when cultured under hypoxic conditions [98]. MIF expression is likely a mechanism by which hypoxia stimulates angiogenesis, providing a reason for why MIF is expressed in wounds, inflamed tissue, and cancer [129].

MIF contributes to the stabilization of hypoxia since it stimulates the release of hypoxia-inducible factor (HIF)-1α via CD74 [130], which is key for the expression of angiogenic proteins such as VEGF, lysyl oxidase (LOX), and connective tissue growth factor (CTGF) [131–133]. CRC is characterized by the regions of variable hypoxia, which is known to be a consequence of vascular disorder [134]. However, there are no studies correlating MIF and hypoxia in CRC. In Fig. 3, we summarize the role of MIF as a promoter of tumor inflammation, proliferation, and growth in CRC.

Anti-MIF antibodies and inhibitors

MIF has become a target for therapy through the use of antibodies or synthetic inhibitors, which have recently been actively researched.

MIF inhibitors can be divided into three classes depending on their mechanism of action: (1) covalent modifiers of the catalytic site at the N-terminal proline residue, (2) molecules that noncovalently bind the catalytic site, and (3) disrupters of the trimeric structure of MIF. Class 1 and 2 inhibitors impede both enzymatic and anti-glucocorticoid MIF activities [135], but class 1 agent modification of the N-terminal proline residue also results in partial loss of the chemotactic recruitment activity of MIF in lungs [136]. However, it is not known whether all class 1 inhibitors induce this effect. The only known member of the third class is the anti-inflammatory drug ebselen, which disrupts MIF trimers and induces aggregation through cysteine residues. Trimer dissociation inhibits the catalytic and ligand-binding activities of MIF, strongly favoring its chemotactic activity [135]. Some inhibitors have been tested in cancer models; some shown promising results, but are still in the clinical testing phase (Table 3).

Imalumab remains the only human MIF-directed monoclonal antibody [161], but other antibodies directed to MIF family receptors exist and can be considered for therapy.

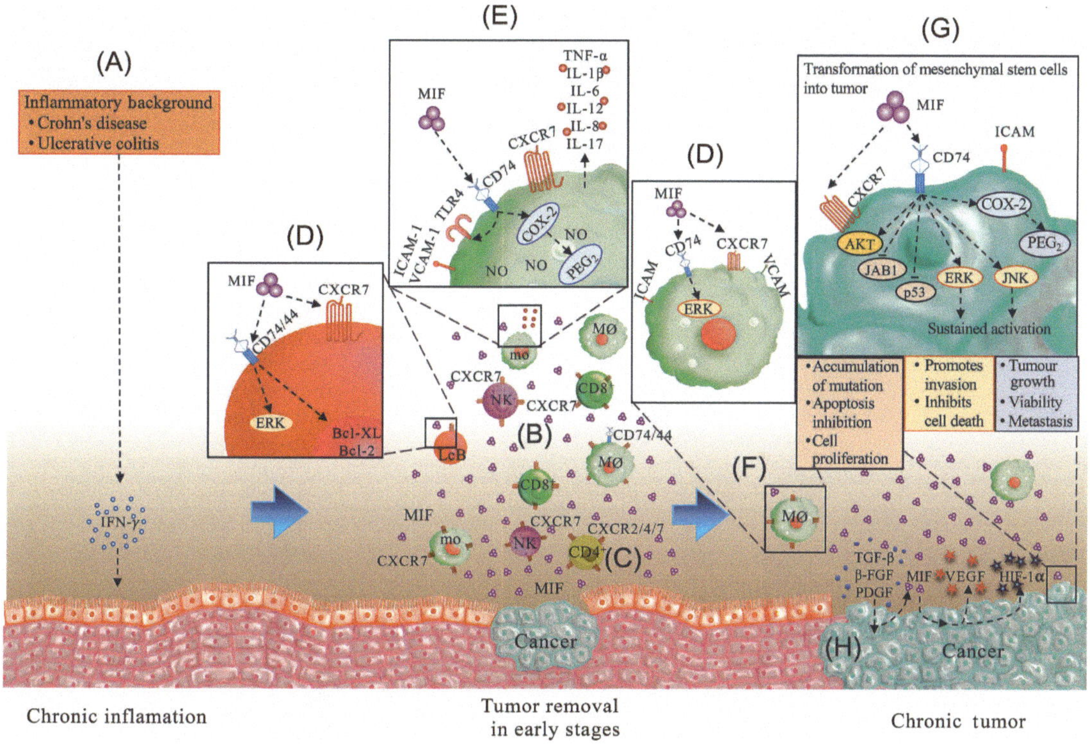

FIG. 3 Role of MIF in the genesis, progression, and metastasis of CRC. (A) Chronic inflammatory diseases such as Crohn's disease and ulcerative colitis promote an inflammatory environment characterized by an increase in IFN-γ favoring the proliferation of malignant cells. (B) The inflammatory environment favors the formation of MIF-producing cancer cells, which favors the recruitment of immune cells such as NK cells, B lymphocytes, CD8+ T cells, monocytes (mo), and macrophages (Mφ) that express CXCR7. (C) $CD4^+$ T cells express CD74/44 and CXCR2, CXCR4, and CXCR7. (D) The binding of MIF to the CXCR7 and CD74/CD44 complexes favors the elevation of cyclin E, and the transcription of the survival genes Bcl-XL and Bcl-2. (E) The binding of MIF to the CD74/CD44 complex and CXCR7 in mo favors the expression of ICAM, VCAM, and TLR-4 in Mφ, in addition to the production of NO, TNF-α, IL-1β, IL-6, IL-12, IL-8, and IL-17 as well as COX-2 activation and PGE2 production. (F) The tumor cells that manage to escape from the immune effector cells proliferate and spread until they lead to the development of cancer. (G) MIF is secreted by tumor cells, is responsible for attracting mesenchymal stem cells, and activates the ERK, JNK, and AKT signaling pathways through the CXCR4 junction. (H) TGF-β, β-FGF, and PDGF favor the release of MIF located in the cytoplasm of tumor cells; finally, MIF promotes the production of VEGF and HIF-1α and consequently angiogenesis.

Milatuzumab (hLL1) is a humanized recombinant antibody directed against CD74; blockade of CD74 by this antibody leads to increased cell death through the PI3K-Akt signaling pathway [162], which is the same mechanism used by MIF to induce cell survival [163]. Milatuzumab has been tested in phase I clinical trials with favorable results; it had a short clearance time and was well tolerated, and the disease became stable for at least 3 months in generally refractory patients [164]. Ulocuplumab (BMS-936564/MDX1338) is an antibody directed against CXCR4 that induces cell death in chronic lymphocytic leukemia through a ROS-dependent pathway [165]. A phase 1 clinical study of ulocuplumab alone and in

TABLE 3 The most common MIF inhibitors that have been tested in different models of inflammatory, autoimmune, and cancer diseases.

Inhibitor	Structure	Mechanism	Observations	Reference
4-IPP (4-iodo-6-phenylpyrimidine)	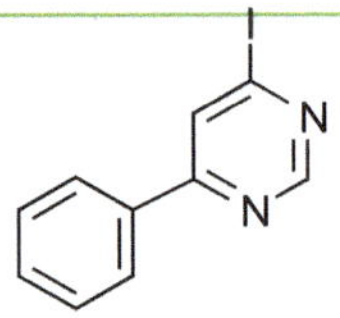	- Blocks internalization of the MIF/CD74 axis, and activation of the JNK pathway - Inhibits proliferation and induces apoptosis and cell death in vitro	- Reduces motility and growth of lung carcinoma cells in vitro - Inhibits the proliferation and invasiveness of squamous cell carcinoma of the head and neck - Induces apoptosis and mitotic death of thyroid carcinomas, and sensitizes multiple myeloma cells to chemotherapy	[137–141]
PEITC (phenethyl isothiocyanate)	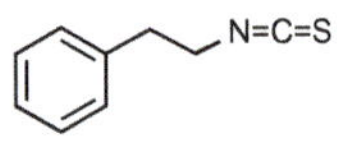	- Inhibits activation of the AKT and NF-κB pathway - Inhibits the synthesis of growth factors EGFR and HER2 - Favors activation of the Rb protein in prostate cells, and attenuates the progression of the cell cycle - Inhibits the synthesis of MMP	- Modulates inflammatory pathways by inhibiting NF-κB - Participates in angiogenesis inhibition, metastasis, suppression, and apoptosis	[138,142–147]
ISO-1 ((*S*,*R*)-3-(4-hydroxyphenyl)-4,5-dihydro-5-isoxazole acetic acid)	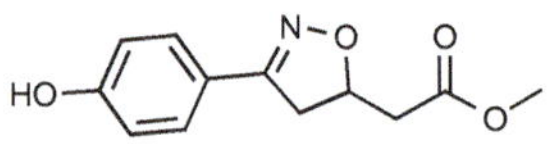	- Inhibits TNF secretion by Mφ after LPS stimulation	- In prostate and colon cancers, there was a decrease in tumor growth and vascularity - Reduces the proliferation of multiform glioblastoma cells in vitro	[119,138,148–152]

TABLE 3 The most common MIF inhibitors that have been tested in different models of inflammatory, autoimmune, and cancer diseases—cont'd

Inhibitor	Structure	Mechanism	Observations	Reference
ISO-66	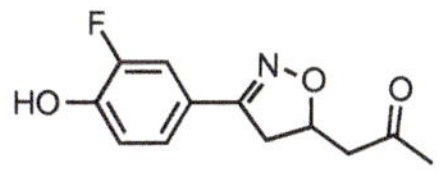	- An ISO-1 analog with enhanced features	- Improves specific and non-specific immune response against cancer - Decreases tumor burden in mice with melanoma or colon cancer	[153]
RPS19 (ribosomal protein S19)		- Inhibits the MIF/CD74 interaction - Suppresses MIF/CXCR2-dependent monocyte adhesion to endothelial cells in vitro	- Blocks MIF signaling - Inhibits macrophage and T cell infiltration and activation	[154,155]
OXIM-11	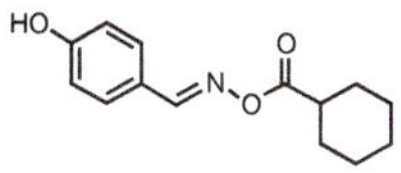	- Inhibits various proinflammatory activities in vitro and in vivo - Reduces inflammation and cellular activation	- Decreases NF-κB activation - Preserves the anti-inflammatory activities of glucocorticoids	[156]
Benzoxazol-2-one	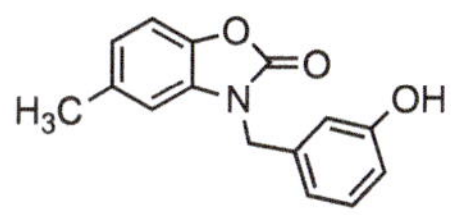	- Avoid MIF/CD74 interaction	- Inhibits ERK1/2 phosphorylation - Reduces collagen synthesis and fibrosis of the pulmonary artery	[157,158]
K-664-1	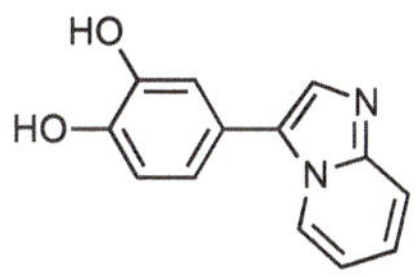	- Prevents the apoptosis of pancreatic β cells, triggered by proinflammatory cytokines - Suppresses the expression of pro-apoptotic ABX and anti-apoptotic Bcl-2 proteins - Suppresses the release of MCP-1 (Mφ chemoattractant)	- Eliminates the effect of MIF in experimental models mediated by inflammatory diseases - Prevents the progression of diabetes	[150,157,158]

Continued

TABLE 3 The most common MIF inhibitors that have been tested in different models of inflammatory, autoimmune, and cancer diseases—cont'd

Inhibitor	Structure	Mechanism	Observations	Reference
K-647-1	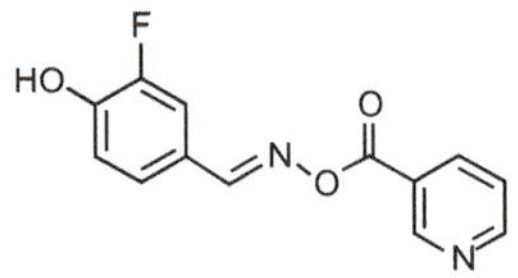	- Prevents β cell apoptosis triggered by proinflammatory cytokines - Suppresses the expression of pro-apoptotic ABX and antiapoptotic Bcl-2 proteins	- Eliminates the effect of MIF in experimental models mediated by inflammatory diseases - Prevents the progression of diabetes	[150]
T-614	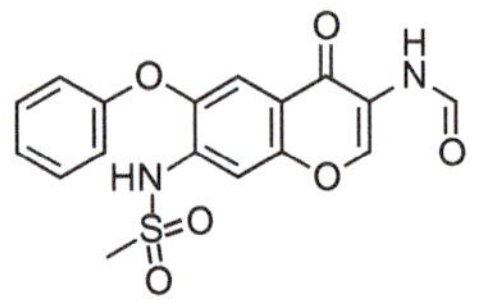	- Inhibits COX-2 activity Inhibits the release and intracellular accumulation of IL-1β in LPS-stimulated human peripheral blood monocytes - Inhibits TNF-α, IL-1β, IL-6, IL-8, and MCP-1 production - Modulates NF-κB activation	- Orally bioavailable drug capable of attenuating joint edema and destruction in arthritis models, in addition to exhibiting analgesic properties - It is useful in multiple sclerosis and cachexia in the context of adenocarcinoma	[157–160]

combination with other treatments has been completed with a general clinical benefit and limited adverse reactions [166].

Conclusion

MIF and its receptors participate in multiple biochemical regulatory mechanisms, including self-regulation. These complex mechanisms have a profound impact on processes typically related to the genesis of CRC and its progression, including (I) inducing inflammation via the MIF/CD74/CD44 interaction; (II) inducing cell proliferation via activation of the MAPK ERK1/2 and AKT pathways, the suppression of JAB1 activity and the inhibition of p53-dependent apoptosis; (III) inducing VEGF production and consequently the angiogenic characteristics of tumor progression and metastasis; and (IV) inducing epithelial to mesenchymal transition. Together, these characteristics of MIF make it an attractive therapeutic and/or diagnostic agent target. However, the characteristics of MIF as chemokine should also be considered since MIF may be an important factor in recruitment immune cells capable of containing CRC onset and malignancy. Therefore, the known regulatory and effective mechanisms of MIF should be further explored, and the regulatory mechanisms already hinted at by the existing data should be confirmed in CRC as in other cancers.

Acknowledgments

This work was supported in part by grants from the "Support Program for Research Projects and Technological Innovation" UNAM-DGAPA-PAPIIT, IN-209718 and National Council of Science and Technology of Mexico (CONACYT) [A1-S-10463]. Tonathiu Rodríguez is a doctoral student from Programa de Doctorado en Ciencias Biomédicas, Universidad Nacional Autónoma de México (UNAM) and received a fellowship (CVU CONACyT No. 5502200). Imelda Juárez-Avelar is a doctoral student from Programa de Doctorado en Ciencias Biológicas, UNAM.

References

[1] Ferlay J, et al. Estimating the global cancer incidence and mortality in 2018: GLOBOCAN sources and methods. Int J Cancer 2019;144(8):1941–53.
[2] World Health Organization. Cáncer, 09-11-2018. Available from: http://www.who.int/es/news-room/fact-sheets/detail/cancer; 2018.
[3] Bray F, et al. Global cancer statistics 2018: GLOBOCAN estimates of incidence and mortality worldwide for 36 cancers in 185 countries. CA Cancer J Clin 2018;68(6):394–424.
[4] Balkwill F. Cancer and the chemokine network. Nat Rev Cancer 2004;4(7):540–50.
[5] David JR. Delayed hypersensitivity in vitro: its mediation by cell-free substances formed by lymphoid cell-antigen interaction. Proc Natl Acad Sci U S A 1966;56(1):72–7.
[6] Bloom BR, Bennett B. Mechanism of a reaction in vitro associated with delayed-type hypersensitivity. Science 1966;153(3731):80–2.
[7] Weiser WY, et al. Molecular cloning of a cDNA encoding a human macrophage migration inhibitory factor. Proc Natl Acad Sci U S A 1989;86(19):7522–6.
[8] Bernhagen J, et al. MIF is a pituitary-derived cytokine that potentiates lethal endotoxaemia. Nature 1993;365 (6448):756–9.
[9] Nishihira J. Macrophage migration inhibitory factor (MIF): its essential role in the immune system and cell growth. J Interferon Cytokine Res 2000;20(9):751–62.
[10] Richard B. MIF: most interesting factor. 332.
[11] Du J, et al. Macrophage migration inhibitory factor (MIF) in Chinese amphioxus as a molecular marker of immune evolution during the transition of invertebrate/vertebrate. Dev Comp Immunol 2004;28(10):961–71.
[12] Bacher M, et al. An essential regulatory role for macrophage migration inhibitory factor in T-cell activation. Proc Natl Acad Sci U S A 1996;93(15):7849–54.
[13] Calandra T, et al. The macrophage is an important and previously unrecognized source of macrophage migration inhibitory factor. J Exp Med 1994;179(6):1895–902.
[14] Calandra T, Roger T. Macrophage migration inhibitory factor: a regulator of innate immunity. Nat Rev Immunol 2003;3(10):791–800.
[15] Imamura K, et al. Identification and immunohistochemical localization of macrophage migration inhibitory factor in human kidney. Biochem Mol Biol Int 1996;40(6):1233–42.
[16] Nishihira J, Koyama Y, Mizue Y. Identification of macrophage migration inhibitory factor (MIF) in human vascular endothelial cells and its induction by lipopolysaccharide. Cytokine 1998;10(3):199–205.
[17] Abe R, et al. Regulation of the CTL response by macrophage migration inhibitory factor. J Immunol 2001;166 (2):747–53.
[18] Waeber G, et al. Insulin secretion is regulated by the glucose-dependent production of islet beta cell macrophage migration inhibitory factor. Proc Natl Acad Sci U S A 1997;94(9):4782–7.

[19] Calandra T, Bucala R. Macrophage migration inhibitory factor (MIF): a glucocorticoid counter-regulator within the immune system. Crit Rev Immunol 2017;37(2–6):359–70.
[20] Choi SS, et al. Human astrocytes: secretome profiles of cytokines and chemokines. PLoS ONE 2014;9(4):e92325.
[21] Suzuki T, et al. Augmented expression of macrophage migration inhibitory factor (MIF) in the telencephalon of the developing rat brain. Brain Res 1999;816(2):457–62.
[22] Koda M, et al. Up-regulation of macrophage migration-inhibitory factor expression after compression-induced spinal cord injury in rats. Acta Neuropathol 2004;108(1):31–6.
[23] Roger T, Glauser MP, Calandra T. Macrophage migration inhibitory factor (MIF) modulates innate immune responses induced by endotoxin and Gram-negative bacteria. J Endotoxin Res 2001;7(6):456–60.
[24] Calandra T. Macrophage migration inhibitory factor and host innate immune responses to microbes. Scand J Infect Dis 2003;35(9):573–6.
[25] Sun HW, et al. Crystal structure at 2.6-A resolution of human macrophage migration inhibitory factor. Proc Natl Acad Sci U S A 1996;93(11):5191–6.
[26] Ruiz-Rosado Jde D, et al. MIF promotes classical activation and conversion of inflammatory Ly6C(high) monocytes into TipDCs during murine toxoplasmosis. Mediators Inflamm 2016;2016:9101762.
[27] Sanchez-Zamora YI, et al. Altered macrophage and dendritic cell response in Mif−/− mice reveals a role of Mif for inflammatory-Th1 response in type 1 diabetes. J Diabetes Res 2016;2016:7053963.
[28] Nathan CF, Remold HG, David JR. Characterization of a lymphocyte factor which alters macrophage functions. J Exp Med 1973;137(2):275–90.
[29] Juttner S, et al. Migration inhibitory factor induces killing of Leishmania major by macrophages: dependence on reactive nitrogen intermediates and endogenous TNF-alpha. J Immunol 1998;161(5):2383–90.
[30] Satoskar AR, et al. Migration-inhibitory factor gene-deficient mice are susceptible to cutaneous Leishmania major infection. Infect Immun 2001;69(2):906–11.
[31] Reyes JL, et al. Macrophage migration inhibitory factor contributes to host defense against acute Trypanosoma cruzi infection. Infect Immun 2006;74(6):3170–9.
[32] Flores M, et al. Macrophage migration inhibitory factor (MIF) is critical for the host resistance against *Toxoplasma gondii*. FASEB J 2008;22(10):3661–71.
[33] Terrazas CA, et al. MIF synergizes with *Trypanosoma cruzi* antigens to promote efficient dendritic cell maturation and IL-12 production via p38 MAPK. Int J Biol Sci 2011;7(9):1298–310.
[34] Terrazas CA, et al. *Toxoplasma gondii*: impaired maturation and pro-inflammatory response of dendritic cells in MIF-deficient mice favors susceptibility to infection. Exp Parasitol 2010;126(3):348–58.
[35] Bernhagen J, et al. An essential role for macrophage migration inhibitory factor in the tuberculin delayed-type hypersensitivity reaction. J Exp Med 1996;183(1):277–82.
[36] Bernhagen J, Calandra T, Bucala R. The emerging role of MIF in septic shock and infection. Biotherapy 1994;8(2):123–7.
[37] Morand EF, Leech M. Macrophage migration inhibitory factor in rheumatoid arthritis. Front Biosci 2005;10:12–22.
[38] Finucane OM, et al. Insights into the role of macrophage migration inhibitory factor in obesity and insulin resistance. Proc Nutr Soc 2012;71(4):622–33.
[39] Sanchez-Zamora YI, Rodriguez-Sosa M. The role of MIF in type 1 and type 2 diabetes mellitus. J Diabetes Res 2014;2014:804519.
[40] Yamaguchi E, et al. Macrophage migration inhibitory factor (MIF) in bronchial asthma. Clin Exp Allergy 2000;30(9):1244–9.
[41] Santos LL, Morand EF. Macrophage migration inhibitory factor: a key cytokine in RA, SLE and atherosclerosis. Clin Chim Acta 2009;399(1–2):1–7.
[42] Denkinger CM, et al. Macrophage migration inhibitory factor and its role in autoimmune diseases. Arch Immunol Ther Exp (Warsz) 2004;52(6):389–400.
[43] Burger-Kentischer A, et al. Expression of macrophage migration inhibitory factor in different stages of human atherosclerosis. Circulation 2002;105(13):1561–6.
[44] Su Y, et al. Macrophage migration inhibitory factor activates inflammatory responses of astrocytes through interaction with CD74 receptor. Oncotarget 2017;8(2):2719–30.
[45] Wraight CJ, et al. Human major histocompatibility complex class II invariant chain is expressed on the cell surface. J Biol Chem 1990;265(10):5787–92.

[46] Leng L, et al. MIF signal transduction initiated by binding to CD74. J Exp Med 2003;197(11):1467–76.
[47] Bucala R, Shachar I. The integral role of CD74 in antigen presentation, MIF signal transduction, and B cell survival and homeostasis. Mini Rev Med Chem 2014;14(14):1132–8.
[48] Amin MA, et al. Migration inhibitory factor up-regulates vascular cell adhesion molecule-1 and intercellular adhesion molecule-1 via Src, PI3 kinase, and NFκB. Blood 2006;107(6):2252–61.
[49] Mitchell RA, et al. Sustained mitogen-activated protein kinase (MAPK) and cytoplasmic phospholipase A2 activation by macrophage migration inhibitory factor (MIF). Regulatory role in cell proliferation and glucocorticoid action. J Biol Chem 1999;274(25):18100–6.
[50] Stojanovic I, et al. Macrophage migration inhibitory factor stimulates interleukin-17 expression and production in lymph node cells. Immunology 2009;126(1):74–83.
[51] Toh ML, et al. Regulation of IL-1 and TNF receptor expression and function by endogenous macrophage migration inhibitory factor. J Immunol 2006;177(7):4818–25.
[52] Calandra T, et al. MIF as a glucocorticoid-induced modulator of cytokine production. Nature 1995;377(6544):68–71.
[53] Mitchell RA, et al. Macrophage migration inhibitory factor (MIF) sustains macrophage proinflammatory function by inhibiting p53: regulatory role in the innate immune response. Proc Natl Acad Sci U S A 2002;99(1):345–50.
[54] Bernhagen J, et al. MIF is a noncognate ligand of CXC chemokine receptors in inflammatory and atherogenic cell recruitment. Nat Med 2007;13(5):587–96.
[55] Dewor M, et al. Macrophage migration inhibitory factor (MIF) promotes fibroblast migration in scratch-wounded monolayers in vitro. FEBS Lett 2007;581(24):4734–42.
[56] Tarnowski M, et al. Macrophage migration inhibitory factor is secreted by rhabdomyosarcoma cells, modulates tumor metastasis by binding to CXCR4 and CXCR7 receptors and inhibits recruitment of cancer-associated fibroblasts. Mol Cancer Res 2010;8(10):1328–43.
[57] Alampour-Rajabi S, et al. MIF interacts with CXCR7 to promote receptor internalization, ERK1/2 and ZAP-70 signaling, and lymphocyte chemotaxis. FASEB J 2015;29(11):4497–511.
[58] Nagarajan P, et al. MIF antagonist (CPSI-1306) protects against UVB-induced squamous cell carcinoma. Mol Cancer Res 2014;12(9):1292–302.
[59] Gordon-Weeks AN, et al. Macrophage migration inhibitory factor: a key cytokine and therapeutic target in colon cancer. Cytokine Growth Factor Rev 2015;26(4):451–61.
[60] Tomiyasu M, et al. Quantification of macrophage migration inhibitory factor mRNA expression in non-small cell lung cancer tissues and its clinical significance. Clin Cancer Res 2002;8(12):3755–60.
[61] Xu X, et al. Overexpression of macrophage migration inhibitory factor induces angiogenesis in human breast cancer. Cancer Lett 2008;261(2):147–57.
[62] Meyer-Siegler KL, Bellino MA, Tannenbaum M. Macrophage migration inhibitory factor evaluation compared with prostate specific antigen as a biomarker in patients with prostate carcinoma. Cancer 2002;94(5):1449–56.
[63] Liao B, et al. Macrophage migration inhibitory factor contributes angiogenesis by up-regulating IL-8 and correlates with poor prognosis of patients with primary nasopharyngeal carcinoma. J Surg Oncol 2010;102(7):844–51.
[64] Munaut C, et al. Macrophage migration inhibitory factor (MIF) expression in human glioblastomas correlates with vascular endothelial growth factor (VEGF) expression. Neuropathol Appl Neurobiol 2002;28(6):452–60.
[65] Kamimura A, et al. Intracellular distribution of macrophage migration inhibitory factor predicts the prognosis of patients with adenocarcinoma of the lung. Cancer 2000;89(2):334–41.
[66] Anzano MA, et al. Growth factor production by human colon carcinoma cell lines. Cancer Res 1989;49(11):2898–904.
[67] Verjans E, et al. Dual role of macrophage migration inhibitory factor (MIF) in human breast cancer. BMC Cancer 2009;9:230.
[68] Bando H, et al. Expression of macrophage migration inhibitory factor in human breast cancer: association with nodal spread. Jpn J Cancer Res 2002;93(4):389–96.
[69] Conroy H, Mawhinney L, Donnelly SC. Inflammation and cancer: macrophage migration inhibitory factor (MIF) – the potential missing link. QJM 2010;103(11):831–6.
[70] Bucala R, Donnelly SC. Macrophage migration inhibitory factor: a probable link between inflammation and cancer. Immunity 2007;26(3):281–5.
[71] Bach JP, et al. Role of MIF in inflammation and tumorigenesis. Oncology 2008;75(3–4):127–33.

[72] Yasasever V, et al. Macrophage migration inhibitory factor in cancer. Cancer Invest 2007;25(8):715–9.
[73] Candido J, Hagemann T. Cancer-related inflammation. J Clin Immunol 2013;33(Suppl 1):S79–84.
[74] Murakami H, et al. Macrophage migration inhibitory factor in the sera and at the colonic mucosa in patients with ulcerative colitis: clinical implications and pathogenic significance. Eur J Clin Invest 2001;31 (4):337–43.
[75] Ohkawara T, et al. Lack of macrophage migration inhibitory factor suppresses innate immune response in murine dextran sulfate sodium-induced colitis. Scand J Gastroenterol 2008;43(12):1497–504.
[76] Leon-Cabrera SA, et al. Lack of STAT6 attenuates inflammation and drives protection against early steps of colitis-associated colon cancer. Cancer Immunol Res 2017;5(5):385–96.
[77] Park CH, Eun CS, Han DS. Intestinal microbiota, chronic inflammation, and colorectal cancer. Intest Res 2018;16 (3):338–45.
[78] Beswick EJ, Reyes VE. CD74 in antigen presentation, inflammation, and cancers of the gastrointestinal tract. World J Gastroenterol 2009;15(23):2855–61.
[79] Mera RM, et al. Dynamics of *Helicobacter pylori* infection as a determinant of progression of gastric precancerous lesions: 16-year follow-up of an eradication trial. Gut 2017;67(7):1239–46.
[80] Chang YW, et al. Combination of *Helicobacter pylori* infection and the interleukin 8 -251 T > A polymorphism, but not the mannose-binding lectin 2 codon 54 G > A polymorphism, might be a risk factor of gastric cancer. BMC Cancer 2017;17(1):388.
[81] Talukdar A, et al. Prevalence of extra-esophageal cancers in patients with Barrett's esophagus and esophageal adenocarcinoma. Trop Gastroenterol 2012;33(3):185–8.
[82] Munigala S, et al. Increased risk of pancreatic adenocarcinoma after acute pancreatitis. Clin Gastroenterol Hepatol 2014;12(7):1143–50 e1.
[83] Ma C, et al. Optimized ROI size on ADC measurements of normal pancreas, pancreatic cancer and mass-forming chronic pancreatitis. Oncotarget 2017;8(58):99085–92.
[84] Rubin DC, Shaker A, Levin MS. Chronic intestinal inflammation: inflammatory bowel disease and colitis-associated colon cancer. Front Immunol 2012;3:107.
[85] Ohkawara T, et al. Amelioration of dextran sulfate sodium–induced colitis by anti-macrophage migration inhibitory factor antibody in mice. Gastroenterology 2002;123(1):256–70.
[86] de Jong YP, et al. Development of chronic colitis is dependent on the cytokine MIF. Nat Immunol 2001;2 (11):1061–6.
[87] Ohkawara T, et al. Transgenic over-expression of macrophage migration inhibitory factor renders mice markedly more susceptible to experimental colitis. Clin Exp Immunol 2005;140(2):241–8.
[88] Mittelbronn M, et al. Macrophage migration inhibitory factor (MIF) expression in human malignant gliomas contributes to immune escape and tumour progression. Acta Neuropathol 2011;122(3):353–65.
[89] Cuthbert RJ, et al. Differential CD74 (major histocompatibility complex Class II invariant chain) expression in mouse and human intestinal adenomas. Eur J Cancer 2009;45(9):1654–63.
[90] Takahashi N, et al. Involvement of macrophage migration inhibitory factor (MIF) in the mechanism of tumor cell growth. Mol Med 1998;4(11):707–14.
[91] Rosenberg DW, Giardina C, Tanaka T. Mouse models for the study of colon carcinogenesis. Carcinogenesis 2009;30(2):183–96.
[92] Ong GL, et al. Cell surface expression and metabolism of major histocompatibility complex class II invariant chain (CD74) by diverse cell lines. Immunology 1999;98(2):296–302.
[93] Maharshak N, et al. CD74 is a survival receptor on colon epithelial cells. World J Gastroenterol 2010; 16(26):3258–66.
[94] Jiang Z, et al. Invariant chain expression in colon neoplasms. Virchows Arch 1999;435(1):32–6.
[95] Hogan NM, et al. Impact of mesenchymal stem cell secreted PAI-1 on colon cancer cell migration and proliferation. Biochem Biophys Res Commun 2013;435(4):574–9.
[96] Shin HN, Moon HH, Ku JL. Stromal cell-derived factor-1alpha and macrophage migration-inhibitory factor induce metastatic behavior in CXCR4-expressing colon cancer cells. Int J Mol Med 2012;30(6):1537–43.
[97] Dessein AF, et al. Autocrine induction of invasive and metastatic phenotypes by the MIF-CXCR4 axis in drug-resistant human colon cancer cells. Cancer Res 2010;70(11):4644–54.
[98] Simons D, et al. Hypoxia-induced endothelial secretion of macrophage migration inhibitory factor and role in endothelial progenitor cell recruitment. J Cell Mol Med 2011;15(3):668–78.

[99] Infantino S, Moepps B, Thelen M. Expression and regulation of the orphan receptor RDC1 and its putative ligand in human dendritic and B cells. J Immunol 2006;176(4):2197–207.
[100] Zabel BA, et al. The novel chemokine receptor CXCR7 regulates trans-endothelial migration of cancer cells. Mol Cancer 2011;10:73.
[101] Deschoolmeester V, et al. Immune cells in colorectal cancer: prognostic relevance and role of MSI. Cancer Microenviron 2011;4(3):377–92.
[102] Galon J, et al. Type, density, and location of immune cells within human colorectal tumors predict clinical outcome. Science 2006;313(5795):1960–4.
[103] Tosolini M, et al. Clinical impact of different classes of infiltrating T cytotoxic and helper cells (Th1, th2, treg, th17) in patients with colorectal cancer. Cancer Res 2011;71(4):1263–71.
[104] Choi S, et al. Role of macrophage migration inhibitory factor in the regulatory T cell response of tumor-bearing mice. J Immunol 2012;189(8):3905–13.
[105] Legendre H, et al. Prognostic values of galectin-3 and the macrophage migration inhibitory factor (MIF) in human colorectal cancers. Mod Pathol 2003;16(5):491–504.
[106] Pacheco-Fernandez T, et al. Macrophage migration inhibitory factor promotes the interaction between the tumor, macrophages, and T cells to regulate the progression of chemically induced colitis-associated colorectal cancer. Mediators Inflamm 2019;2019:2056085.
[107] Nakasaki T, et al. Expression of tissue factor and vascular endothelial growth factor is associated with angiogenesis in colorectal cancer. Am J Hematol 2002;69(4):247–54.
[108] Plate KH, et al. Vascular endothelial growth factor is a potential tumour angiogenesis factor in human gliomas in vivo. Nature 1992;359(6398):845–8.
[109] Chesney J, et al. An essential role for macrophage migration inhibitory factor (MIF) in angiogenesis and the growth of a murine lymphoma. Mol Med 1999;5(3):181–91.
[110] Amin MA, et al. Migration inhibitory factor mediates angiogenesis via mitogen-activated protein kinase and phosphatidylinositol kinase. Circ Res 2003;93(4):321–9.
[111] Warren RS, et al. Regulation by vascular endothelial growth factor of human colon cancer tumorigenesis in a mouse model of experimental liver metastasis. J Clin Invest 1995;95(4):1789–97.
[112] Ogawa H, et al. An antibody for macrophage migration inhibitory factor suppresses tumour growth and inhibits tumour-associated angiogenesis. Cytokine 2000;12(4):309–14.
[113] Sun B, et al. Macrophage migration inhibitory factor promotes tumor invasion and metastasis via the Rho-dependent pathway. Clin Cancer Res 2005;11(3):1050–8.
[114] Han I, et al. Expression of macrophage migration inhibitory factor relates to survival in high-grade osteosarcoma. Clin Orthop Relat Res 2008;466(9):2107–13.
[115] Meyer-Siegler K, Hudson PB. Enhanced expression of macrophage migration inhibitory factor in prostatic adenocarcinoma metastases. Urology 1996;48(3):448–52.
[116] Bini L, et al. Protein expression profiles in human breast ductal carcinoma and histologically normal tissue. Electrophoresis 1997;18(15):2832–41.
[117] Ren Y, et al. Macrophage migration inhibitory factor: roles in regulating tumor cell migration and expression of angiogenic factors in hepatocellular carcinoma. Int J Cancer 2003;107(1):22–9.
[118] Wilson JM, et al. Macrophage migration inhibitory factor promotes intestinal tumorigenesis. Gastroenterology 2005;129(5):1485–503.
[119] He X-X, et al. Macrophage migration inhibitory factor promotes colorectal cancer. Mol Med 2009;15(1–2):1–10.
[120] Lee H, et al. Macrophage migration inhibitory factor may be used as an early diagnostic marker in colorectal carcinomas. Am J Clin Pathol 2008;129(5):772–9.
[121] Morris KT, et al. Chronic macrophage migration inhibitory factor exposure induces mesenchymal epithelial transition and promotes gastric and colon cancers. PLoS ONE 2014;9(6)e98656.
[122] Li Z, et al. Macrophage migration inhibitory factor enhances neoplastic cell invasion by inducing the expression of matrix metalloproteinase 9 and interleukin-8 in nasopharyngeal carcinoma cell lines. Chin Med J (Engl) 2004;117(1):107–14.
[123] Werner JA, Rathcke IO, Mandic R. The role of matrix metalloproteinases in squamous cell carcinomas of the head and neck. Clin Exp Metastasis 2002;19(4):275–82.
[124] Hagemann T, et al. Macrophages induce invasiveness of epithelial cancer cells via NF-κB and JNK. J Immunol 2005;175(2):1197–205.

[125] Zeng ZS, Cohen AM, Guillem JG. Loss of basement membrane type IV collagen is associated with increased expression of metalloproteinases 2 and 9 (MMP-2 and MMP-9) during human colorectal tumorigenesis. Carcinogenesis 1999;20(5):749–55.
[126] Moran A, et al. Clinical relevance of MMP-9, MMP-2, TIMP-1 and TIMP-2 in colorectal cancer. Oncol Rep 2005;13(1):115–20.
[127] Li BH, et al. Matrix metalloproteinase-2 and tissue inhibitor of metallo-proteinase-2 in colorectal carcinoma invasion and metastasis. World J Gastroenterol 2005;11(20):3046–50.
[128] Lunt SJ, Chaudary N, Hill RP. The tumor microenvironment and metastatic disease. Clin Exp Metastasis 2009;26(1):19–34.
[129] Maity A, Koumenis C. HIF and MIF – a nifty way to delay senescence? Genes Dev 2006;20(24):3337–41.
[130] Gaber T, et al. Macrophage migration inhibitory factor counterregulates dexamethasone-mediated suppression of hypoxia-inducible factor-1 alpha function and differentially influences human CD4+ T cell proliferation under hypoxia. J Immunol 2011;186(2):764–74.
[131] Winner M, et al. Amplification of tumor hypoxic responses by macrophage migration inhibitory factor-dependent hypoxia-inducible factor stabilization. Cancer Res 2007;67(1):186–93.
[132] Oda S, et al. Macrophage migration inhibitory factor activates hypoxia-inducible factor in a p53-dependent manner. PLoS ONE 2008;3(5)e2215.
[133] No YR, et al. HIF1alpha-induced by lysophosphatidic acid is stabilized via interaction with MIF and CSN5. PLoS ONE 2015;10(9)e0137513.
[134] Moulder JE, Rockwell S. Hypoxic fractions of solid tumors: experimental techniques, methods of analysis, and a survey of existing data. Int J Radiat Oncol Biol Phys 1984;10(5):695–712.
[135] Ouertatani-Sakouhi H, et al. Identification and characterization of novel classes of macrophage migration inhibitory factor (MIF) inhibitors with distinct mechanisms of action. J Biol Chem 2010;285(34):26581–98.
[136] Rajasekaran D, et al. Targeting distinct tautomerase sites of D-DT and MIF with a single molecule for inhibition of neutrophil lung recruitment. FASEB J 2014;28(11):4961–71.
[137] Schmohl KA, et al. Thyroid hormones and tetrac: new regulators of tumour stroma formation via integrin avß3. Endocr Relat Cancer 2015;22:941–52.
[138] O'Reilly C, et al. Targeting MIF in cancer: therapeutic strategies, current developments, and future opportunities. Med Res Rev 2016;36(3):440–60.
[139] Gutiérrez-González A, et al. Evaluation of the potential therapeutic benefits of macrophage reprogramming in multiple myeloma. Blood 2016;128(18):2241–52.
[140] Rajasekaran D, et al. Targeting distinct tautomerase sites of D-DT and MIF with a single molecule for inhibition of neutrophil lung recruitment. FASEB J 2014;28(11):4961–71.
[141] Zheng L, et al. Macrophage migration inhibitory factor (MIF) inhibitor 4-IPP suppresses osteoclast formation and promotes osteoblast differentiation through the inhibition of the NF-κB signaling pathway. FASEB J 2019;33 (6):7667–83.
[142] Gupta P, et al. Phenethyl isothiocyanate: a comprehensive review of anti-cancer mechanisms. Biochim Biophys Acta 2014;1846(2):405–24.
[143] Brown KK, et al. Direct modification of the proinflammatory cytokine macrophage migration inhibitory factor by dietary isothiocyanates. J Biol Chem 2009;284(47):32425–33.
[144] Tyndall JD, et al. Macrophage migration inhibitory factor covalently complexed with phenethyl isothiocyanate. Acta Crystallogr Sect F Struct Biol Cryst Commun 2012;68(9):999–1002.
[145] Gao N, et al. Phenethyl isothiocyanate exhibits antileukemic activity in vitro and in vivo by inactivation of Akt and activation of JNK pathways. Cell Death Dis 2011;2(4):e140.
[146] Loganathan S, et al. Inhibition of EGFR-AKT axis results in the suppression of ovarian tumors in vitro and in preclinical mouse model. PLoS ONE 2012;7(8):e43577.
[147] Okubo T, Washida K, Murakami A. Phenethyl isothiocyanate suppresses nitric oxide production via inhibition of phosphoinositide 3-kinase/Akt-induced IFN-γ secretion in LPS-activated peritoneal macrophages. Mol Nutr Food Res 2010;54(9):1351–60.
[148] Xu L, et al. Current developments of macrophage migration inhibitory factor (MIF) inhibitors. Drug Discov Today 2013;18(11–12):592–600.
[149] Al-Abed Y, et al. ISO-1 binding to the tautomerase active site of MIF inhibits its pro-inflammatory activity and increases survival in severe sepsis. J Biol Chem 2005;280(44):36541–4.

[150] Vujicic M, et al. Novel inhibitors of macrophage migration inhibitory factor prevent cytokine-induced beta cell death. Eur J Pharmacol 2014;740:683–9.
[151] Baron N, et al. Role of macrophage migration inhibitory factor in primary glioblastoma multiforme cells. J Neurosci Res 2011;89(5):711–7.
[152] Meyer-Siegler KL, et al. Inhibition of macrophage migration inhibitory factor or its receptor (CD74) attenuates growth and invasion of DU-145 prostate cancer cells. J Immunol 2006;177(12):8730–9.
[153] Ioannou K, et al. ISO-66, a novel inhibitor of macrophage migration, shows efficacy in melanoma and colon cancer models. Int J Oncol 2014;45(4):1457–68.
[154] Filip A-M, et al. Ribosomal protein S19 interacts with macrophage migration inhibitory factor and attenuates its pro-inflammatory function. J Biol Chem 2009;284(12):7977–85.
[155] Li J, et al. Blocking macrophage migration inhibitory factor protects against cisplatin-induced acute kidney injury in mice. Mol Ther 2018;26(10):2523–32.
[156] Crichlow GV, et al. Alternative chemical modifications reverse the binding orientation of a pharmacophore scaffold in the active site of macrophage migration inhibitory factor. J Biol Chem 2007;282(32):23089–95.
[157] Kok T, et al. Small-molecule inhibitors of macrophage migration inhibitory factor (MIF) as an emerging class of therapeutics for immune disorders. Drug Discov Today 2018;23(11):1910–8.
[158] Bloom J, et al. Identification of iguratimod as an inhibitor of macrophage migration inhibitory factor (MIF) with steroid-sparing potential. J Biol Chem 2016;291(51):26502–14.
[159] Aikawa Y, et al. A new anti-rheumatic drug, T-614, effectively suppresses the development of autoimmune encephalomyelitis. J Neuroimmunol 1998;89(1–2):35–42.
[160] Tanaka K, et al. Effect of iguratimod and other anti-rheumatic drugs on adenocarcinoma colon 26-induced cachexia in mice. Inflamm Res 2007;56(1):17–23.
[161] Mahalingam D, et al. Phase I study of imalumab (BAX69), a fully human recombinant antioxidized macrophage migration inhibitory factor antibody in advanced solid tumours. Br J Clin Pharmacol 2020;.
[162] Smith MR, Jin F, Joshi I. Milatuzumab and veltuzumab induce apoptosis through JNK signalling in an NF-κB dependent human transformed follicular lymphoma cell line. Br J Haematol 2014;165(1):151–3.
[163] Lue H, et al. Macrophage migration inhibitory factor (MIF) promotes cell survival by activation of the Akt pathway and role for CSN5/JAB1 in the control of autocrine MIF activity. Oncogene 2007;26(35):5046–59.
[164] Kaufman JL, et al. Phase I, multicentre, dose-escalation trial of monotherapy with milatuzumab (humanized anti-CD74 monoclonal antibody) in relapsed or refractory multiple myeloma. Br J Haematol 2013;163(4):478–86.
[165] Kashyap MK, et al. Ulocuplumab (BMS-936564/MDX1338): a fully human anti-CXCR4 antibody induces cell death in chronic lymphocytic leukemia mediated through a reactive oxygen species-dependent pathway. Oncotarget 2016;7(3):2809–22.
[166] Ghobrial IM, et al. A phase Ib/II trial of the first-in-class anti-CXCR4 antibody ulocuplumab in combination with lenalidomide or bortezomib plus dexamethasone in relapsed multiple myeloma. Clin Cancer Res 2020;26(2):344–53.

CHAPTER

11

Fcγ receptors—Master regulators of antibody therapy

Stephen A. Beers[a] *and Björn Frendéus*[b]

[a]**Antibody and Vaccine Group, Centre for Cancer Immunology, University of Southampton Faculty of Medicine, Southampton, United Kingdom** [b]**BioInvent International AB, Lund, Sweden**

Abstract

Antibody therapeutics have been at the heart of transformative cancer treatment for more than 20 years: first with antibodies directly targeting the tumor and more recently with immune checkpoint blockade. Despite their impact and widespread utilization, antibody therapy mechanisms of action and factors governing response or resistance in patients are still poorly understood. One aspect that has emerged as important for all clinically developed antibodies is antibody Fc interactions with Fcγ receptors. Antibody Fc acts to connect the specificity of antibody to the power of the innate immune system and ensuing adaptive immunity. What has become clear is that in the context of both direct tumor-targeting and ICB mAbs, these interactions can be pivotal to therapeutic activity and survival. Through improved understanding and evolving strategies of Fc engineering, FcγR blockade, and pharmacological modulation of immune effector cell FcγR expression, we are at the dawn of harnessing the power of already clinically validated and new classes of antibody-based cancer immunotherapeutics. Current understanding including the impact and consequence of Fc:Fcγ receptor interactions in defining antibody efficacy and resistance is discussed here for developed antibody classes together with strategies to enhance these to increase patient responses and overcome resistance.

Abbreviations

Ab	antibody
Ag	antigen
ADCC	antibody-dependent cellular cytotoxicity
ADCP	antibody-dependent cellular phagocytosis
A:I	activatory to inhibitory
BCR	B-cell receptor
CDC	complement-dependent cytotoxicity
CNV	copy-number variation
CTLA-4	cytotoxic T-lymphocyte-associated protein 4
DLBCL	diffuse large B-cell lymphoma
EMA	European Medicines Agency

Immunotherapy in Resistant Cancer:
From the Lab Bench Work to Its Clinical Perspectives
https://doi.org/10.1016/B978-0-12-822028-3.00014-5

Fc	fragment crystallizable
FcR	Fc receptor
FcγR	Fcγ receptor
FDA	Food and Drug Administration
H	heavy chain
HER2	human epidermal growth factor receptor 2
ICB	immune checkpoint blocking
ISA	immune-stimulatory agonist
Ig	immunoglobulin
IgG	Gamma immunoglobulin
ITAM	immune tyrosine activation motif
ITIM	immune tyrosine inhibitory motif
mAb	monoclonal antibody
MCL	mantle cell lymphoma
PD-1	programmed cell death 1
PD-L1	programmed cell death ligand 1
PDX	patient-derived xenograft
SNP	single nucleotide polymorphism
STING	stimulator of interferon genes
TAM	tumor-associated macrophage
Tg	transgenic
TLR	toll-like receptor
TME	tumor microenvironment
TNFRSF	tumor necrosis factor receptor superfamily
Treg	regulatory T cell

Conflict of interest

BF is an employee of and holds stock in BioInvent (www.BioInvent.com), a company developing antibody-based cancer immunotherapeutics, including anti-FcγRIIb antibodies. S.A.B. has received institutional support from BioInvent for grants and patents.

Introduction

Monoclonal antibody (mAb) drug development for immuno-oncology represents one of the most active and dynamic areas of biomedical research spanning academia and industry. Indeed, since rituximab received FDA approval to treat non-Hodgkin's lymphoma in 1997, becoming the first mAb approved for use in malignant disease, more than 30 mAbs have been approved in a range of hematological and solid cancers transforming patient outcomes [1]. Although the number of approved mAbs is expanding rapidly, the number of successful mAb targets to treat cancer patients is more limited with success largely restricted to mAbs that directly target cancer cells for killing or those that target immune inhibitory receptors to release antitumor immunity (Fig. 1). Despite notable successes and patient impacts, all mAb treatments to date have been limited by resistance to treatment or relapse [2–5]. Investigating what characteristics and mechanistic requirements are needed for success in the clinic with different mAb targets and the mAbs targeting them are some of the most active areas of pursuit in the field. Understanding these requirements will expand therapeutic responses with existing mAbs and serve to inform new drug development, allowing more patients to be treated successfully.

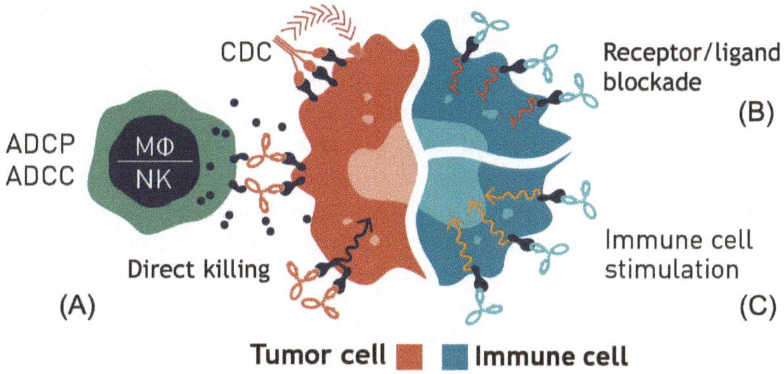

FIG. 1 Anticancer antibody classes. (A) Direct-targeting/cytotoxic mAbs bind to tumor-associated antigen and inducing; cell death or growth arrest, recruiting complement leading to complement-dependent cytotoxicity, or recruiting FcγR-expressing effector cells to induce antibody-dependent cellular cytotoxicity or phagocytosis. (B) Immune checkpoint blocking mAbs bind to inhibitory receptors on immune cells and stop inhibitory signaling to "release the brakes" on antitumor immunity. (C) Immune-stimulatory agonist mAbs bind to costimulatory receptors and induce activatory signaling to "step on the gas" and evoke antitumor immunity.

Cancer-cell surface receptors

Monoclonal antibodies targeting cell surface receptors dominate the field of immuno-oncology and can broadly be divided into three classes according to their respective mechanisms of action (Fig. 1): firstly, as exemplified by anti-CD20 are mAbs that directly target cancer cells to induce cell death or growth arrest and importantly tag them for clearance through innate effector mechanisms [6]; secondly, immune checkpoint blocking (ICB) mAbs, as represented by anti-PD-1, that block immune inhibitory signaling to "release the brakes" on antitumor immune responses [7]; and finally, the immune-stimulatory agonists (ISAs), which are less well-developed clinically and are represented by anti-CD40, that activate immune costimulatory receptors to "step on the gas" and provoke antitumor immune responses [8]. In keeping with their different target receptors and their respective cellular expression, these mAbs have been demonstrated, in preclinical models with supporting clinical correlation studies, to require different mAb characteristics. The study and understanding of what requirements are necessary for optimal clinical performance of these different classes of mAbs drive a great deal of the research into mAb development.

Despite the marked differences in the mechanisms of action of these classes of mAb drugs, a common finding of preclinical studies and available clinical data is the importance played by the fragment crystallizable (Fc), part of the antibody constant region, on function and efficacy [9–12]. Natural antibodies (Abs), also termed immunoglobulins (Ig), are produced by B cells as part of antipathogen responses. This arsenal of specialized soluble proteins serves to link the specificity of humoral adaptive immunity to the power of innate immune effector function to clear pathogens. Antibodies constitute variable bivalent F(ab) regions that bind to target antigen (Ag) and a constant Fc region that links to appropriate immune effector function. B cells are able to produce a range of Ig isotypes—IgA, IgD, IgE, IgG, and IgM—with each specialized to recruit and engage the most suitable innate effector functions to neutralize and induce clearance of different forms of pathogens.

In the context of cancer therapy, direct-targeting mAbs, which are analogous to those antibodies targeting pathogens for opsonization or the induction of soluble and cellular cytotoxicity, require inflammatory effector function recruitment to elicit beneficial therapeutic responses [13]. For mAbs that are used to block inhibitory immune receptors expressed on immune cells, then a lack of Fc-mediated immune recruitment is frequently optimal for therapy [10], and finally, for immune-stimulatory mAbs, an Fc that is able to mediate target receptor cross-linking and clustering to induce target cell signaling has been shown to be required [12]. So, although the Fc is commonly critical to effective therapeutic effects in preclinical models and therefore likely clinical effects in patients, the requirements for this Fc vary dramatically.

The Fc portion has also been shown to directly impact on mAb toxicity, so an understanding of the properties and the immune effectors engaged to elicit therapy or toxicity will be critical to harnessing the full power of current mAb drugs and to inform the rational design and combination of future mAbs and approaches [14,15]. Antibody Fcs are capable of interacting with a range of immune mediators from complement to Fc receptors (Fig. 1), and the relative roles of these immune mediators to mAb activity are still the subject of debate and conjecture. However, what is clear from preclinical models and more limited clinical studies with therapeutic mAbs is that Fc interactions with Fc receptors (FcRs) can critically affect mAb efficacy. These FcR interactions can, by their nature, be either positive or negative and, due to the complexity of tumor immune environment, will represent a combination of these effects with the net outcome the patient responses observed. Given the clear role of FcRs in dictating mAb responses across preclinical models and from recent clinical studies, the consideration of their role in mAb therapy efficacy and resistance will constitute the focus of the rest of this review.

FcR family

Fc receptors represent a broad family of largely type I membrane receptors belonging to the immunoglobulin gene superfamily which have evolved to bind specific Ab Fc classes and isotypes. The binding of Ab Fcs to their cognate receptors enables the recruitment of appropriate, mainly innate, immune effector function as part of coordinated innate and adaptive antipathogen responses.

There are five major Fc receptor classes: FcαRs that are expressed largely on myeloid cells and bind IgA that is secreted at high levels at mucosal surfaces and systemically [16]; FcδRs that bind IgD and are yet to be cloned but have been shown to be expressed on basophils and other granulocytes [17], some fibroblast-like synoviocytes [18], and a small percentage of T cells [19]; FcεRs that are expressed on myeloid cells and bind IgE found predominantly in tissues [20]; FcγRs that are the most diverse having multiple classical and non-classical members, are expressed on all leukocytes and some non-hematopoietic cells, and bind to IgG, the most prevalent Ig isotype in circulation [21]; and finally, FcμRs that were discovered relatively recently and are expressed on B, T, and NK cells and bind to IgM—however, their role is still poorly defined [22]. Given that all therapeutic mAbs developed to date for anticancer use are of IgG isotype, the remainder of this review will focus on FcγRs and their role in mAb efficacy and resistance mechanisms.

Fcγ receptors

Effective immune responses require controlled activation and inhibition to elicit protective immunity without excessive tissue damage or the induction of autoimmunity. Reflective of this and common to several immune receptor families, the classical FcγRs have evolved to have activatory and inhibitory members that induce proinflammatory and suppressive effects, respectively, in response to engagement by Abs/Ig. FcγRs have wide cellular expression profiles, among FcRs second only to the neonatal FcR (FcRn). FcRn is a non-classical FcγR, which is widely expressed on stromal cells as well as leukocytes and has a ubiquitous role in Ab recycling and transcytosis of IgG from the blood to the tissues [23]. Other non-classical receptors binding IgG include the intracellular tripartite motif-containing protein 21 (TRIM21) [24] involved in antiviral responses and the Fc receptor-like family [25], but these have not been demonstrated to impact on the efficacy of mAb therapeutics in preclinical or clinical studies to date.

The wide cellular expression of the classical FcγRs, with evidence supporting the presence of members in all leukocytes and a range of other non-hematopoietic cell types involved in immune cell and pathogen interactions at the interface of circulatory fluids and tissues, means that to harness the full potential of these receptors therapeutically requires detailed understanding of FcγR biology.

In humans, the classical FcγR family consists of six members: five activatory FcγRs: FcγRI (CD64), FcγRIIa (CD32a), FcγRIIc (CD32c), FcγRIIIa (CD16a), and FcγRIIIb (CD16b); and one inhibitory FcγR: FcγRIIb (CD32b). The activatory receptors (with the notable exception of FcγRIIIb, which is unique among FcR being GPI linked and without signaling capacity) signal via an immune tyrosine activation motif (ITAM) contained either within their alpha chain or through association with the FcR gamma chain. The single inhibitory member of the family, FcγRIIb, contains an immune tyrosine inhibitory motif (ITIM) within its cytoplasmic domain and acts to counteract the effects of IgG-induced ITAM signaling, the latter triggered through engagement of activatory FcγRs or the B-cell receptor [26]. A further layer of complexity is brought to the biology of FcγR family through variations in the forms of a range of single nucleotide polymorphisms (SNPs), copy-number variations (CNVs), and splice variants present in several members that alter IgG affinity, receptor clustering, trafficking, and expression levels (Fig. 2) [27].

FcγRI is the sole high-affinity FcγR with a kD in the nanomolar range and is capable of binding monomeric IgG [28]. Its high affinity for IgG might be expected to mean that this receptor would be continually occupied given normal circulating Ab levels and therefore of little consequence for mAb therapy considerations. However, FcγRI is quite poorly expressed under resting conditions only becoming upregulated in response to proinflammatory signals [29–31]. Under these conditions, immune complexes are able to displace monomeric IgG enabling this receptor to mediate effector function. FcγRI can be induced on monocytes, macrophages, neutrophils, dendritic cells, and mast cells facilitating their role in early immune responses and the efficacy of mAb therapeutics targeting cells for innate destruction.

The remaining activatory and inhibitory FcγRs are of low affinity with kD in the micromolar ranges and bind multimeric arrayed Fcs as soluble or membrane receptor-associated immune complexes [28]. The formation of immune complexes increases the avidity of interactions between Fc and FcγR sufficiently to allow productive binding and cellular

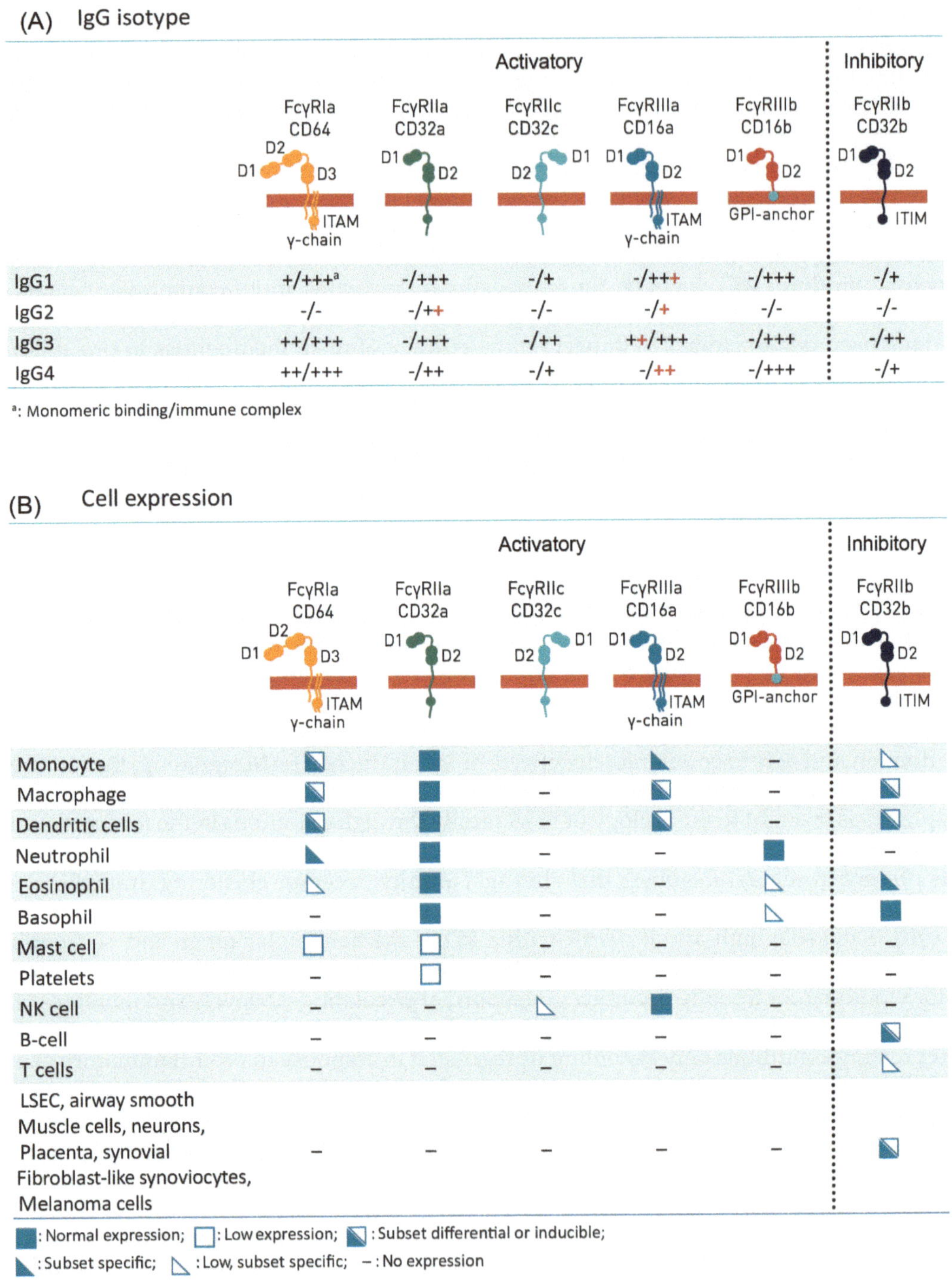

FIG. 2 Fcγ receptor IgG isotype binding and cellular expression. (A) IgG isotype binding to classical FcγRs. Monomeric/immune complex binding. – indicates no binding; + indicates binding, with the number indicating relative strength of binding; + indicates SNP-dependent binding. (B) FcγR expression on major hematopoietic and non-hematopoietic cell populations.

responses. The requirement for multimerization of both Fc and FcγR presents a potential for variation in constituent partners and increases the complexity of these interactions and the understanding of their impact on therapeutic mAb efficacy and resistance.

FcγRIIa has the broadest expression of the FcγR but is found predominantly on phagocytic cells and is involved in the processes of ADCP, immune complex clearance, and antigen presentation. The role of FcγRIIa in the clinical efficacy of several direct-targeting mAbs has been inferred from SNP association studies where the presence of a SNP at position 131 that impacts affinity to IgG correlates with responses [32–34].

FcγRIIc is an enigmatic receptor that, due to a SNP that encodes for either an early stop codon or a functioning protein, is reported to be restricted to low-level expression in NK and B cells [35] in approximately 20% of individuals. This receptor is the product of gene duplication and recombination which resulted in a receptor with the extracellular domains of FcγRIIb fused to the ITAM-containing intracellular tail of FcγRIIa [36]. Whether it is really expressed, on what cells it is expressed, and how it is relevant to mAb therapy are still the subject of debate.

FcγRIIIa is expressed on NK cells and mononuclear phagocytes and is involved in mediating ADCC and ADCP. Clinical association studies have shown a correlation with SNPs at position 158 of the receptor which alter affinity for IgG Fc and response to a number of mAbs [37,38]. Although many of the studies reporting these effects have small patient numbers and some are contradictory, these findings have helped drive numerous Fc engineering strategies to enhance Fc:FcγR interactions with the aim to boost effector capacity of mAb therapeutics.

FcγRIIIb is restricted to neutrophils and has variously been categorized as an activatory or decoy receptor, and its function remains debated, although a role in modulating the effects of FcγRIIa through neutrophil degranulation and reaction oxygen species production has been demonstrated [39].

FcγRIIb is the sole inhibitory FcγR and has two main splice variants; FcγRIIb1, which is found on B cells where it associates with the BCR signaling complex and acts to counteract signaling from antigen-antibody complexes bound to the BCR; and FcγRIIb2, which is found on innate effector cells including phagocytes and granulocytes and acts to counteract signaling through activatory FcγRs [40]. FcγRIIb2 is also found on non-hematopoietic cells such as liver sinusoidal endothelial cells, airway smooth muscle cells, neurons, placenta, synovial fibroblast-like synoviocytes, and melanoma tumor cells where, due to its ability to rapidly internalize, it can act as a scavenger receptor of IgG immune complexes [41].

FcγRs have been variously reported on T cells, but the data are conflicting and any function is difficult to elucidate due to a paucity of highly specific reagents to detect the closely related low-affinity FcγR. Recent developments and the validation of panels of mAbs highly specific to individual FcγRs should enable this to be clarified [42].

Nimmerjahn and Ravetch were the first to describe the activatory-to-inhibitory (A:I) engagement ratio of mAbs and thereby provided a clear rationale for why some mAb isotypes were more effective for direct tumor targeting compared with others [13]. These findings have been recapitulated for human IgG and their receptors [43], and these A:I requirements underlie isotype selection of tumor-targeting mAbs. FcγR expression, particularly on myeloid cells that express both activatory and inhibitory FcγRs, adds another layer of complexity to this isotype-dependent potency of such mAbs as their differential cellular expression and the absence or presence of such effector cells in therapeutically relevant compartments, e.g., within tumors and in tumor-draining lymph nodes, will impact on the efficacy of effector function. This effect is mediated through two means: the different affinities these receptors possess for various IgG isotypes (Fig. 2) and the fact that the levels of these receptors are exquisitely sensitive to the activation status of the myeloid cells themselves. These factors together impact on relative engagement and the expression of the activatory versus the inhibitory FcγR and lead to cellular effectors with variable activatory-to-inhibitory receptor expression profiles. The host immune FcγR-expressing cell landscape can therefore impact the physiological A:I ratio of mAbs through receptor availability and has been found to underlie local resistance in tumors and associated niches to mAb therapy in vivo, and is likely to significantly contribute to FcγR-mediated resistance to mAb therapeutics clinically.

Anticancer mAb mechanisms of action and resistance

Direct-targeting/cytotoxic mAbs

Direct-targeting mAbs, also termed cytotoxic mAbs, bind to target cell surface receptors and mark the cell for innate immune destruction. There are three main mechanisms through which these immune effects may be mediated, and all are dependent upon mAb Fc. The first of these is complement-dependent cytotoxicity (CDC) mediated through activation of the classical or alternative complement pathways, a process dependent upon a series of serum proteins and cell surface receptors. The second is antibody-dependent cellular cytotoxicity (ADCC) largely mediated by natural killer cells and FcγRIIIa. The final one is antibody-dependent cellular phagocytosis (ADCP), a process that is mediated by a range of innate immune cells, primarily myeloid cells of the monocyte/macrophage lineage, and is capable of being carried out by all ITAM-containing FcγRs and inhibited by engagement with the inhibitory FcγRIIb. This latter observation makes the process of ADCP dependent upon both the A:I FcγR engagement ratio of the mAb Fc in question and the A:I FcγR expression of the effector cells themselves. These factors make this process particularly susceptible to inhibition and consequently contribution to resistance mechanisms in the context of suppressive tumor microenvironments, but also as a consequence, an attractive target mechanism for manipulation to enhance therapeutic activity. Although the role for FcγRs in the activity of direct-targeting/cytotoxic mAbs has been established for some time [9], the impact of differential FcγR availability on mAb activity in tumor settings is still relatively poorly understood, particularly in the context of human disease.

Rituximab is arguably the best-studied antibody used in cancer treatment, providing a prime example of the complex mechanisms that govern efficacy and resistance to direct tumor-cell-targeting antibodies. Among multiple mechanisms proposed to contribute to and underlie rituximab therapeutic activity, including the above-mentioned induction of apoptosis and triggering of complement-dependent cell-mediated cytotoxicity [6,44], the strongest preclinical and clinical evidence points to FcγR-dependent mechanisms [45–48].

In accordance with the activating FcγRs acting in concert to provide immune effector cell activation, and controlling the cytotoxic activity of cell-depleting antibodies such as rituximab, polymorphic variants with higher affinity for antibody constant domain induce stronger immune effector cell activation and accordingly are associated with survival advantage compared to lower affinity allelic variants. Thus, independent retrospective studies have demonstrated a correlation between one or more activating FcγRs and rituximab efficacy in different types of lymphomas. Patients homozygous for high-affinity allelic variants of FcγRIIIa or FcγRIIa show improved responses and survival in response to rituximab therapy compared to patients carrying one or more lower affinity alleles [32,37].

In keeping with FcγRs controlling activity of antibody (drugs) as a class, similar findings have been made for other direct-targeting antibodies in additional cancers. Breast cancer and colorectal cancer patients expressing high-affinity allelic variants of activating FcγRs show improved survival when treated with herceptin (anti-HER2) and cetuximab (anti-EGFR), respectively [49,50], compared to those expressing lower affinity activating FcγRs.

In summary, therefore, increasing binding to activating FcγRs is associated with stronger antibody-mediated target cell depletion, and accordingly improved therapeutic activity and survival in response to direct cancer-cell-targeting mAbs.

Analogously, and in keeping with the A:I theorem, i.e., that it is the ratio of engaging activating FcγRs relative to inhibitory FcγRs which determines antibody-mediated target cell depletion, one might predict that cancer patients expressing higher affinity variants of the inhibitory FcγRIIb would show impaired survival to antibody-based therapy, compared to those expressing lower affinity variants. While such clinical data remain to be generated, genetic deletion of the sole inhibitory antibody checkpoint FcγRIIb in preclinical models has been shown to enhance in vivo therapeutic activity of direct cancer-cell-targeting antibodies, including those specific for CD20, HER2, and EGFR, i.e., clinically validated targets in therapy of hematologic malignancy as well as solid cancer [9].

Interestingly, and in further support of FcγRIIb being a tractable target in resistance to cancer immunotherapy, recent data have demonstrated that inhibitory FcγRIIb limits therapeutic antibody efficacy and promotes antibody drug resistance by additional mechanisms, distinct from inhibitory signaling in immune effector cells, when expressed on tumor B cells [51]. Lim et al. found that FcγRIIb expressed on tumor B cells promoted internalization of rituximab antibody molecules from the tumor B-cell surface [52]. Rituximab internalization, which correlated with tumor FcγRIIb expression levels across several different lymphoma subtypes studied, increased antibody consumption in vivo and left fewer rituximab molecules to engage critical FcγR-dependent effector cell-mediated antitumor activity, e.g., ADCP [53]. Highest and most homogenous expression of FcγRIIb is observed in chronic lymphocytic leukemia (CLL), mantle cell lymphoma (MCL), and marginal zone lymphoma, although a fraction of follicular lymphoma (FL) and diffuse large B-cell lymphoma shows exceptionally high FcγRIIb expression [52,54]. Further, consistent with tumor B-cell-expressed FcγRIIb limiting antibody therapeutic efficacy and promoting antibody resistance, retrospective clinical studies of MCL and FL patients treated with rituximab-containing therapy showed decreased survival of patients with higher FcγRIIb expression on tumor cells [52,55]. Tumor cell-expressed FcγRIIb appears to be a general mechanism limiting antibody therapeutic efficacy and promoting antibody drug resistance in the tumor microenvironment. Using a humanized model of treatment refractory B-cell leukemia, and the CD52-specific antibody alemtuzumab, Pallasch et al. found that FcγRIIb is highly overexpressed on leukemic tumor cells in such antibody drug-resistant tumor microenvironments and that shRNA-mediated knockdown of tumor cell FcγRIIb restored responsiveness to therapeutic antibody, resulting in animal cure [56]. Finally, high expression of FcγRIIb in B-cell malignancy may indicate that immunocompetent antibodies to FcγRIIb could have single-agent therapeutic activity in this setting [51,57].

As discussed below, these and additional observations collectively provided the rationale to develop antagonistic anti-FcγRIIb antibodies that block FcγRIIb-mediated antibody internalization for combination immunotherapy of B-cell cancer with direct-targeting antibodies, e.g., rituximab [51,58].

Immune checkpoint blockers

Immune checkpoint blocking mAbs have made a tremendous impact on a variety of previously hard-to-treat cancers including metastatic melanoma and lung cancer. Indeed, this class of mAbs can largely be credited with confirming to many that the immune system can be harnessed with therapeutic mAbs to eradicate cancers [59]. In this respect, they have therefore served to reinvigorate drug development efforts to find more ways to treat patients using mAbs. ICB mAbs were generated against critical immune inhibitory receptors, largely involved in limiting T-cell-mediated immunity subsequent to T-cell receptor-dependent immune recognition and activation. These inhibitory receptors thereby act to limit antitumor immune responses, and their blockade can be said to "release the brakes" on antitumor immunity. In this respect, ICB mAbs have been viewed through the prism of blocking mAbs that have the simplest mAb requirements: strong target binding, long half-life, and no effector function. This has led to a range of ICB mAbs with IgG formats where little concern about Fc and effector engagement was considered or to mAb development featuring isotypes or engineered Fc with low immune effector engagement characteristics. Recent preclinical and clinical correlation studies indicate that the "blocking of the brakes" analogy is oversimplified and that therapeutic activity of ICB mAbs—much like direct-targeting mAbs—can be regulated by activating and inhibitory FcγRs.

Intriguingly, despite CTLA-4 and the PD-1/PD-L1 axes acting in common to counteract overt immune activation, mAbs to these immune checkpoints appear to be regulated quite differently by FcγRs (Figs. 3 and 4). As discussed in detail below, antibody Fc:FcγR interactions may promote or compromise therapeutic activity and cancer resistance to different immune checkpoint blocking antibodies. Consistent with the molecular and cellular mechanism of FcγRs being to sense, and respond by elimination, antibody-coated target receptor-expressing cells, how ICB mAbs to CTLA-4, PD-1, or PD-L1 are modulated by FcγRs appears to be determined by these targets' expression levels on antitumor immune effector cells, e.g., CD8+ T cells versus immune suppressor cells Tregs and TAMs. As discussed below, this and how strongly individual ICB mAbs bind activating compared with inhibitory FcγRs, i.e., what isotype they are, determine whether ICB mAb engagement of FcγRs will enhance or compromise therapy efficacy. Of critical relevance to cancer patients, these factors are also likely to determine if, how, and to what extent resistance may develop.

CTLA-4: "Releasing the brakes and eliminating the culprit"

In pioneering studies, James Allison established the scientific rationale to harness the potential of immune checkpoint blockade to target the immune system, rather than tumor cells directly, with antibodies to combat cancer [60]. Following successful translation to the clinic, the CTLA-4-specific antibody ipilimumab in 2011 became the first ICB mAb to be approved for cancer immunotherapy. Twenty-two years later, Arce Vargas et al., in a landmark paper for the first time in human subjects, demonstrated a link between FcγRs and clinical response to ICB mAb therapy [11]. Similar to what had been observed for rituximab in the context of the tumor B-cell-depleting mAb rituximab, melanoma patients carrying a high-affinity allele of the activating FcγRIIIa (V158) showed improved survival in response to treatment with ipilimumab compared to patients carrying a lower affinity FcγRIIIa (F158) allele. While a role for FcγRs underlying efficacy of ICB mAbs had not been considered in the development of ipilimumab, Arce Vargas's findings to many were not unexpected. Following approval of ipilimumab, anti-CTLA-4 therapy in the mouse had been shown to critically depend on FcγR-mediated depletion of intratumoral regulatory T cells [61–64], which express CTLA-4 at higher levels compared with effector T cells in the tumor microenvironment [11]. Consistent with coordinate regulation of anti-CTLA-4 therapeutic efficacy by activating and inhibitory FcγRs according to the A:I ratio theorem (Fig. 3), in FcγR-humanized mice, mAb variants engineered for enhanced binding to activating FcγRs showed enhanced therapeutic activity [11]. As expected, and conversely, mAb variants with diminished binding to activating FcγRs failed to induce protective immunity against cancer.

Despite its life-changing potential for cancer patients—when used alone or in combination with PD-1 ICB mAb—ipilimumab use is limited by tolerability issues, which are frequent and can be severe [65–67]. While ipilimumab efficacy and side effects appear to go hand in hand and are both dose-dependent, recent observations indicate that anti-CTLA-4 mAb efficacy and tolerability potentially can be separated. The anti-CTLA-4 mAb tremelimumab (IgG2) is a similarly potent blocker of CTLA-4:B7 ligand interactions as ipilimumab (IgG1) and has therapeutically relevant pharmacokinetics, but, unlike ipilimumab, is a poor engager of FcγR-dependent Treg depletion owing to its human IgG2 isotype. In this respect, it is noteworthy that despite similar and significant clinical development efforts, tremelimumab, unlike ipilimumab, has not shown efficacy in clinical trials and has yet to be approved for use in cancer immunotherapy. Moreover, despite different apparent efficacy profiles, both mAbs are associated with similar tolerability issues and side-effect profiles [68,69]. In light of these observations, it is tempting to speculate that CTLA-4:B7 blockade and FcγR-dependent Treg depletion mechanisms representing "release of the brakes" and "elimination of culprit immune cells," respectively, may differently contribute to anti-CTLA-4 efficacy and side effects. If so, fine-tailoring these mechanisms may prove clinically and therapeutically important. A caveat is that ipilimumab and tremelimumab antibodies have not been trialed head-to-head in randomized controlled studies, so apparent differences could well be attributed to uncontrolled factors. Further studies are needed to indicate the potential therapeutic relevance of Fc:FcγR-modulating strategies to improve on anti-CTLA-4.

Anti-PD-1: "Releasing the brakes—but take care to spare the effectors"

While links between FcγR polymorphisms and survival to anti-PD-1/PD-L1 mAb-treated cancer patients remain to be established, preclinical studies indicate that therapeutic efficacy and cancer resistance to antibodies specific for both targets may be regulated by FcγRs. Interestingly, however, these data indicate differential FcγR regulation for anti-PD-1 and anti-PD-L1 antibodies. Dahan et al. reported that anti-PD-L1 antibodies' therapeutic efficacy, similar to what had been observed for anti-CTLA-4, was enhanced with antibody isotypes that preferentially engage activating over inhibitory FcγRs [10]. Conversely, anti-PD-1 antibody variants that did not engage FcγRs showed greatest therapeutic activity, and FcγR-engaging mAb activity decreased with increasing A:I ratios. Similarly, Pittet and coworkers found that the in vitro and in vivo efficacy of clinically approved anti-PD-1 antibodies nivolumab and pembrolizumab, and a murine surrogate antibody variant with claimed similar FcγR engagement in the mouse to that of the human mAbs in human cancer patients, were compromised by FcγR engagement [70]. Deglycosylation of antibodies with EndoS rendering them incapable of engaging FcγRs, or antibody-mediated FcγR blockade, significantly improved anti-PD-1 antibody therapeutic activity. This demonstrates that FcγRs can negatively regulate anti-PD-1 antibody efficacy.

So, how can differential FcγR regulation of mAbs to two immune checkpoints belonging to the same axis (PD-1/PL-1) be explained and reconciled with a principal function of FcγRs such as depletion of antibody-coated target cells? As for rituximab (Fig. 5) and ipilimumab (Fig. 3), the answer appears to lie in how PD-1 (Fig. 4) and PD-L1 are expressed on different cell subpopulations. While the targets for rituximab (CD20) and ipilimumab (CTLA-4) are most highly expressed on B cells and intratumoral Tregs, respectively, PD-1 expression is highly expressed on both intratumoral CD8+ T cells and Tregs. From this and the molecular and cellular principles of FcγR function, we would expect antibodies to PD-1 to be capable of

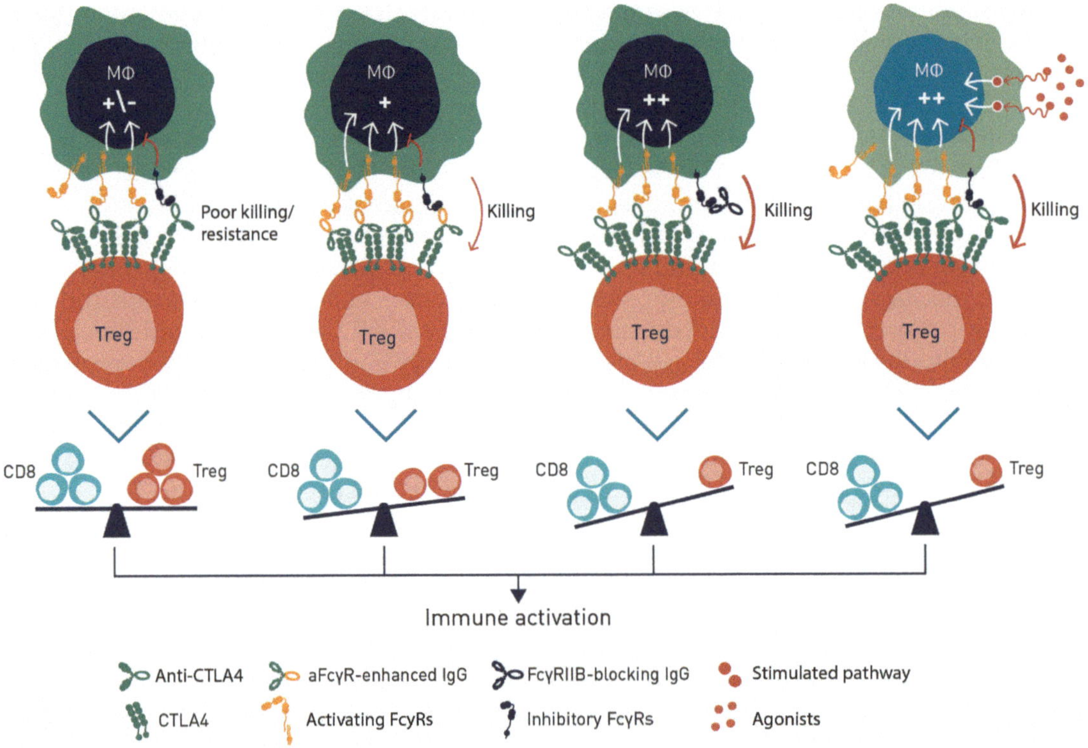

FIG. 3 Fcγ receptor-dependent resistance to anti-CTLA-4 antibodies. *First/left panel*: Anti-CTLA-4 immune checkpoint blocking mAbs, such as anti-HER2 tumor-direct-targeting mAbs, show limited FcγR-dependent killing of Tregs in intratumoral compartments owing to upregulated inhibitory FcγRIIB and balanced activatory FcγR expression on myeloid effector cells. *Second panel*: Fc-engineered anti-CTLA-4 mAb; *third panel*: FcγRIIB blockade; and *fourth/right panel*: pharmacological modulation (e.g., by TLR or STING agonists) inducing altered FcγR expression and improved A:I ratios; all result in improved A:I ratios and enhanced anti-CTLA-4 antibody, myeloid effector cell-dependent, killing of Tregs, and improved antitumor immunity.

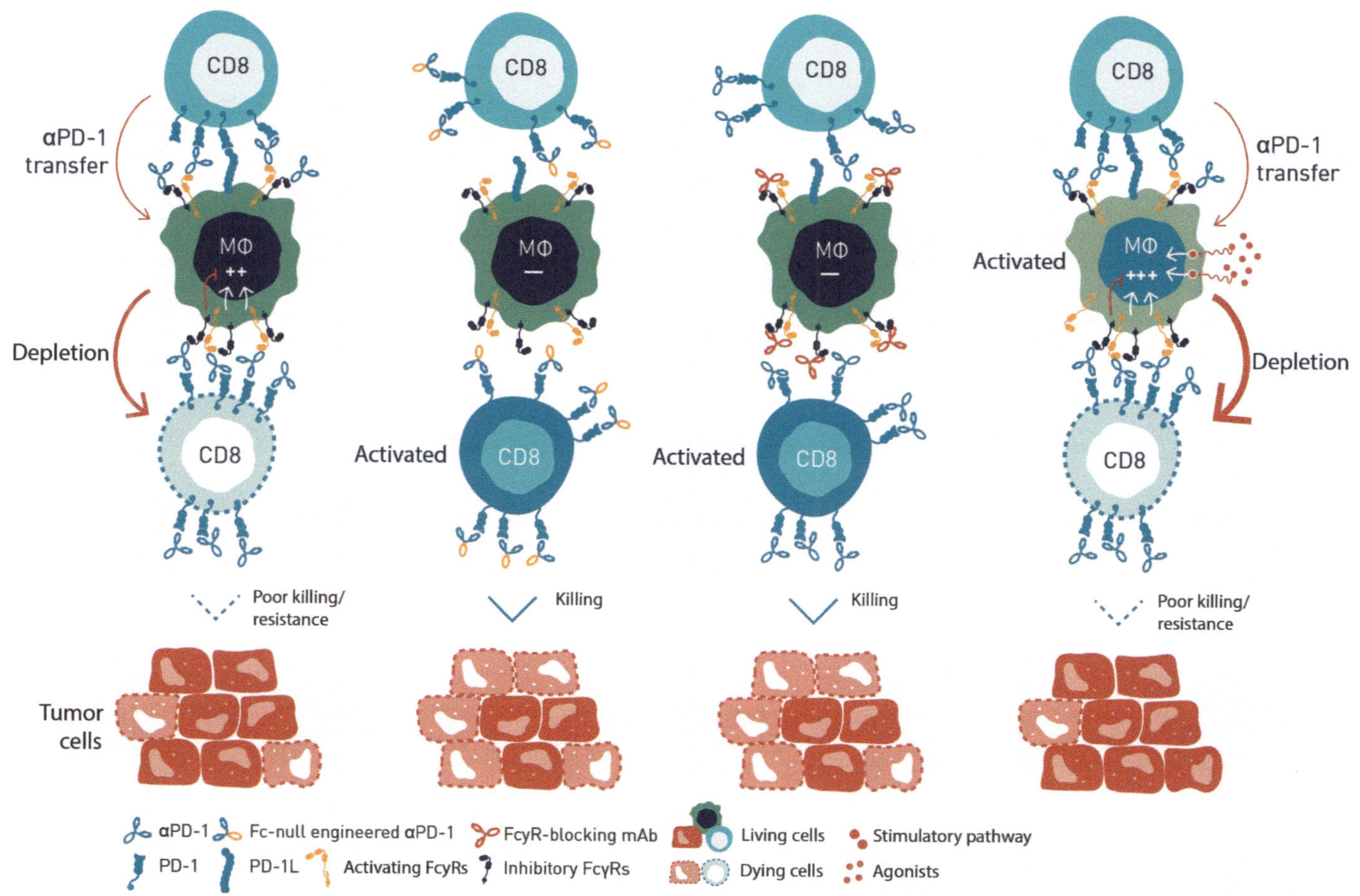

FIG. 4 Fcγ receptor-dependent resistance to anti-PD-1 antibodies. *First/left panel*: The efficacy of anti-PD-1 immune checkpoint blocking mAbs, unlike anti-CTLA-4 (Fig. 3) and anti-HER2 mAbs (Fig. 5), is compromised by FcγR binding and engagement. FcγRs promote resistance to anti-PD-1 mAb by at least two mechanisms, both acting on CD8+ T cells expressing PD-1 at high density. *Left panel*: FcγRs promote anti-PD-1 resistance by transfer of anti-PD-1 antibody molecules from anti-PD-1-coated CD8+ T cells to FcγR-expressing macrophages, reducing PD-1/PD-L1 blockade and reducing relief of T-cell exhaustion. Further, FcγRs promote myeloid effector cell-mediated killing of densely anti-PD-1-coated PD-1 high-expressing antitumor CD8+ T cells. *Second panel*: Fc-engineered FcγR-null anti-PD-1 mAbs are left coated on PD-1-expressing CD8+ T cells and are spared from FcγR-expressing myeloid cell killing, resulting in full "release of the brakes" on antitumor immunity. *Third panel*: FcγR blockade, similar to the second panel, results in full "release of the brakes" on antitumor immunity. *Fourth/right panel*: Pharmacological modulation (e.g., by TLR or STING agonists) resulting in improved A:I ratios and enhanced FcγR-dependent antibody killing, compromises anti-PD-1 antibody therapeutic activity by promoting trogocytosis and killing of PD-1-expressing antitumor CD8+ T cells.

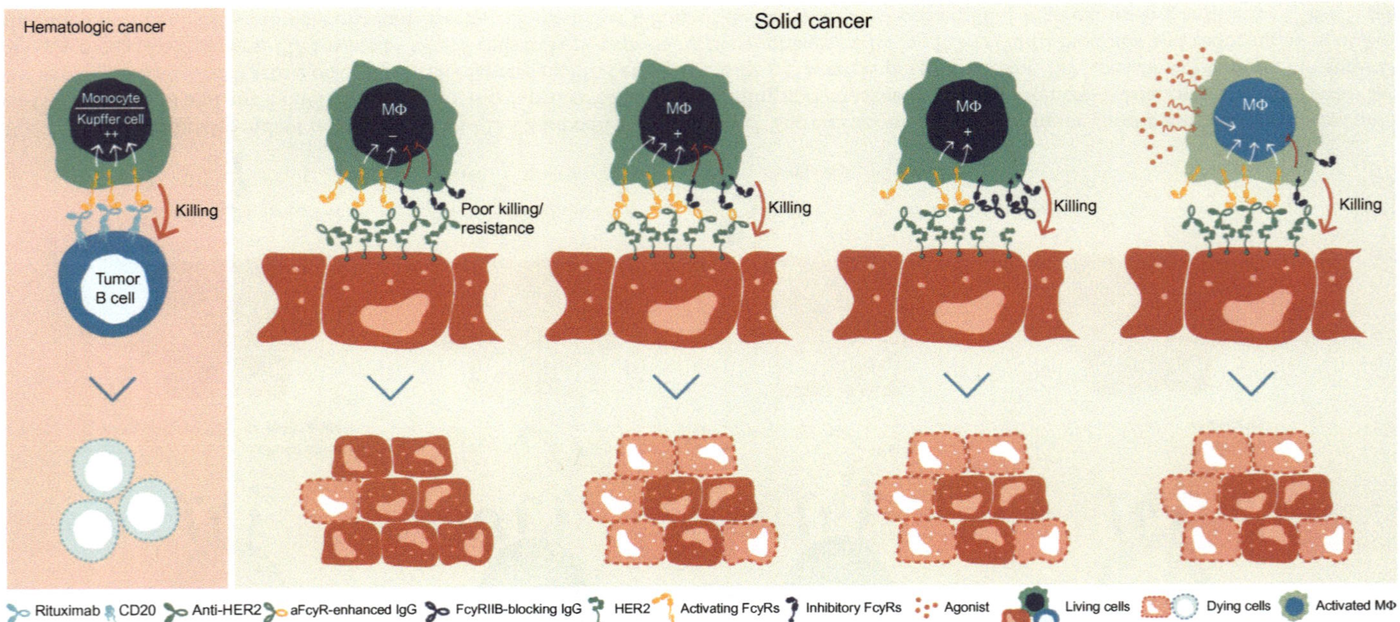

FIG. 5 Fcγ receptor-dependent resistance to direct tumor-targeting antibodies. *Left panel* (rituximab killing of leukemic tumor cell in blood compartment): Leukemic tumor cells are efficaciously killed by peripheral blood NK cells, monocytes, and liver Kupffer cells, which express activatory FcγRs only and no inhibitory FcγRs. *Second panel*: Trastuzumab anti-HER2 IgG1 antibody has limited FcγR-dependent killing capacity of epithelial breast cancer cells owing to upregulated inhibitory FcγRIIB expression and balanced activatory FcγR expression on intratumoral myeloid cells and limited infiltration of NK cell effectors. *Third/center panel*: Fc-engineered anti-HER2 antibody shows enhanced killing of breast cancer cells owing to its stronger binding and engagement of activatory compared with inhibitory FcγRs and stronger activation by FcγRIIIa-expressing effectors. *Fourth panel*: Antibody efficacy is enhanced by "release of the antibody brake" on multiple intratumoral myeloid effector cells, unleashing killing of antibody-coated breast cancer cells. *Fifth/right panel*: Pharmacological modulation (e.g., by TLR or STING agonists) induces altered FcγR expression and improved A:I ratios, resulting in enhanced killing by anti-HER2 antibody following engagement of greater numbers of activating compared with inhibitory FcγRIIB. Anti-HER2 antibody HER2-signal-blocking effects, which are known to contribute to antitumor activity, are not illustrated for the sake of clarity. Blockade of HER2 signaling is known to contribute to anti-HER2 antibody antitumor activity, alongside FcγR-dependent ADCC and ADCP, albeit in an antibody Fv-dependent manner (at similar magnitude in all strategies illustrated in panels 2–5).

depleting not only culprit Tregs but equally efficaciously CD8+ T cells—one of the strongest predictors of good clinical outcome across cancer types [71–73]. Accordingly, Dahan and coworkers found reduced numbers of CD8+ T cells following treatment with anti-PD-1 variants capable of binding FcγRs (mouse IgG2a isotype) compared with anti-PD-1 antibody variants engineered for impaired FcγR binding (N297A) [10]. It has since been reported that FcγRs may decrease efficacy and promote resistance to anti-PD-1 mAb by additional mechanisms. Pittet and coworkers demonstrated that FcγR-expressing macrophages could transfer anti-PD-1 mAb from coated T cells—a process commonly referred to as trogocytosis [70]. Trogocytosis occurred in vivo in the same mouse tumor models where FcγR engagement was shown to compromise anti-PD-1 mAb therapeutic efficacy, and in vitro in cocultures of human T cells and macrophages with clinically approved nivolumab and pembrolizumab anti-PD-1 mAbs. Additional mechanisms may contribute to a detrimental role for FcγRs in anti-PD-1 antibodies' efficacy. A recent report by Lo Russo et al. indicated that FcγR-mediated cross-linking of T-cell-bound anti-PD-1 mAb may result in agonism of inhibitory signaling in PD-1-expressing T cells [74]. However, the therapeutic relevance of the latter is pending translation to therapeutically relevant immune-competent settings, e.g., syngeneic mouse tumor models.

In summary, although a clinical (inverse) correlation between FcγR engagement and cancer patient survival in response to anti-PD-1 treatment remains to be established, preclinical observations by independent groups indicate that FcγR engagement compromises anti-PD-1 therapeutic activity through potentially several different resistance mechanisms (Fig. 4).

The limited data that have been presented on a role for FcγRs in anti-PD-L1 therapy suggest that for mAb to PD-L1, which may be variably expressed on cancer cells and both antitumor and protumor types of myeloid cells, the effects will be dependent on the tumor immune landscape and relative and absolute PD-L1 expression on these different types of cells.

Immune-stimulatory agonists

This class of mAb therapeutics has thus far failed to translate to clinical utility and success in the way both direct targeting and ICB have achieved. Recent studies investigating tumor necrosis factor receptor superfamily (TNFRSF) member-targeting ISA mAbs have elucidated a potential reason for some of these clinical failures and have served to highlight once more the critical role that effective FcγR engagement plays in the efficacy or resistance to mAb therapy [12]. Arguably, ISAs have demonstrated the greatest potential of the three mAb classes in preclinical models, frequently being able to induce effective and durable antitumor immunity in large established tumors where direct-targeting mAbs and ICB mAbs prove ineffective. However, when initial attempts were made to develop these mAbs for clinical use, they failed to induce durable responses in patients with their activity limited by two factors: a lack of agonistic activity or, where agonism was evident, dose-limiting toxicities. Importantly, these early attempts to translate ISAs were carried out before the Fc requirement for this mAb class was fully understood. The lack of efficacy observed to date is likely, in part, a reflection of this. While ISA mAbs were failing in clinical trial studies, a striking difference in requirements for ISA agonistic function compared to other mAbs was established in mouse models [75,76]. ISA mAbs were found to be dependent upon the inhibitory FcγRIIb for their agonism. In this context, engagement of FcγRIIb was demonstrated to be optimal for agonism and the activatory

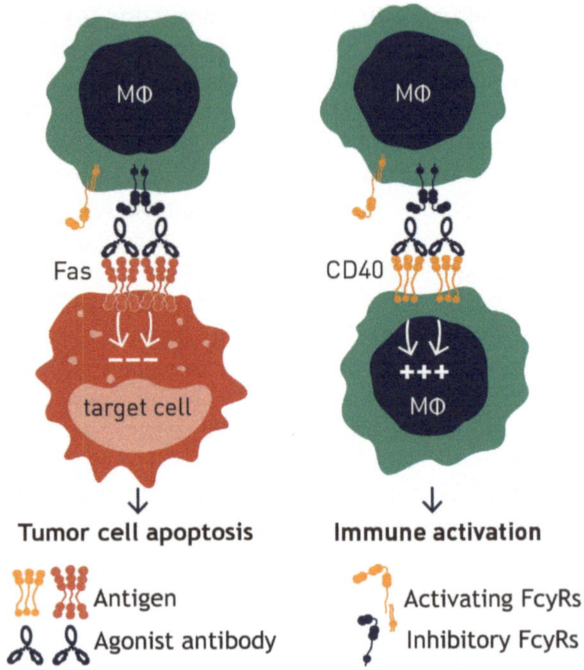

FIG. 6 Fcγ receptor dependence of immune-stimulatory agonist antibodies. FcγRIIB promotes clustering and signaling by binding to Fc domains of antibodies that have bound (specifically in an antibody Fv-dependent manner) to targeted cell surface receptors. Antibody-induced cellular responses are determined by the antibody-targeted receptor's biological function, e.g., macrophage costimulation (anti-CD40 mAb) or tumor cell apoptosis (anti-Fas mAb).

FcγR to be largely ineffective or even detrimental. Evidence supported a mechanism whereby FcγRIIb provided cross-linking for the agonistic mAbs likely inducing target receptor clustering and mimicking interactions with the TNFRSF members' natural trimeric ligands and inducing signaling leading to agonism and cell activation (Fig. 6). These requirements have since been confirmed in a range of human in vitro and human FcγR Tg in vivo studies from several research groups. One noteworthy fact about these observations is that the original murine studies utilized mAbs of mIgG1 isotype, an isotype with a low A:I binding characteristic which does not have a representative human orthologue, meaning that with wild-type human mAbs, the efficacy induced in murine systems cannot be directly translated. Attempts to overcome this problem have been addressed in two ways: Fc engineering to enhance FcγRIIb binding [77] and the generation of FcγR-independent agonists dependent on other means for their cross-linking requirements [78–80] (Fig. 6).

Therapeutic interventions to overcome FcγR-mediated resistance

Since the very early demonstration of human anti-mouse antibody (HAMA) responses against Orthoclone OKT3, the first mAb to obtain FDA approval [81], and the subsequent impact on patient outcomes by engineered mAbs such as rituximab in non-Hodgkin's lymphoma and trastuzumab in HER2+ breast cancer, drug developers have sought to enhance mAb Fc to maximize activity and reduce toxicity and resistance [82]. A large proportion of these efforts have centered on engineering the Fc portion of mAbs given its established importance in determining the mechanism of action and resistance. These approaches have largely focused on manipulation of Fc glycosylation and amino acid substitution in human IgG1 to enhance antibody binding to activating FcγRs. More recently, the alternative approach of increasing antibody A:I ratios, i.e., through blockade of the inhibitory FcγRIIb, is receiving increasing attention, and clinical trials are ongoing evaluating FcγRIIb-blocking mAb ability to overcome resistance to direct tumor-targeting mAbs and ICB mAbs in hematologic and solid cancers.

Alternative mAb isotypes and formats, which do not depend upon FcγR interactions, constitute separate developmental efforts. Given that these latter strategies do not depend on FcγRs for mechanisms of action and are only now reaching early clinical trial, these will not be discussed further here. Reviews for these approaches are available elsewhere [83,84].

Fc engineering

Glycoengineering

Fc:FcγR binding requires extensive, structurally sensitive interactions between the mAb and the receptor. The Fc portion of all IgG has a conserved, complex, N-linked oligosaccharide at position 297 of the Ab heavy chain (HC), and this acts to open the HC structure enabling Fc:FcγR loop interactions and binding [85]. Alterations to the sugars within the complex glycan structure can markedly impact binding affinity. The removal of fucose has been shown in a number of studies to have the most dramatic impact leading to an approximately 50-fold enhancement of interactions with the activatory FcγRIIIa and FcγRIIIb [86], thereby enhancing ADCC [87] and ADCP [88]. Clinical data can also be interpreted to support these findings given the dramatically enhanced effect the glycoengineered anti-CD20 mAb obinutuzumab has over rituximab in chronic lymphocytic leukemia [89]. There are a number of afucosylated mAbs in clinical trials or recently approved (inebilizumab, anti-CD19 [90]; ublituximab, anti-CD20 [91]; mogamulizumab, anti-CCR4 [92]), and as these mAbs mature in the clinic, the potential of glycan manipulation to enhance mAb efficacy across a range of target antigens and malignancies should become clearer. Hypergalactosylation [93] and reduced sialylation [94] have also recently emerged as a means to enhance FcγRIII interactions and ADCC; however, how much of these effects is due directly to the changes in these other glycans as opposed to associated changes to fucose content or antigen binding is yet to be fully determined. Notably, although manipulation of specific sugar residues can enhance FcγR interactions, the removal of the entire oligosaccharide through a mutation to replace the asparagine, typically with an alanine, at position 297 leads to a loss of nearly all FcγR-binding capability and innate effector recruitment, creating what is sometimes referred to as an Fc-null or Fc-disabled variant. Although the N297A Fc mutation has been incorporated into the clinically approved anti-PD-L1 mAb atezolizumab to remove effector function [95], this mutation has been demonstrated to still be able to bind FcγRI when complexed [96] and mediate ADCP, although this is impacted in the presence of serum [97]. Therefore, clinically these aglycosylated mAbs may still induce undesired Fc:FcγR binding in settings of high antigen expression, membrane clustering, or immune complex formation. As such, these interactions could impact mAb efficacy, and it will be interesting to observe how such aglycosylated mAbs compare against other Fc engineering approaches to abrogate unwanted Fc:FcγR interactions.

Amino acid manipulations

Amino acid mutations to the protein backbone of mAb Fc permit more nuanced manipulation of Fc:FcγR interactions compared to glycoengineering. Here, in addition to enhancing activatory FcγRIII interactions and reducing all FcγR binding, it is possible to increase binding more specifically to selected activatory and inhibitory FcγRs. In 2001, a seminal work from Shields et al. detailed the critical residues for human IgG1 Fc:FcγR interactions and thereby laid the foundation stone for much of the Fc amino acid engineering carried out since [86]. The number of different mutants designed to influence Fc:FcγR binding in various stages of development stands at nearly 30 and can be grouped into those that enhance binding to activatory FcγRs or the inhibitory FcγRIIb or disable FcγR binding all together. The majority of these mutations are specific to IgG1 although a few do pertain to the other clinically relevant isotypes—IgG2 and IgG4. Tafasitamab is an IgG1 anti-CD19 mAb currently in clinical trials containing a double mutation (S329D/I332E) that increases affinity to both SNP variants of FcγRIIIa and to a lesser extent binding to the inhibitory FcγRIIb [98] and has demonstrated enhanced ADCC and ADCP [99]. Tafasitamab was granted FDA Breakthrough Therapy and Fast Track designation as well as being designated an Orphan Drug by FDA and EMA for relapsed/refractory DLBCL in late 2019 [100]. Margetuximab is an IgG1 anti-HER2 mAb currently in trial containing five mutations (L235V/F243L/R292P/Y300L/P396L) that have the double effect to both enhance FcγRIIIa interactions and inhibit FcγRIIb interactions [101]. This engineered mAb demonstrated enhanced ADCC in vitro activity and in vivo activity compared with an unmodified control and trastuzumab. Finally, ocaratuzumab, an IgG1 anti-CD20 mAb engineered with a double mutation (P247I/A339Q) that increases binding to both variants of FcγRIIIa and has enhanced ADCC [102], has also entered trial.

While Fc engineering has proven an effective strategy to enhance mAb and FcγR interactions, improving patient outcomes compared to non-engineered counterparts targeting the same receptors [89] and more mAbs in clinical development to follow [100], Fc engineering has thus far failed to generate two types of reagents that potentially could help transform antibody cancer immunotherapy—Fcs that are null for binding to the inhibitory FcγRIIb and simultaneously enhance binding to all activating FcγRs, and Fcs that, like the mouse IgG1 Fcs, show reduced or no binding to activating FcγRs and robust engagement of the inhibitory FcγRIIb.

To date, no mAbs with Fc-bearing amino acid mutations that enhance binding to the inhibitory FcγRIIb or remove FcγR binding completely have moved to clinical trial. This lack of development to date is reflected by the lack of effective ISA mAbs, which, as discussed above in preclinical models, have been shown to be dependent on FcγRIIb, and for ICB, the utilization of IgG2 and IgG4 isotypes with reduced but significant retained Fc:FcγR interactions for all ICB mAbs except atezolizumab. Notably, these isotypes do not ablate all FcγR interactions but only reduce them (Fig. 2), and the potential impact of these interactions is becoming clear and will likely drive harder the push to generate truly Fc-silent ICB mAbs, or alternatively and pending demonstrated efficacy and tolerability combined treatment with FcγR-blocking mAbs.

Immune modulation/interventions to change FcγR expression profiles

However potent engineered mAbs are with regard to maximizing appropriate Fc:FcγR interactions, responses may be limited in patients where there is a lack of appropriate FcγR expression or there is tumor microenvironment-induced FcγR deregulation. For example, clinically validated afucosylated antibodies, e.g., obinutuzumab, selectively enhance binding to and activity of FcγRIIIa-expressing effector cells, e.g., NK cells and patrolling monocytes. While NK cells are readily abundant in blood, they are generally scarce in human tumors compared to myeloid cells and macrophages, which may constitute up to half of the tumor mass. Depending on environmental cues and ontogeny, myeloid cells show varying expression of activating FcγRs with FcγRIIa predominating, and as discussed in tumor and tumor-associated compartments, resistance-promoting inhibitory FcγRIIb is often upregulated and highly expressed. In these contexts, a different but complementary approach may be pursued using interventions that aim to modulate immune cell activation profiles to alter FcγR expression and thereby enhance mAb efficacy. These combination approaches have, to date, largely been focused on innate immune agonists due to the expression of both activatory and inhibitory FcγRs on myeloid cells, their predominance in inflammatory infiltrates in tumors, and their demonstrated role in mAb potency in murine and in vitro human preclinical studies. Myeloid cells are exquisitely sensitive to their microenvironment having evolved to respond appropriately to a myriad of pathogenic threats with very different requirements for their effective neutralization and disposal and yet also to be capable of responding to host-derived stimuli in order to support normal tissue homeostasis and wound healing [103]. This has imbued myeloid cells with the ability to change their activation state and functional capabilities rapidly and profoundly when stimulated with appropriate cues. These activation states represent a continuum of differing phenotypes from the extremes of classically activated proinflammatory "M1-like" to alternatively activated "M2-like" macrophage activation states [104].

Tumor microenvironments are largely immunosuppressive with infiltrates frequently dominated by macrophages, termed tumor-associated macrophages (TAMs) in this context. Indeed, TAMs can form up to as much as 50% of the infiltrating cells in tumors, and their presence is one of the strongest immune correlates for poor response across tumor types [72]. TAM phenotypes are complex and heterogeneous representing the assimilation of a range of tumor- and stroma-derived cytokines and factors such as IL-4, IL-10, and IL-13, CSF1, TGF-β, and PGE2, and metabolic challenges such as hypoxia, low pH, and competition for nutrients [105,106]. In this complex milieu of stimulatory factors, TAMs are generally accepted to exhibit a range of tumor-promoting [107] and immune-suppressive properties [105]. This has led to strategies designed to delete TAMs with a view to improving patient responses [108,109] or to reprogram them to an antitumor proinflammatory phenotype [110,111]. The majority of strategies aimed at reprogramming TAMs, understandably, have focused on reversing these protumoral and immune-suppressive properties to enhance adaptive antitumor immunity and frequently have overlooked the impact these approaches may have on antibody combinations through changes induced in FcγR expression directly.

As effector molecules, FcγR expression is intimately linked to myeloid cell activation status. Proinflammatory macrophages upregulate activatory FcγRs and downregulate the inhibitory FcγRIIb to maximize their cytotoxic capability against pathogens targeted for destruction by Abs. In contrast, macrophages involved in wound healing and tissue remodeling (analogous to established tumor microenvironments) downregulate activatory FcγRs and upregulate the inhibitory receptor to prevent tissue-damaging inflammatory responses to IC deposition during these processes.

Given their prevalence within the tumor microenvironment, the established role of macrophages as mAb effectors [112], and their sensitivity to inflammatory cues, it is clear that TAMs have the potential to be exploited to provide more potent mAb-mediated antitumor activity [113]. In support of this capacity, TAMs were demonstrated as the effector cells responsible for deleting Tregs within certain tumors [61–63] as well as tumor cells directly in lung and skin cancer models [114]. Despite these observations, studies indicate that TAM FcγR expression profiles are far from optimal for mAb-dependent effector function, with tumor inducing high expression of inhibitory FcγRIIb and reduced expression of activatory FcγRs in preclinical models [11,111,115], impairing effector capacity. Therefore, strategies to alter TAMs from an immunosuppressive to an activated phenotype may be an effective strategy to enhance therapeutic mAbs. This approach is supported by reports demonstrating the plasticity of myeloid cells [104,116,117] and our own recent data showing that these profiles are reversible and can be manipulated in mouse and human models [111]. Moreover, these observations clearly validate the concept that FcγR expression patterns can be specifically, rapidly, and profoundly altered to mediate desired mAb activity and may contribute to the poor efficacy of mAbs, such as rituximab, in resistant anatomical niches.

Small molecules

There are an array of small molecules being investigated for their ability to target the reprogramming of TAMs, and these have been reviewed recently at length elsewhere [118]. These agents fall into two broad categories: those that provide stimulation of pathogen or danger-associated molecular pattern detection pathways to induce infectious and noninfectious inflammatory responses respectively and those, in contrast, that seek to inhibit anti-inflammatory pathways induced in the tumor microenvironment.

Toll-like receptor (TLR) agonists have been the most exhaustively investigated in the context of combination with direct tumor-targeting/cytotoxic mAbs, and these studies have ranged from preclinical studies through to phase III trials. TLR agonism seems ideally suited to augmenting mAb effector function given the demonstrated synergy between TLR and FcγR pathways in innate pathogen recognition and uptake and antigen presentation [119,120]. Despite this demonstrated cooperation, the links between pathways and responses still need to be fully elucidated. TLR activation leads to NFκB, STAT1, and IRF-3 signaling to induce a range of inflammatory gene expression including FcγR. Although a number of preclinical studies do support the potential, at least in principle, of these approaches to reprogram TAMs and augment mAb efficacy, few have examined directly the FcγR mediating effector function. Again, as discussed above, this in part is a reflection of the fact that until recently, panels of highly specific reagents to measure mouse and human FcγRs were not widely available [42]. Despite this, Butchar and colleagues demonstrated that the TLR7/8 ligand resiquimod was able to upregulate activatory FcγRs and downregulate inhibitory FcγRIIb as detected by

Western blot in peripheral blood monocytes and further that these changes led to enhanced ADCC in vitro. They also showed enhancement of antitumor activity in combination with mAbs in tumor models but did not directly assess FcγRs in this context [110]. Recently, we have also investigated the potential for a range of TLR ligands to induce phenotypic changes in macrophages reminiscent of an "M1-like" phenotype and demonstrated increased activatory FcγRs and reduced inhibitory FcγRIIb that in both human and mouse in vitro models led to enhanced mAb-mediated phagocytosis of target cells [111]. Notably, in the context of a suppressive tumor microenvironment where mAb effector function was poor, TLRs were unable in our hands to overcome suppression and enhance mAb efficacy.

Another attractive but relatively recently discovered innate sensing pathway for macrophage reprogramming is the simulator of interferon genes (STING) pathway. STING is an innate immune sensor that detects intracellular double-stranded DNA in the form of cyclic dinucleotides and mediates its effects through type I interferons (IFNs) [121,122]. Studies have shown that stimulating this pathway has powerful immunoregulatory and anticancer effects [123]. We have shown that STING agonists can overcome immunosuppression in the tumor microenvironment (TME), reversing TAM inhibitory FcγR profiles to produce strong adjuvant effects for antibody immunotherapy [111]. Importantly, primary human macrophage FcγR and ADCP responses mirrored those of mouse, supporting the translational potential of these findings.

These studies demonstrate an important proof of concept and validate the potential of these approaches, but clearly, a major hurdle that must be overcome is that, to date, immune modulatory agents have struggled in trial to demonstrate sufficiently high benefit–risk ratio to justify combination approvals [124,125]. Indeed, STING agonists have recently joined the ranks of innate immune activators with underwhelming clinical responses, although a number of trials are ongoing and yet to report [126]. Notably, these trials do not seek to directly augment direct-targeting/cytotoxic mAbs and so these investigations would still be merited even in the absence of single-agent or ICB combination efficacy.

Recently, Roghanian et al. demonstrated that existing therapeutics may be "hiding" novel, poorly understood combination mechanisms that can be exploited if better understood. In this study, tumor-bearing bone marrow macrophage FcγR profiles were shown to be suppressed by bone marrow intrinsic as well as the tumor extrinsic effects leading to an upregulation of inhibitory FcγRIIb in both murine models and lymphoma patient biopsies [127]. Notably, this suppressed A:I FcγR expression profile could be reversed using the long-established chemotherapy drug cyclophosphamide in combination with rituximab serving to further demonstrating the potential of such strategies.

Future studies with more in-depth investigation of FcγR-specific effects and mechanisms may serve to better guide the clinical translation of mAb combinations with novel innate agonist and potentially reawaken interest in some old friends.

Antibodies

More recently, mAbs have been generated to target key surface receptors on TAMs with a view to shutting down anti-inflammatory signaling and inducing macrophages to become more proinflammatory and subsequently antitumor.

One of the furthest developed of these approaches targets phosphatidylserine, a lipid exposed on apoptotic cells as a marker for non-inflammatory removal by phagocytes. PS targeting has been reported to reprogram TAMs to an inflammatory phenotype [128] but has disappointed in a phase III trial [129]. Another approach, targeting macrophage receptor with collagenous structure (MARCO) a scavenger receptor, has been shown to potently reprogram TAMs. Interestingly, mAb activity was dependent upon the inhibitory FcγRIIb, supporting the role of agonistic signaling in this process [130] as demonstrated for TNFRSF members [75,76]. The leukocyte immunoglobulin-like receptor B (LILRB) family of inhibitory receptors are widely expressed on TAMs and involved in their negative regulation. Chen et al. demonstrated utility of mAbs targeting LILRB2 in reprogramming TAMs and enhancing ICB responses [131]. In a very recent development, Zhou et al. demonstrated the promise of blocking tumor cell efferocytosis by TAMs to induce STING activation and type I IFN responses leading to productive antitumor immunity and combination with ICB in a preclinical model [132]. This study serves to demonstrate how a strategy targeting one facet of TAM biology can seemingly unlock multiple antitumoral factors leading to profound responses and serves to support the investigation of other TAM reprogramming targets.

Compared to small-molecule approaches, mAb targeting of TAMs is still relatively underexplored but holds great promise due to the potential for better on-target activation and reduced systemic toxicity profiles. However, careful consideration will likely be required to the optimal mAb characteristics required when targeting FcγR-expressing TAMs with regard to the target receptor, its biology, and the potential interactions of mAbs bearing intact Fcs with both activatory and inhibitory FcγRs to maximize potential utility of these approaches.

These various immunomodulatory approaches hold the potential for wider therapeutic activity and impact beyond enhancing TAM FcγR effector function. TAMs can have reduced antigen-presenting capabilities [133], and repolarizing them to a more inflammatory phenotype offers the potential to enhance mAb- and FcγR-mediated tumor-associated antigen presentation, thereby inducing adaptive antitumor immune responses and the potential of long-term antitumor memory. FcγRIIa-dependent vaccinal effects have been demonstrated in dendritic cells in cancer models [134], and given the potential to upregulate activatory FcγRs including FcγRIIa in macrophages with agonists and their potential to be induced to present antigen in tumor microenvironments [135], such approaches merit further investigation.

Another FcγR-mediated mechanism closely related to phagocytosis that may be associated with potentially context-dependent efficacy or resistance is trogocytosis. This is the removal and internalization of target antigen and cell membrane by FcγR-expressing phagocytes. This controversial mechanism has been shown to induce cell death in breast cancer cell lines [136] and conflictingly tumor escape and resistance in B-cell malignancies [137,138]. Although the mechanism at play here is the same, the outcome is proposed to be very different, perhaps reflecting the underlying properties of the targeted cells or antigens. We have reported that polarizing macrophages to a more activated proinflammatory phenotype using classic immunomodulatory agents enhances trogocytosis [139], and therefore, immunomodulatory strategies to induce these changes in TAMs may well have a direct impact increasing trogocytosis, the outcome of which may be efficacy-enhancing or efficacy-reducing.

Although innate immune modulation is a promising approach to augment mAb activity through FcγR manipulation given the very different FcγR requirements of the mAb classes described above and the complexity of the tumor immune microenvironment, these approaches need to be considered carefully to ensure that they do not have unexpected detrimental effects or lead to increased toxicity limiting their utility.

FcγR blockade

The final approach to be discussed here is FcγR blockade where, analogous to ICB, mAbs targeting FcγRs are used to block detrimental Fc:FcγR interaction. Here, FcγR-blocking mAbs are used in combination with other mAb approaches where inappropriate interactions limit efficacy. As discussed above, clinically validated Fc engineering approaches are largely restricted to enhancing antibody affinity for one (FcγRIIIa) of several activating FcγRs. This is suboptimal since activating FcγR expression in intratumoral and tumor-associated compartments is highly heterogeneous and since emerging data indicate high expression and functional significance of the inhibitory FcγRIIb in conferring tumor resistance to mAb therapy. Additional benefits of the FcγR-blocking approach, as for innate immune modulatory approaches described above, include that this strategy allows existing mAbs to be improved, thereby expanding patient responses with approved drugs and, potentially, widening use to indications where activity of the existing mAbs alone is insufficiently powerful. An additional advantage of the mAb combination strategy is that it allows for variable dosing, with respect to dose, timing, frequency, and potentially route of administration. As such, the same blocking mAbs may be used in different dosing regimens optimized to enhance or deepen responses, or overcome resistance, of multiple approved mAbs used in distinct cancer subtypes and individual patients. Given that drug development and approval are expensive, slow, and arduous tasks, the potential of one mAb to expand the utility of many has clear cost benefits.

The position of FcγRIIb as the solitary inhibitory FcγR, its upregulation in the tumor microenvironment [115], and its documented role in conferring resistance to antibody-based therapy in this niche [51,56,115] through mechanisms operating on both innate immune effector cells and tumor cells indicate the importance of targeting FcγRIIb to boost therapeutic antibody efficacy and overcome cancer resistance. The strategy does, however, put exquisite requirements on a therapeutic antibody candidate, from both target receptor specificity and function-modulating perspectives. The extracellular, antibody-accessible domain of the inhibitory FcγRIIb is ~93% homologous with the activating FcγRIIa. Nevertheless, probing of a highly diversified human recombinant antibody library [51], or immunization of mice transgenic for human FcγRIIa [140], generated diverse pools of highly specific antibodies that selectively bound to FcγRIIb, and not to FcγRIIa, and which in a dose-dependent manner blocked immune complex binding to cell surface-expressed FcγRIIb. Functional screening revealed that only a minority of the highly FcγRIIb-specific human recombinant antibodies were able to block antibody-induced FcγRIIb inhibitory signaling [51]. Based on observations that FcγRIIb limits antibody efficacy and promotes tumor cell resistance by dual mechanisms in B-cell malignancy, acting at the level of both immune effector cells and tumor B cells, the authors have further characterized the therapeutic potential of FcγRIIb-blocking antibodies to boost efficacy and overcome resistance to antibody therapy in vivo focusing initially on this setting. A lead human antagonistic anti-FcγRIIb IgG1 antibody (6G11 or BI-1206), which synergistically enhanced rituximab B-cell depletion in FcγRIIb- and CD20-humanized mice and overcame refractoriness of primary leukemic B cells to anti-CD20-based antibody therapy in vivo, is currently in early phase clinical testing [51]. While it is early days and optimal biological doses have yet to be administered, preliminary findings indicate that FcγRIIb blockade may indeed help overcome antibody drug resistance also in patients. In an ongoing study of cancer patients with relapsed/refractory B-cell lymphoma that had previously progressed on rituximab-containing therapy (NCT02933320), two follicular lymphoma patients were reported to have stable disease and complete response, respectively, and a patient with leukemic mantle cell lymphoma (MCL) experienced complete loss of circulating tumor cells following combined treatment with FcγRIIb-blocking antibody and rituximab. Despite recent development of highly potent targeted therapies, e.g., inhibitors of Bruton's tyrosine kinase (BTKi) and Bcl-2, MCL is characterized by a significant unmet medical need, where patients that initially respond but later acquire resistance—so-called "double-hit" patients—have a particularly grave prognosis. We recently confirmed a generally high and homogenous FcγRIIb expression on MCL tumor cells and demonstrated that in vivo administration of the FcγRIIb-blocking antibody BI-1206 generated single-agent antitumor activity and enhanced rituximab therapeutic efficacy alone or on the back of clinically relevant chemotherapies or targeted therapies [141]. Impressively, BI-1206 had antitumor activity in PDX models derived from double-hit MCL patients resistant to ibrutinib and venetoclax, and the addition of this FcγRIIb blocker to rituximab and venetoclax significantly enhanced survival compared to V + R treatment in a PDX model of ibrutinib- and rituximab-resistant MCL. These findings are consistent with and extend our previous reports [51] that BI-1206 is able to suppress and overcome FcγRIIb-mediated rituximab resistance in diverse clinically relevant in vivo settings.

Based on the apparent potential of FcγRIIb blockade to enhance therapeutic efficacy, and potentially overcome resistance, also of immune checkpoint blocking antibodies, our current efforts aim at translating observations of FcγRIIb-regulated antitumor immunity to the most appropriate antibody drug combinations and expanding development to the solid cancer clinical setting. A phase 1/2a clinical trial of BI-1206 in combination with pembrolizumab in subjects with advanced solid tumors previously treated with anti-PD-1 or anti-PD-L1 antibodies is now open and will soon be enrolling patients (NCT04219254).

Besides affording efficacy, therapeutic targeting of Fc-gamma receptors, whether by blocking antibodies or Fc engineering, must be safe and associated with therapeutically relevant pharmacokinetics. In addition to its high expression on B cells and certain macrophage/dendritic cells, FcγRIIb has been reported to be highly expressed in mouse and rat liver sinusoidal endothelial cells (LSECs) [142], where they have been implicated in removal of circulating small immune complexes [41]. These observations raise potential safety concerns of undesirably cytotoxic activity with therapeutic antibodies targeting FcγRIIb. However, our recent observations of human and mouse liver indicate lower LSEC expression in humans [42], and dosing of FcγRIIb-humanized mice with therapeutically relevant doses of anti-human FcγRIIb IgG1 antibody 6G11 showed no apparent acute or chronic treatment-related adverse effects [42,51]. Ultimately, the safety and efficacy of targeting FcγRIIb need to be assessed in human subjects. Available clinical data support a tolerable profile of FcγRIIb blockade (NCT03571568 and NCT02933320). While it is early days, doses achieving transient receptor saturation (~2 days) have been safely administered indicating that complete FcγRIIb saturation can be safely achieved.

Conclusion

Antibody drugs are established first-line treatment modalities for a range of cancer types, and yet we still understand relatively little of why they work in some patients and fail in others. FcγR interactions have emerged as one component that should be considered when seeking to understand mechanisms of action and resistance with antibody drugs.

The requirement for activatory FcγR engagement and the detrimental effects of inhibitory FcγRIIb binding in the activity of direct tumor-targeting mAbs has been established for some time. Consequently, enhancing the activatory-to-inhibitory engagement ratio in this setting is an attractive approach to increase patient responses. In principle, this enhancement of proinflammatory character can be achieved through a number of means: Fc engineering, immune modulation, and FcγR blockade. These strategies are being investigated both preclinically and clinically and should not necessarily be seen in isolation but be perhaps better considered when carefully applied as complementary strategies that could synergize to enhance effects. Inevitably, with mAb therapeutics caution needs to be taken with these approaches as it has been demonstrated that enhancing Fc:FcγR interactions can also compromise therapeutic antibody activity as illustrated by ICB to PD-1. Through careful characterization of antibody mechanism of action as a function of target biology, subcellular expression, and engagement of activating and inhibitory FcγR, FcγR interactions can be tailored through Fc engineering, FcγR blockade, or through pharmacological modulation of myeloid FcγR expression. Further studies will be required to understand if and how we might best apply these approaches to enhance mAb activity.

The recent findings that immune checkpoint blockers can be critically, and differentially dependent upon Fc:FcγR interactions, are scientifically intriguing, but at the same time indicate an imminent need to improve our understanding of how FcγRs regulate therapeutic efficacy and cancer resistance to individual targets and mAbs. The factors determining these outcomes are complex but largely reflect whether the antigen is more highly expressed on critical immune effector populations, immune-suppressive cells, or the tumor and its wider stromal compartment. The presence of antigen on cells with very different contributions to antitumor immunity explains the variable Fc requirements demonstrated in model systems. Here, to engage activatory FcγRs could lead to depletion of key effectors, suppressor cells or tumor with negative or positive impacts accordingly. The expression variance seen with ICB targets can be a reflection of tumor type, inflammatory state, and host genetics, and outcomes may be enhanced if expression can be determined to ensure ICB mAbs are administered to patients able to respond. Here, for example, immune modulation to enhance activatory FcγR expression could be coupled with Fc engineering to enhance activatory FcγR binding and/or FcγRIIb blockade to enhance anti-CTLA-4-mediated deletion of intratumor-suppressive Tregs. Contrarily, to enhance on anti-PD-1 mAbs, Fc engineering or blocking antibody approaches to blunt anti-PD-1 antibody engagement with FcγRs are warranted and in clinical development. These opposing strategies simply reflect the different intratumoral cellular expression profiles of these immune inhibitory receptors.

It is worth considering that FcγRs have evolved to play critical roles linking innate, humoral, and cellular adaptive immunity and that using this biology rather than seeking to disregard it might allow mAb therapeutics to better induce long-term adaptive antitumor immunity. Perhaps, improved understanding of the complex mechanisms and biology of Fc: FcγR interactions may allow already transformative antibody drugs to be harnessed to even greater therapeutic potential, and help guide the development of next-generation mAbs and pathways that by complementing currently available great drugs help achieve pan-cancer responses and overcome resistance to cancer on the whole.

Acknowledgments

We thank Joost Bakker (www.scicomvisuals.com) for help generating schematic figures.

References

[1] The Antibody Society. Antibody therapeutics approved or in regulatory review in the EU or US. https://www.antibodysociety.org/resources/approved-antibodies/; 2020.
[2] Sharma P, et al. Primary, adaptive, and acquired resistance to cancer immunotherapy. Cell 2017;168(4):707–23.
[3] Goede V, et al. Obinutuzumab plus chlorambucil in patients with CLL and coexisting conditions. N Engl J Med 2014;370(12):1101–10.
[4] Baselga J, et al. Pertuzumab plus trastuzumab plus docetaxel for metastatic breast cancer. N Engl J Med 2012;366(2):109–19.
[5] Gopal AK, et al. PI3Kdelta inhibition by idelalisib in patients with relapsed indolent lymphoma. N Engl J Med 2014;370(11):1008–18.
[6] Maloney DG. Anti-CD20 antibody therapy for B-cell lymphomas. N Engl J Med 2012;366(21):2008–16.
[7] Wei SC, Duffy CR, Allison JP. Fundamental mechanisms of immune checkpoint blockade therapy. Cancer Discov 2018;8(9):1069–86.
[8] Garber K. Immune agonist antibodies face critical test. Nat Rev Drug Discov 2020;19(1):3–5.
[9] Clynes RA, et al. Inhibitory Fc receptors modulate in vivo cytotoxicity against tumor targets. Nat Med 2000;6 (4):443–6.
[10] Dahan R, et al. FcgammaRs modulate the anti-tumor activity of antibodies targeting the PD-1/PD-L1 axis. Cancer Cell 2015;28(3):285–95.
[11] Arce Vargas F, et al. Fc effector function contributes to the activity of human anti-CTLA-4 antibodies. Cancer Cell 2018;33(4):649–63 e4.
[12] White AL, Beers SA, Cragg MS. Fc gamma RIIB as a key determinant of agonistic antibody efficacy. Curr Top Microbiol Immunol 2014;382:355–72.
[13] Nimmerjahn F, Ravetch JV. Divergent immunoglobulin g subclass activity through selective Fc receptor binding. Science 2005;310(5753):1510–2.
[14] Hussain K, et al. Upregulation of FcgammaRIIb on monocytes is necessary to promote the superagonist activity of TGN1412. Blood 2015;125(1):102–10.
[15] Knorr DA, Dahan R, Ravetch JV. Toxicity of an Fc-engineered anti-CD40 antibody is abrogated by intratumoral injection and results in durable antitumor immunity. Proc Natl Acad Sci U S A 2018;115(43):11048–53.
[16] Bakema JE, van Egmond M. The human immunoglobulin A Fc receptor FcalphaRI: a multifaceted regulator of mucosal immunity. Mucosal Immunol 2011;4(6):612–24.
[17] Chen K, et al. Immunoglobulin D enhances immune surveillance by activating antimicrobial, proinflammatory and B cell-stimulating programs in basophils. Nat Immunol 2009;10(8):889–98.
[18] Wu YJ, et al. The immunoglobulin D Fc receptor expressed on fibroblast-like synoviocytes from patients with rheumatoid arthritis contributes to the cell activation. Acta Pharmacol Sin 2017;38(11):1466–74.
[19] Coico RF, et al. IgD-receptor-positive human T lymphocytes. I. Modulation of receptor expression by oligomeric IgD and lymphokines. J Immunol 1990;145(11):3556–61.

[20] Wu LC, Zarrin AA. The production and regulation of IgE by the immune system. Nat Rev Immunol 2014;14 (4):247–59.
[21] Nimmerjahn F, Ravetch JV. Fcgamma receptors as regulators of immune responses. Nat Rev Immunol 2008;8 (1):34–47.
[22] Liu J, et al. Role of the IgM Fc receptor in immunity and tolerance. Front Immunol 2019;10:529.
[23] Roopenian DC, Akilesh S. FcRn: the neonatal Fc receptor comes of age. Nat Rev Immunol 2007;7(9):715–25.
[24] Keeble AH, et al. TRIM21 is an IgG receptor that is structurally, thermodynamically, and kinetically conserved. Proc Natl Acad Sci U S A 2008;105(16):6045–50.
[25] Rostamzadeh D, et al. Update on Fc receptor-like (FCRL) family: new immunoregulatory players in health and diseases. Expert Opin Ther Targets 2018;22(6):487–502.
[26] Vivier E, Daeron M. Immunoreceptor tyrosine-based inhibition motifs. Immunol Today 1997;18(6):286–91.
[27] Nagelkerke SQ, et al. Genetic variation in low-to-medium-affinity Fcgamma receptors: functional consequences, disease associations, and opportunities for personalized medicine. Front Immunol 2019;10:2237.
[28] Bruhns P, et al. Specificity and affinity of human Fcgamma receptors and their polymorphic variants for human IgG subclasses. Blood 2009;113(16):3716–25.
[29] Lu J, Sun PD. Structural mechanism of high affinity FcγRI recognition of immunoglobulin G. Immunol Rev 2015;268(1):192–200.
[30] Chenoweth AM, et al. The high-affinity receptor for IgG, FcγRI, of humans and non-human primates. Immunol Rev 2015;268(1):175–91.
[31] Swisher JF, Feldman GM. The many faces of FcγRI: implications for therapeutic antibody function. Immunol Rev 2015;268(1):160–74.
[32] Weng WK, Levy R. Two immunoglobulin G fragment C receptor polymorphisms independently predict response to rituximab in patients with follicular lymphoma. J Clin Oncol 2003;21(21):3940–7.
[33] Cheung NK, et al. FCGR2A polymorphism is correlated with clinical outcome after immunotherapy of neuroblastoma with anti-GD2 antibody and granulocyte macrophage colony-stimulating factor. J Clin Oncol 2006;24 (18):2885–90.
[34] Tamura K, et al. FcgammaR2A and 3A polymorphisms predict clinical outcome of trastuzumab in both neoadjuvant and metastatic settings in patients with HER2-positive breast cancer. Ann Oncol 2011;22(6):1302–7.
[35] Li X, et al. Allelic-dependent expression of an activating Fc receptor on B cells enhances humoral immune responses. Sci Transl Med 2013;5(216)216ra175.
[36] Lejeune J, Brachet G, Watier H. Evolutionary story of the low/medium-affinity IgG Fc receptor gene cluster. Front Immunol 2019;10:1297.
[37] Cartron G, et al. Therapeutic activity of humanized anti-CD20 monoclonal antibody and polymorphism in IgG Fc receptor FcgammaRIIIa gene. Blood 2002;99(3):754–8.
[38] Musolino A, et al. Immunoglobulin G fragment C receptor polymorphisms and clinical efficacy of trastuzumab-based therapy in patients with HER-2/neu-positive metastatic breast cancer. J Clin Oncol 2008;26(11):1789–96.
[39] Salmon JE, et al. Fc gamma receptor IIIb enhances Fc gamma receptor IIa function in an oxidant-dependent and allele-sensitive manner. J Clin Invest 1995;95(6):2877–85.
[40] Daeron M. Fc receptor biology. Annu Rev Immunol 1997;15:203–34.
[41] Ganesan LP, et al. FcgammaRIIb on liver sinusoidal endothelium clears small immune complexes. J Immunol 2012;189(10):4981–8.
[42] Tutt AL, et al. Development and characterization of monoclonal antibodies specific for mouse and human Fcgamma receptors. J Immunol 2015;195(11):5503–16.
[43] Schwab I, Lux A, Nimmerjahn F. Pathways responsible for human autoantibody and therapeutic intravenous IgG activity in humanized mice. Cell Rep 2015;13(3):610–20.
[44] Wang SY, et al. Depletion of the C3 component of complement enhances the ability of rituximab-coated target cells to activate human NK cells and improves the efficacy of monoclonal antibody therapy in an in vivo model. Blood 2009;114(26):5322–30.
[45] Biburger M, et al. Monocyte subsets responsible for immunoglobulin G-dependent effector functions in vivo. Immunity 2011;35(6):932–44.
[46] Montalvao F, et al. The mechanism of anti-CD20-mediated B cell depletion revealed by intravital imaging. J Clin Invest 2013;123(12):5098–103.
[47] Uchida J, et al. The innate mononuclear phagocyte network depletes B lymphocytes through Fc receptor-dependent mechanisms during anti-CD20 antibody immunotherapy. J Exp Med 2004;199(12):1659–69.

[48] Biburger M, Lux A, Nimmerjahn F. How immunoglobulin G antibodies kill target cells: revisiting an old paradigm. Adv Immunol 2014;124:67–94.
[49] Zhang W, et al. FCGR2A and FCGR3A polymorphisms associated with clinical outcome of epidermal growth factor receptor expressing metastatic colorectal cancer patients treated with single-agent cetuximab. J Clin Oncol 2007;25(24):3712–8.
[50] Mellor JD, et al. A critical review of the role of Fc gamma receptor polymorphisms in the response to monoclonal antibodies in cancer. J Hematol Oncol 2013;6:1.
[51] Roghanian A, et al. Antagonistic human FcgammaRIIB (CD32B) antibodies have anti-tumor activity and overcome resistance to antibody therapy in vivo. Cancer Cell 2015;27(4):473–88.
[52] Lim SH, et al. Fc gamma receptor IIb on target B cells promotes rituximab internalization and reduces clinical efficacy. Blood 2011;118(9):2530–40.
[53] Beers SA, et al. Antigenic modulation limits the efficacy of anti-CD20 antibodies: implications for antibody selection. Blood 2010;115(25):5191–201.
[54] Camilleri-Broet S, et al. FcgammaRIIB is differentially expressed during B cell maturation and in B-cell lymphomas. Br J Haematol 2004;124(1):55–62.
[55] Lee CS, et al. Expression of the inhibitory Fc gamma receptor IIB (FCGR2B, CD32B) on follicular lymphoma cells lowers the response rate to rituximab monotherapy (SAKK 35/98). Br J Haematol 2015;168(1):145–8.
[56] Pallasch CP, et al. Sensitizing protective tumor microenvironments to antibody-mediated therapy. Cell 2014;156(3):590–602.
[57] Rankin CT, et al. CD32B, the human inhibitory Fc-gamma receptor IIB, as a target for monoclonal antibody therapy of B-cell lymphoma. Blood 2006;108(7):2384–91.
[58] Vaughan AT, et al. Inhibitory FcgammaRIIb (CD32b) becomes activated by therapeutic mAb in both cis and trans and drives internalization according to antibody specificity. Blood 2014;123(5):669–77.
[59] Couzin-Frankel J. Breakthrough of the year 2013. Cancer immunotherapy. Science 2013;342(6165):1432–3.
[60] Leach DR, Krummel MF, Allison JP. Enhancement of antitumor immunity by CTLA-4 blockade. Science 1996;271(5256):1734–6.
[61] Selby MJ, et al. Anti-CTLA-4 antibodies of IgG2a isotype enhance antitumor activity through reduction of intratumoral regulatory T cells. Cancer Immunol Res 2013;1(1):32–42.
[62] Simpson TR, et al. Fc-dependent depletion of tumor-infiltrating regulatory T cells co-defines the efficacy of anti-CTLA-4 therapy against melanoma. J Exp Med 2013;210(9):1695–710.
[63] Bulliard Y, et al. Activating Fc gamma receptors contribute to the antitumor activities of immunoregulatory receptor-targeting antibodies. J Exp Med 2013;210(9):1685–93.
[64] Peggs KS, et al. Blockade of CTLA-4 on both effector and regulatory T cell compartments contributes to the antitumor activity of anti-CTLA-4 antibodies. J Exp Med 2009;206(8):1717–25.
[65] Larkin J, et al. Combined nivolumab and ipilimumab or monotherapy in untreated melanoma. N Engl J Med 2015;373(1):23–34.
[66] Sharma P, Allison JP. The future of immune checkpoint therapy. Science 2015;348(6230):56–61.
[67] Topalian SL, Drake CG, Pardoll DM. Immune checkpoint blockade: a common denominator approach to cancer therapy. Cancer Cell 2015;27(4):450–61.
[68] Ribas A, et al. Antitumor activity in melanoma and anti-self responses in a phase I trial with the anti-cytotoxic T lymphocyte-associated antigen 4 monoclonal antibody CP-675,206. J Clin Oncol 2005;23(35):8968–77.
[69] Bertrand A, et al. Immune related adverse events associated with anti-CTLA-4 antibodies: systematic review and meta-analysis. BMC Med 2015;13:211.
[70] Arlauckas SP, et al. In vivo imaging reveals a tumor-associated macrophage-mediated resistance pathway in anti-PD-1 therapy. Sci Transl Med 2017;9(389):eaal3604.
[71] Galon J, et al. Type, density, and location of immune cells within human colorectal tumors predict clinical outcome. Science 2006;313(5795):1960–4.
[72] Rooney MS, et al. Molecular and genetic properties of tumors associated with local immune cytolytic activity. Cell 2015;160(1–2):48–61.
[73] Gentles AJ, et al. The prognostic landscape of genes and infiltrating immune cells across human cancers. Nat Med 2015;21(8):938–45.
[74] Lo Russo G, et al. Antibody-Fc/FcR interaction on macrophages as a mechanism for hyperprogressive disease in non-small cell lung cancer subsequent to PD-1/PD-L1 blockade. Clin Cancer Res 2019;25(3):989–99.

[75] White AL, et al. Essential cross-linking role for FcgRIIB in the in vivo activity of anti-CD40 monoclonal antibody. Immunology 2011;135:189.
[76] Li FB, Ravetch JV. Inhibitory Fc gamma receptor engagement drives adjuvant and anti-tumor activities of agonistic CD40 antibodies. Science 2011;333(6045):1030–4.
[77] Dahan R, et al. Therapeutic activity of agonistic, human anti-CD40 monoclonal antibodies requires selective FcgammaR engagement. Cancer Cell 2016;29(6):820–31.
[78] White AL, et al. Fcgamma receptor dependency of agonistic CD40 antibody in lymphoma therapy can be overcome through antibody multimerization. J Immunol 2014;193(4):1828–35.
[79] White AL, et al. Conformation of the human immunoglobulin G2 hinge imparts superagonistic properties to immunostimulatory anticancer antibodies. Cancer Cell 2015;27(1):138–48.
[80] Claus C, et al. Tumor-targeted 4-1BB agonists for combination with T cell bispecific antibodies as off-the-shelf therapy. Sci Transl Med 2019;11(496): eaav5989.
[81] Chatenoud L, et al. Restriction of the human in vivo immune response against the mouse monoclonal antibody OKT3. J Immunol 1986;137(3):830–8.
[82] Strohl WR. Current progress in innovative engineered antibodies. Protein Cell 2018;9(1):86–120.
[83] Spiess C, Zhai QT, Carter PJ. Alternative molecular formats and therapeutic applications for bispecific antibodies. Mol Immunol 2015;67(2):95–106.
[84] Labrijn AF, et al. Bispecific antibodies: a mechanistic review of the pipeline. Nat Rev Drug Discov 2019;18 (8):585–608.
[85] Arnold JN, et al. The impact of glycosylation on the biological function and structure of human immunoglobulins. Annu Rev Immunol 2007;25:21–50.
[86] Shields RL, et al. High resolution mapping of the binding site on human IgG1 for Fc gamma RI, Fc gamma RII, Fc gamma RIII, and FcRn and design of IgG1 variants with improved binding to the Fc gamma R. J Biol Chem 2001;276(9):6591–604.
[87] Mossner E, et al. Increasing the efficacy of CD20 antibody therapy through the engineering of a new type II anti-CD20 antibody with enhanced direct and immune effector cell-mediated B-cell cytotoxicity. Blood 2010;115 (22):4393–402.
[88] Herter S, et al. Glycoengineering of therapeutic antibodies enhances monocyte/macrophage-mediated phagocytosis and cytotoxicity. J Immunol 2014;192(5):2252–60.
[89] Goede V, et al. Obinutuzumab as frontline treatment of chronic lymphocytic leukemia: updated results of the CLL11 study. Leukemia 2015;29(7):1602–4.
[90] Ohmachi K, et al. A multicenter phase I study of inebilizumab, a humanized anti-CD19 monoclonal antibody, in Japanese patients with relapsed or refractory B-cell lymphoma and multiple myeloma. Int J Hematol 2019;109 (6):657–64.
[91] Lunning M, et al. Ublituximab and umbralisib in relapsed/refractory B-cell non-Hodgkin lymphoma and chronic lymphocytic leukemia. Blood 2019;134(21):1811–20.
[92] Mogamulizumab tops standard of care for CTCL. Cancer Discov 2018;8(2):OF1.
[93] Thomann M, et al. Fc-galactosylation modulates antibody-dependent cellular cytotoxicity of therapeutic antibodies. Mol Immunol 2016;73:69–75.
[94] Scallon BJ, et al. Higher levels of sialylated Fc glycans in immunoglobulin G molecules can adversely impact functionality. Mol Immunol 2007;44(7):1524–34.
[95] Herbst RS, et al. Predictive correlates of response to the anti-PD-L1 antibody MPDL3280A in cancer patients. Nature 2014;515(7528):563–7.
[96] Lux A, et al. Impact of immune complex size and glycosylation on IgG binding to human FcgammaRs. J Immunol 2013;190(8):4315–23.
[97] Nesspor TC, et al. Avidity confers FcgammaR binding and immune effector function to aglycosylated immunoglobulin G1. J Mol Recognit 2012;25(3):147–54.
[98] Lazar GA, et al. Engineered antibody Fc variants with enhanced effector function. Proc Natl Acad Sci U S A 2006;103(11):4005–10.
[99] Horton HM, et al. Potent in vitro and in vivo activity of an Fc-engineered anti-CD19 monoclonal antibody against lymphoma and leukemia. Cancer Res 2008;68(19):8049–57.
[100] Kaplon H, et al. Antibodies to watch in 2020. MAbs 2020;12(1):1703531.
[101] Nordstrom JL, et al. Anti-tumor activity and toxicokinetics analysis of MGAH22, an anti-HER2 monoclonal antibody with enhanced Fcgamma receptor binding properties. Breast Cancer Res 2011;13(6):R123.

[102] Cheney CM, et al. Ocaratuzumab, an Fc-engineered antibody demonstrates enhanced antibody-dependent cell-mediated cytotoxicity in chronic lymphocytic leukemia. MAbs 2014;6(3):749–55.
[103] Sica A, Mantovani A. Macrophage plasticity and polarization: in vivo veritas. J Clin Invest 2012;122(3):787–95.
[104] Xue J, et al. Transcriptome-based network analysis reveals a spectrum model of human macrophage activation. Immunity 2014;40(2):274–88.
[105] Qian BZ, Pollard JW. Macrophage diversity enhances tumor progression and metastasis. Cell 2010;141(1):39–51.
[106] Mazzone M, Menga A, Castegna A. Metabolism and TAM functions-it takes two to tango. FEBS J 2018;285 (4):700–16.
[107] Sica A, et al. Tumour-associated macrophages are a distinct M2 polarised population promoting tumour progression: potential targets of anti-cancer therapy. Eur J Cancer 2006;42(6):717–27.
[108] Zhu Y, et al. CSF1/CSF1R blockade reprograms tumor-infiltrating macrophages and improves response to T-cell checkpoint immunotherapy in pancreatic cancer models. Cancer Res 2014;74(18):5057–69.
[109] Ries CH, et al. Targeting tumor-associated macrophages with anti-CSF-1R antibody reveals a strategy for cancer therapy. Cancer Cell 2014;25(6):846–59.
[110] Butchar JP, et al. Reciprocal regulation of activating and inhibitory Fc{gamma} receptors by TLR7/8 activation: implications for tumor immunotherapy. Clin Cancer Res 2010;16(7):2065–75.
[111] Dahal LN, et al. STING activation reverses lymphoma-mediated resistance to antibody immunotherapy. Cancer Res 2017;77(13):3619–31.
[112] Weiskopf K, Weissman IL. Macrophages are critical effectors of antibody therapies for cancer. MAbs 2015;7 (2):303–10.
[113] Biswas SK, Allavena P, Mantovani A. Tumor-associated macrophages: functional diversity, clinical significance, and open questions. Semin Immunopathol 2013;35(5):585–600.
[114] Lehmann B, et al. Tumor location determines tissue-specific recruitment of tumor-associated macrophages and antibody-dependent immunotherapy response. Sci Immunol 2017;2(7): eaah6413.
[115] Arce Vargas F, et al. Fc-optimized anti-CD25 depletes tumor-infiltrating regulatory T cells and synergizes with PD-1 blockade to eradicate established tumors. Immunity 2017;46(4):577–86.
[116] Zhao JL, et al. Forced activation of notch in macrophages represses tumor growth by upregulating miR-125a and disabling tumor-associated macrophages. Cancer Res 2016;76(6):1403–15.
[117] Zhao Y, et al. Bladder cancer cells re-educate TAMs through lactate shuttling in the microfluidic cancer microenvironment. Oncotarget 2015;6(36):39196–210.
[118] van Dalen FJ, et al. Molecular repolarisation of tumour-associated macrophages. Molecules 2018;24(1):9.
[119] van Egmond M, Vidarsson G, Bakema JE. Cross-talk between pathogen recognizing toll-like receptors and immunoglobulin Fc receptors in immunity. Immunol Rev 2015;268(1):311–27.
[120] Lennartz M, Drake J. Molecular mechanisms of macrophage Toll-like receptor-Fc receptor synergy. F1000Res 2018;7:21.
[121] Ishikawa H, Barber GN. STING is an endoplasmic reticulum adaptor that facilitates innate immune signalling. Nature 2008;455(7213):674–8.
[122] Barber GN. STING: infection, inflammation and cancer. Nat Rev Immunol 2015;15(12):760–70.
[123] Corrales L, et al. Direct activation of STING in the tumor microenvironment leads to potent and systemic tumor regression and immunity. Cell Rep 2015;11(7):1018–30.
[124] Kimby E, et al. Two courses of four weekly infusions of rituximab with or without interferon-alpha2a: final results from a randomized phase III study in symptomatic indolent B-cell lymphomas. Leuk Lymphoma 2015;56(9):2598–607.
[125] Friedberg JW, et al. Phase II study of a TLR-9 agonist (1018 ISS) with rituximab in patients with relapsed or refractory follicular lymphoma. Br J Haematol 2009;146(3):282–91.
[126] Ding C, et al. Small molecules targeting the innate immune cGAS–STING–TBK1 signaling pathway. Acta Pharm Sin B 2020. https://doi.org/10.1016/j.apsb.2020.03.001.
[127] Roghanian A, et al. Cyclophosphamide enhances cancer antibody immunotherapy in the resistant bone marrow niche by modulating macrophage FcgammaR expression. Cancer Immunol Res 2019;7(11):1876–90.
[128] Yin Y, et al. Phosphatidylserine-targeting antibody induces M1 macrophage polarization and promotes myeloid-derived suppressor cell differentiation. Cancer Immunol Res 2013;1(4):256–68.
[129] Gerber DE, et al. Randomized phase III study of docetaxel plus bavituximab in previously treated advanced non-squamous non-small-cell lung cancer. Ann Oncol 2018;29(7):1548–53.

[130] Georgoudaki AM, et al. Reprogramming tumor-associated macrophages by antibody targeting inhibits cancer progression and metastasis. Cell Rep 2016;15(9):2000–11.
[131] Chen HM, et al. Blocking immunoinhibitory receptor LILRB2 reprograms tumor-associated myeloid cells and promotes antitumor immunity. J Clin Invest 2018;128(12):5647–62.
[132] Zhou Y, et al. Blockade of the phagocytic receptor MerTK on tumor-associated macrophages enhances P2X7R-dependent STING activation by tumor-derived cGAMP. Immunity 2020;52(2):357–73. e9.
[133] Gabrilovich DI, Ostrand-Rosenberg S, Bronte V. Coordinated regulation of myeloid cells by tumours. Nat Rev Immunol 2012;12(4):253–68.
[134] DiLillo DJ, Ravetch JV. Differential Fc-receptor engagement drives an anti-tumor vaccinal effect. Cell 2015;161 (5):1035–45.
[135] Muraoka D, et al. Antigen delivery targeted to tumor-associated macrophages overcomes tumor immune resistance. J Clin Invest 2019;129(3):1278–94.
[136] Velmurugan R, et al. Macrophage-mediated trogocytosis leads to death of antibody-opsonized tumor cells. Mol Cancer Ther 2016;15(8):1879–89.
[137] Williams ME, et al. Thrice-weekly low-dose rituximab decreases CD20 loss via shaving and promotes enhanced targeting in chronic lymphocytic leukemia. J Immunol 2006;177(10):7435–43.
[138] Beum PV, et al. The shaving reaction: rituximab/CD20 complexes are removed from mantle cell lymphoma and chronic lymphocytic leukemia cells by THP-1 monocytes. J Immunol 2006;176(4):2600–9.
[139] Dahal LN, et al. Shaving is an epiphenomenon of type I and II anti-CD20-mediated phagocytosis, whereas antigenic modulation limits type I monoclonal antibody efficacy. J Immunol 2018;201(4):1211–21.
[140] Veri MC, et al. Monoclonal antibodies capable of discriminating the human inhibitory Fcgamma-receptor IIB (CD32B) from the activating Fcgamma-receptor IIA (CD32A): biochemical, biological and functional characterization. Immunology 2007;121(3):392–404.
[141] Jiang C, et al. BI-1206, a monoclonal antibody against Fcγriib, showed superior anti-tumor activity in an ibrutinib-venetoclax dual resistant PDX model in mantle cell lymphoma. Blood 2019;134(Suppl. 1):2863.
[142] Berntzen G, et al. Identification of a high affinity FcgammaRIIA-binding peptide that distinguishes FcgammaRIIA from FcgammaRIIB and exploits FcgammaRIIA-mediated phagocytosis and degradation. J Biol Chem 2009;284(2):1126–35.

CHAPTER

12

Novel immunotherapy strategies involving matrix metalloproteinase (MMP) family

Claudia A. Garay-Canales[a], *Laura Díaz-Alvarez*[b], *and Georgina I. Lopez-Cortes*[a]

[a]**Departamento de Inmunología, Instituto de Investigaciones Biomédicas, Universidad Nacional Autónoma de México, Ciudad de México, Mexico** [b]**Posgrado en Ciencias Biológicas, Universidad Nacional Autónoma de México, Ciudad Universitaria, Ciudad de México, México**

Abstract

Common cancer therapy includes surgery, radiotherapy, and chemotherapy; these techniques have some disadvantages, such as lack of specificity causing side effects and the possibility of relapse. Novel design of specific, nontoxic therapies for cancer requires deep knowledge of the biology and the molecules involved, which would result in a better outcome for patients. The matrix metalloproteinases (MMPs) are a family of zinc-dependent endopeptidases; some of the major substrates are the components of the extracellular matrix (ECM). MMPs are responsible for regulating numerous physiological and pathological events including bone development, wound repair, and different stages in carcinogenesis. Therefore, these enzymes represent highly relevant targets for cancer therapy. Overexpression of MMPs is well documented in most types of cancers; upregulated expression is present not only in cancer cells but also in stromal cells, which modify and regulate the tumor microenvironment. To break the cancer tolerance, here we present the evolution of some strategies: first efforts using MMPs were addressed to inhibit the catalytic site or the binding zinc domain; although several chemical inhibitors were synthesized, poor specificity to target only cancer cells were achieved, and none of them passed human clinical trials. Then, studies of crystal structure, phage display, and production of monoclonal antibodies revealed specific epitopes or domains within the MMPs; these cryptic sites are helpful to design new targets. Following, one novel strategy is the drug delivery systems (DDS); this strategy uses the upregulated expression of MMPs in cancer cells, and it is based on the specificity of sensitive MMP substrates or monoclonal antibodies attached to a highly toxic drug that only will be "activated or released" in the presence of the selected MMP. Altogether, knowing the features of the MMPs, it is possible to design more specific and less toxic therapies for several types of cancer, and combining these strategies with standard therapies, we could achieve the goal of better survival rate for cancer patients.

Immunotherapy in Resistant Cancer: From the Lab Bench Work to Its Clinical Perspectives
https://doi.org/10.1016/B978-0-12-822028-3.00015-7

Abbreviations

ECM	extracellular matrix
MMP	matrix metalloproteinase
MMPI	MMP inhibitor
MT-MMP	membrane-type matrix metalloproteinase
TIMP	tissue inhibitor of metalloproteinase
mAb	monoclonal antibody
scFv	single-chain variable fragment
BM	basal membrane
TGF-α or β	transforming growth factor-α or β
IL-2Rα	interleukin-2 receptor α
DDS	drug delivery system

Conflict of interest

No potential conflicts of interest were disclosed by the authors.

Introduction

The extracellular matrix (ECM) has a fundamental role in the adhesion of cells and communicates signals through cell surface adhesion receptors. The ECM contains collagen, noncollageneous glycoproteins, and proteoglycans. Other components such as tenascin, fibronectin, and different isoforms of laminin are found in tumors and might facilitate cancer progression. The basal membrane (BM) is a specialized ECM that physically separates the epithelial cells from the underlying stroma, thereby proving the first barrier against the invasion of carcinomas [1]. The matrix metalloproteinases (MMPs) are a broad family of zinc-dependent proteolytic enzymes with multiple roles in tissue remodeling and degradation of various proteins in the extracellular matrix (ECM). Their mechanism of action—degradation of proteins—regulates numerous physiological and pathological processes [2]. MMPs promote degradation of several different proteins: (i) extracellular matrix proteins such as collagen, gelatin, entactin, fibronectin, and laminin; (ii) cytokines, chemokines, and some of their receptors; and (iii) MMPs, once activated they can act over other MMPs, leading to a cascade of activation [3].

The resulting modifications can lead to changes in the cell and their microenvironment. Physiological processes include trophoblast implantation, embryogenesis, bone growth, angiogenesis, cell cycle, cellular proliferation, cell adhesion, other MMPs' activation, phagocytosis, and among others [4]. On the contrary, MMPs have been involved in the occurrence of some diseases, such as cancer, cardiovascular disorders, arthritis, inflammatory diseases, etc. [5–7].

There are many studies that involve MMPs in initial and late steps in cancer development since ECM degradation stimulates or inhibit angiogenesis, cell migration, invasion, metastasis and release activated cytokines stored in ECM [8]. Overexpression and increased activity of MMPs are reported in almost every type of cancer; therefore, there is an ongoing effort to use MMPs as therapeutic targets. Surprisingly, the use of the first chemical inhibitors gave disappointing results, since it was not considered the complex systemic effects affected and other activities of the molecule [2].

Many MMPs have been described as moonlighting proteins, which means they have different functions, aside from their enzymatic activity, due to other domains located along the protein [9]. Distinct domains in the protein perform independent enzymatic processes, such as cell adhesion, phagocytosis, angiogenesis, and receptors for some coronaviruses [10]. An overall view of these molecules will aid the understanding of dependent and independent enzymatic activity functions of these MMPs and would allow a better design to achieve a successful therapy.

History

In 1962, Jerome Gross and Charles Lapiere described for the first time the in vitro collagenolytic activity of a diffusible factor from amphibian tissue; they observed how the collagenase enables the distribution of large amounts of collagen in the tail of tadpoles during metamorphosis, which in turn facilitates the bodily transformation into the adult specimen [11]. This was the milestone for the metalloproteinase's study field, since, from the beginning, Gross and Lapiere's theoretical focus was on tissue remodeling and its importance for physiological growth and development. Soon, MMPs would be related not only to physiological phenomena, such as postpartum regeneration, but also to pathophysiological processes, like periodontal disease [12]. As early as 1967, Ukraine's Balitskiĭ and Pridato published a report on the perceived changes in collagen and its impact on cancer development, showing that the link between cancer and extracellular matrix (ECM)-degrading enzymes is only a scratch away from the surface [13].

Structure

Matrix metalloproteinases (MMPs) are an extensive family of metzincins. Their members are multidomain, highly homologous, zinc-dependent, calcium-containing endopeptidases. The MMPs are expressed as zymogens (pro-MMPs, since they have to be tightly regulated); they can be activated by other enzymes or free radicals through the cysteine-switch mechanism [14]. Aside a few exceptions, the basic structure of an MMP comprises three domains:

- Pro-peptide. Around 85 amino acids in length, this domain contains a thiol-forming cysteine essential for maintaining the inactive form of the enzyme, until otherwise needed.
- Catalytic domain. It spans approximately 170 amino acids, which contains both structural and catalytic metal ions, the most important is one molecule of zinc coordinated by three histidines, the heart of the catalytic site. Between the catalytic and hemopexin-like domains, there is a variable-length linker region, also known as "hinge region."
- Hemopexin-like domain. Named after a soluble protein with which it shares significant homology, this nearly 200-amino acid domain is key for substrate and natural inhibitors binding [15] (Fig. 1).

Classification

MMPs are classified as enzymes according to the Nomenclature Committee of the International Union of Biochemistry and Biology, falling under 3.4.24.X classification, which indicates they are hydrolases (1st number), acting on peptide bonds (peptidases, 2nd number), and grouping as metalloendopeptidases (3rd number). The 4th number is left as "X" since

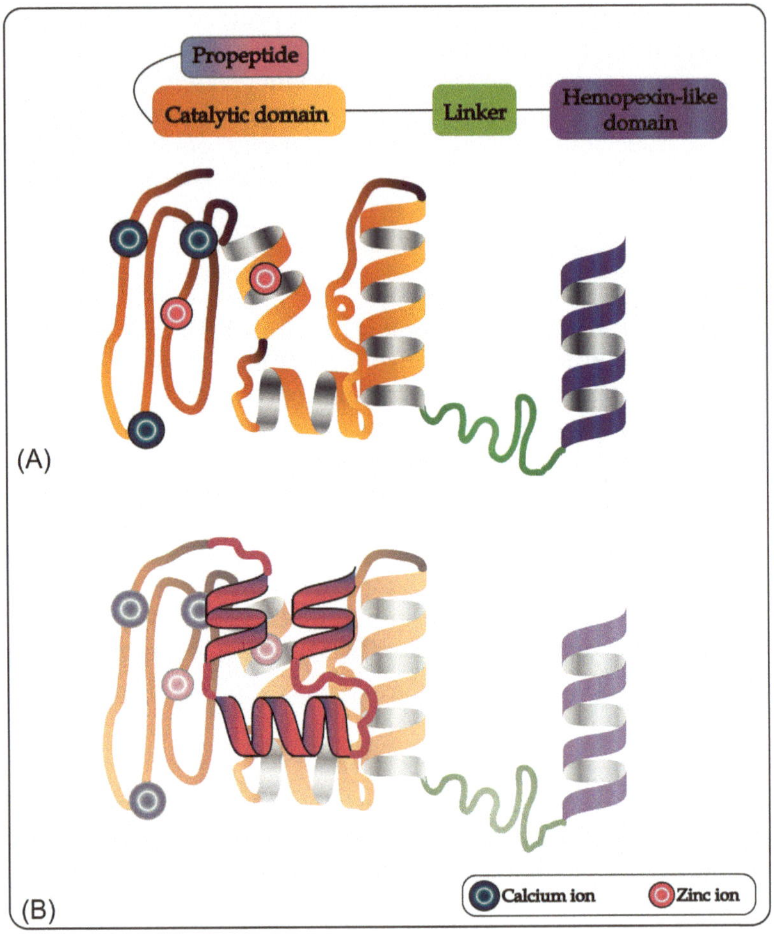

FIG. 1 General structure of matrix metalloproteinases, schematic representation. Domains are color-coded as shown on the top part. (A) would be the active MMP, whereas (B) means to illustrate the inactive MMP with the pro-peptide still attached and making contact with the catalytic zinc ion.

it is the individual identifier of each enzyme. For example, MMP-1 EC number is 3.4.24.7 [16,17]. Nowadays, members of the MMP family are known to cleave virtually any component of the ECM. At first, MMPs were classified based on their specificity for the ECM components; thus, we had collagenases, gelatinases, stromelysins, matrilysins, etc. [2].

1. Collagenases. The main substrates of these enzymes, as its name specifies, are fibrillar collagens; however, they are also known to cleave other proteins, such as growth factors. They are central to tissue regeneration. MMP-1, MMP-8, and MMP-13 are representative members of this family [18].
2. Gelatinases. These enzymes are capable of hydrolyzing components of the basal lamina such as type IV collagen, elastin, fibronectin, and laminin. Giving them a role in processes

like muscular fiber development and integrity maintenance. MMP-2 and MMP-9 are examples of gelatinases [19].

3. Stromelysins. They have a wide range of substrates, from the expected ECM proteins like aggrecan and tenascins, to cytokines like TNF-α and IL-1β, and precursors of other MMPs, thus participating in immune and stress responses. MMP-3 is a member of this family [20].
4. Matrilysins. These peptidases were originally related to cancer from their discovery, for instance, MMP-26 was cloned from an endometrial tumor cDNA library. Among their substrates, there are gelatin, laminin, elastin, and integrins, like E-cadherin. Matrilysins are not highly expressed in a constitutive manner; rather, they augment physiologically, during tissue repairing or pathophysiologically, in tumor invasion and angiogenesis [21,22].
5. Membrane-type (MT). Depending on the enzyme, MT-MMPs are anchored to the plasma membrane either through glycosylphosphatidylinositol or a transmembrane domain. The members of this family cleave a large number of molecules, such as the inactive forms of other MT-MMPs (e.g., pro-MMP-2) and cytokines, integrins (e.g., ICAM-1), and, of course, ECM components (fibronectin, vitronectin, type I, II, and III collagens). There are 6 known members of this family in humans, precisely named from 1 to 6 [23].
6. Other enzymes. These include some well-studied members like MMP-12, MMP-19 and MMP-20, with well-known substrates (type I and IV collagens and amelogenin). Regarding their functions, there is evidence that macrophage elastase (MMP-12) is involved in the recruitment of inflammatory cells and that enamelysin (MMP-20) takes part in odontogenesis [24,25].

Since the substrates for the MMPs continued growing, the new classification consists in a sequential numbering system, starting MMP-1 and ending with MMP-28 (excluding MMP-4, MMP-6 and MMP-22, since these enzymes were discovered simultaneously by different groups); there are 23 members reported in humans [3]. Now, the MMPs are grouped according to their structure, there are eight different structural classes: five are secreted and three are membrane-type MMPs (MT-MMPs) (Table 1).

The MMPs can regulate cell behavior by pericellular proteolysis; MT-MMPs, which are covalently attached to the membrane, are not the only ones whose activity is performed on the cell membrane; the secreted MMPs can be found on the cell surface by binding to integrins [26], CD44 [27], or heparan sulfate proteoglycans, type IV collagen, etc. [28]. MMPs have also been located in the cytoplasm, and some of them even inside the nucleus (Fig. 2).

More than half of human MMPs are reported on more than one location, which correlates with their role in specific scenarios. For instance, high serum levels of MMP-1 are related to poor prognosis in patients with nonsmall cell lung cancer. However, this enzyme is also known to associate with mitochondria and the nucleus, thus conferring resistance to apoptotic degradation [29,30]. It has also been reported that MMPs are released by using an extracellular vesicle strategy such as exosomes and microvesicles [31].

Functions and regulation

As mentioned earlier, MMPs have an important role in physiological processes like morphogenesis, cell proliferation, bone embryonic development, bone differentiation, tissue

TABLE 1 MMPs classified in structural families and location.

Structural family	Localization	Members
Simple hemopexin-domain-containing MMPs	Secreted	MMP-1, MMP-3, MMP-8, MMP-10, MMP-12, MMP-13, MMP-18, MMP-19, MMP-20, MMP-22, MMP-27
Gelatin-binding MMPs	Secreted	MMP-2, MMP-9,
Furin-activated secreted MMPs	Secreted	MMP-11, MMP-28
Minimal-domain MMPs	Secreted	MMP-7, MMP-26
Vitronectin-like insert MMPs	Secreted	MMP-21
Transmembrane MMPs	Membrane-type	MMP-14, MMP-15, MMP-16, MMP-24
GPI-anchored MMPs	Membrane-type	MMP-17, MMP-25
Type II transmembrane MMPs	Membrane-type	MMP-23

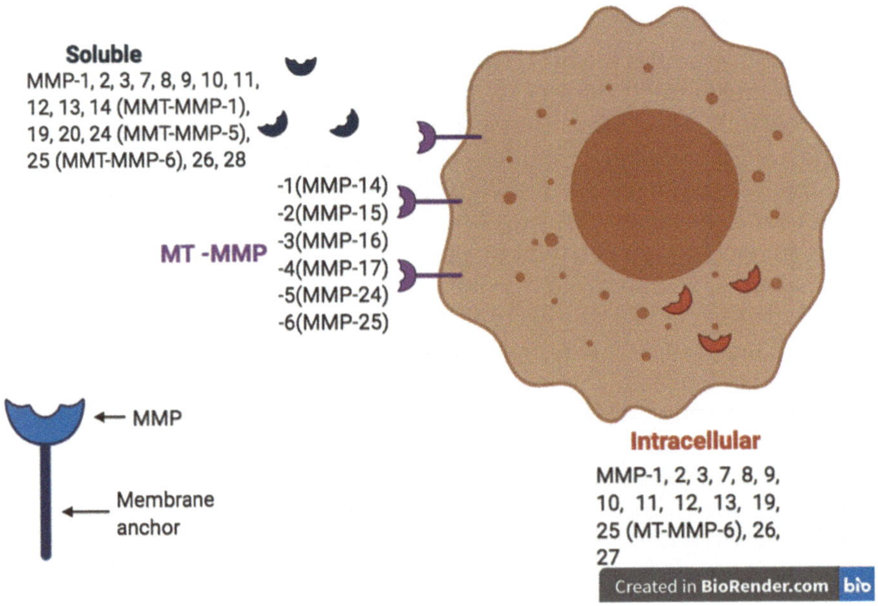

FIG. 2 Different cellular locations of MMP. MMP can be found inside the cell, in the cytoplasm, and a few of them in the nucleus (not shown); anchored to the membrane, either through a transmembrane domain or through a glycosylphosphatidylinositol moiety, or secreted into the extracellular space, where they can sometimes reach circulation.

repair, survival, migration, apoptosis, angiogenesis, tissue regeneration, and immune response. Regarding the enzymatic activity, the cleavage of ECM proteins can generate fragments with new functions: cleavage of laminin-5 and type IV collagen results in exposure of cryptic sites that promote migration; ADAMs (a disintegrin and MMP) can release cell-bound precursor forms including transforming growth factor-α (TGF-α); MMP-2 and MMP-9 can release TGF-β from an inactive extracellular complex [2] (Fig. 3).

As previously described, these MMPs are moonlighting proteins, besides the enzymatic activity, they contain independent domains from the catalytic activity, responsible for different types of processes, such as adhesion, cell-cell interactions, polarization of immune cells, and phagocytosis, among others [9,10]. Conversely, MMPs have been involved in pathological progression of several different human diseases in nervous system, cardiovascular disorders, osteoarthritis, osteoporosis, pathogenesis of coronaviruses (as host receptor), and a major role in carcinogenesis [32].

The expression of MMPs' genes is observed mainly in the connective tissue, pro-inflammatory, and uteroplacental cells, including fibroblast, osteoblast, hematopoietic progenitor cells, neutrophils, monocytes, macrophages, dendritic cells, mast cells, lymphocytes,

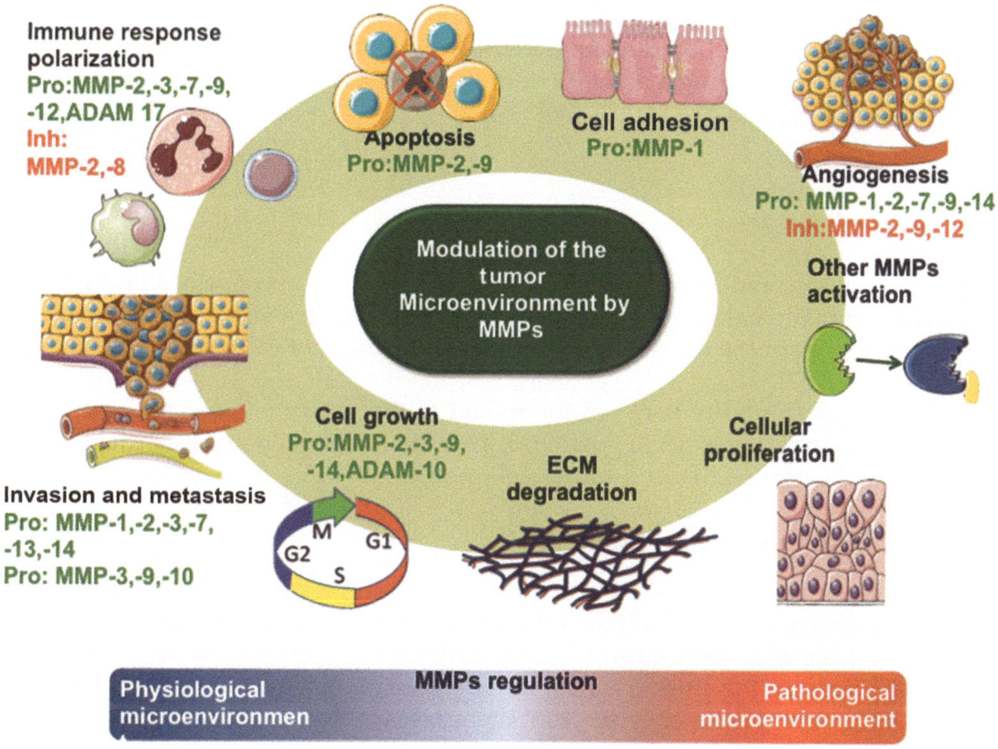

FIG. 3 Processes modulated by the different MMPs. Physiologically, MMPs play different roles for maintaining the tissues' homeostasis by promoting cellular proliferation and cell growth but also avoiding apoptosis. In case of injury, MMPs degrade damaged tissue, cell adhesion molecules, cytokines, and other substrates that may enhance angiogenesis. When MMPs' activity is deregulated, they contribute to enhance a pathological microenvironment breaking down homeostasis. Figure constructed with SMART images. https://smart.servier.com/

vascular smooth muscle (VSM), and endothelial cells (Fig. 3). MMPs are highly active enzymes whose control must be thoroughly regulated. These proteins are synthesized as inactive zymogens, and they are activated by proteinase cleavage. There are several levels of regulation: gene transcription, the MMP mRNA half-life modulation, secretion, compartmentalization, proenzyme activation, endocytosis, and inhibition of enzyme activity [33].

The function of MMPs in vivo depends on the local balance between them and their physiological inhibitors. There are endogenous and exogenous inhibitors of MMPs: (A) α2-macroglobulin and thrombospondin-2 are endogenous inhibitors that form a complex with the MMPs like MMP-2 and then bind to a scavenger receptor and are irreversibly cleared by endocytosis, while thrombospondin-1 binds to pro-MMP-2 or 9 and directly inhibits their activation. (B) Some exogenous inhibitors are tissue inhibitors of MMPs (TIMPs), which are widely distributed in the body and may inhibit more than one MMP with absolute or relative changes. TIMP-1 and TIMP-2 can block MMP effects in promoting tumor cell proliferation and migration, but also can inhibit angiogenesis and apoptosis. On the opposite site, TIM-3 can induce apoptosis through TNF-α receptor stabilization or by overexpression of TIM-4, depending on the region involved [34].

Some MMPs are constitutively expressed in normal tissues; meanwhile, other MMPs appear only in pathological processes like inflammation, tissue repair or remodeling, cancer, etc. [35]. As mentioned earlier, the MMPs are expressed as an inactive zymogen form (pro-MMPs), but once activated they can subsequently activate other MMPs, which may lead to a cascade of activation, creating a network of interactions between MMPs resulting in potentiate a pathological process like cancer [36].

MMPs in cancer

During cell transformation and carcinogenesis, the interactions of cells with ECM modify the microenvironment; malignant cancer cells are embedded in vasculature and surrounded by a dynamic tumor stroma consisting of various nonmalignant cells, such as fibroblast and myeloid cells (Table 2).

There is a strong similarity between the milieu of the tumor microenvironment and the inflammatory response in a healing wound; both promote angiogenesis, degradation of the extracellular matrix (ECM), and tumor cell motility [37].

Role of MMPs in cancer development

Cell migration or angiogenesis can be regulated by the cleavage of some ECM proteins like fibronectin, thrombospondin-1, laminin, and osteopontin. Previously, MMPs were associated only with metastasis because it was only considered the role of the enzymes breaking down the physical barriers. At this time, it is clear that they are involved in all stages of cancer, from initiation to outgrowth of clinically relevant metastases [38].

(a) Dysregulation of expression of MMP in cancer

Special attention has been focused on the dysregulation of MMP expression at the protein and RNA levels in many cancer types; there are extensive literatures

TABLE 2 Activators and inhibitors of different immune cell lineages.

Stromal cells			
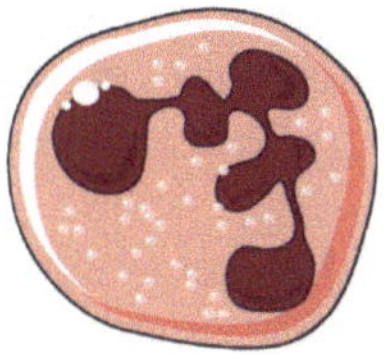	**Neutrophils** **Proteinase** MMP-8, MMP-9 ADAM-8, ADAM-17 ADAMTS-1 **Inhibitors** TIMP-1	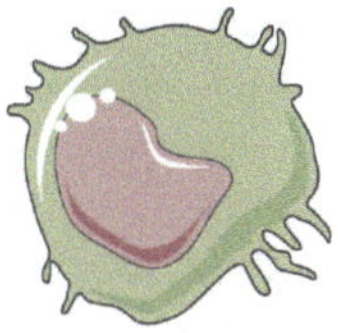	**Macrophages** **Proteinase** MMP-1, MMP-2, MMP-7, MMP-9, MMP-12, MMP-14 ADAM-9, ADAM-15, ADAM-17 ADAMTS-4 **Inhibitors** TIMP-1
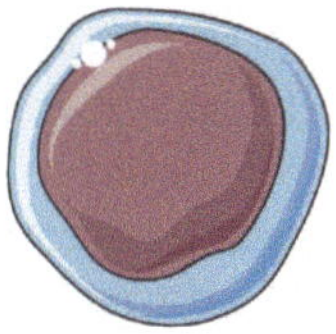	**Lymphocytes** **Proteinase** MMP-3, MMP-9 ADAM-17, ADAM-28 **Inhibitors** TIMP- 1	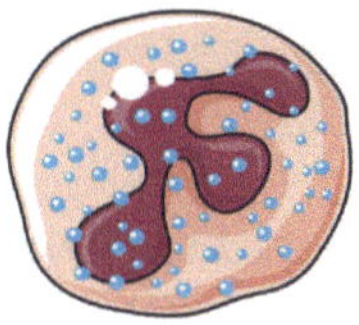	**Mast cells** **Proteinase** MMP-2, MMP-9 Chymase Tryptase **Inhibitors** TIMP-1
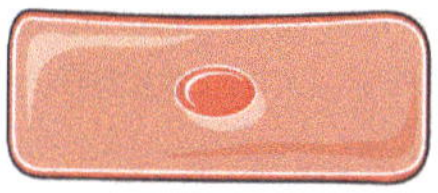	**Endothelial cells** **Proteinase** MMP-2, MMP-3, MMP-7, MMP-14, MMP-19 ADAM-15, ADAM-17 **Inhibitors** TIMP-1, TIMP-2	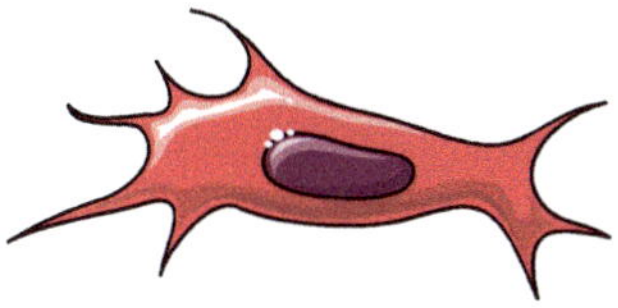	**Fibroblast** **Proteinase** MMP-1, MMP-2, MMP-3, MMP-9, MMP-11, MMP-13, MMP-14, MMP-19 ADAMTS-5 **Inhibitors** TIMP-1, TIMP-2, TIMP-3
	**Dendritic cells** **Proteinase** MMP-1, MMP-2, MMP-3, MMP-9, MMP-19 ADAM-15, ADAM-17 **Inhibitors** TIMP- 1, TIMP-2	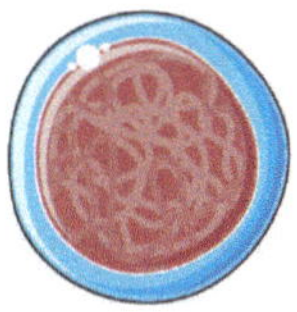	**Hematopoietic progenitor cells** **Proteinase** MMP-1, MMP-2, MMP-3, MMP-9, MMP-11, MMP-13, MMP-14, MMP-19 ADAMTS-5 **Inhibitors** TIMP-1, TIMP-2, TIMP-3
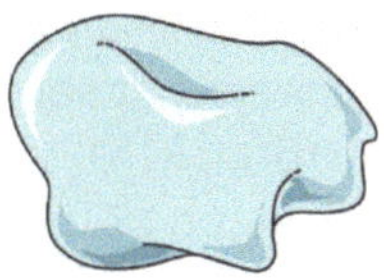	**Platelets** **Proteinase** MMP-1, MMP-2, MMP-3, MMP-14 **Inhibitors** TIMP-1, TIMP-2, TIMP-3, TIMP-4	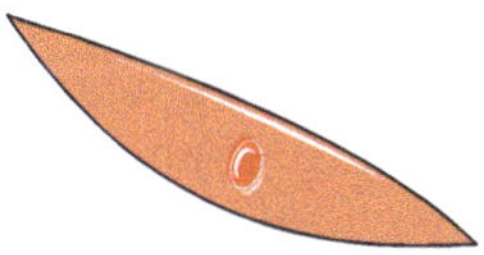	**Vascular smooth muscle cells (VSM)** **Proteinase** MMP-1, MMP-2, MMP-3, MMP-7, MMP-8, MMP-9, MMP-12, MMP-13 **Inhibitors** TIMP-1, TIMP-3

Adapted from Kessenbrock K, Plack V, Werb Z. Matrix metalloproteinases: regulator of the tumor microenvironment. Cell 2015;141(1):52-67.

regarding how the difference in the expression of MMP can be associated with poor prognostic outcome for the cancer patients, especially noticed in breast, ovarian, prostate, and colon cancers. Cancer cells are not the only responsible for changes in the tumor microenvironment and MMP-7 are upregulated in cancer cells, but MMP-2 and MMP-9 are synthesized by the tumor stromal cells, including fibroblast, myofibroblast, inflammatory, and endothelial cells. Cancer cells might stimulate tumor stromal cells to synthesize MMPs in a paracrine manner through secretion of interleukins, interferons, emmprin, and growth factors [28].

A study of differences in gene expression of 24 MMP family members in 15 different cancer types showed mostly upregulation of the majority of collagenases, matrilysins, metalloelastases, and stromelysins, although with a large degree of heterogeneity across cancers; MMP-14 showed strong upregulation; meanwhile, MMP-23B, MMP-27, and MMP-28 were downregulated. MMP-11 was the stromelysin most significantly upregulated in all cancer types, presumably with the fact that this MMP can help evading the immune surveillance by desensitizing NK cells; MMP-3 and MMP-10 were upregulated in 7 and 10 cancer types, respectively. Among collagenases, MMP-1, and MMP-13 were significantly upregulated in 11 and 12 cancer types, respectively.

In colon cancer, MMP-7 upregulated the gene expression 141-fold change; compared with normal tissue, this is consistent with numerous studies of colon cancer, indicating that MMP-7 could predict a more aggressive phenotype of colon cancer and a poor patient survival; although the upregulation of MMP-7 was found on other cancer types such as esophageal, stomach, thyroid and head/neck, the more aggressive phenotype could be explained to the intrinsic expression of MMP-7 in colon. Only lung squamous and uterine corpus endometrial were significantly downregulated [39].

(b) Regulation by oncogenes

Oncogenes, c-erbB-2 or c-Ha-ras, elevate the MMP-2 expression in epithelial mammary cells, and the overexpression became even higher when both oncogenes were active. On the contrary, the same cells decreased the mRNA levels of the metalloproteinase inhibitor TIMP-2. This overexpression of MMP-2 and downregulation of TIMP-2 seemed to facilitate the cancer cells to establish in mice lungs [40]. In a similar way, MMP-9 expression was induced by the transcription factor E1AF in a noninvasive cell line, which conferred an invasive phenotype. Moreover, the downregulation of the transcription factor with an anti-sense RNA caused not only less protein levels of MMP-9 but also of MMP-1 and MMP-3 [41,42].

(c) Activation of MMPs

MMPs' activation is reached by removing the pro-domain of the chain; this pro-domain interacts with the catalytic site impeding the substrate entrance; MMPs are activated by other enzymes in a redundant fashion. It is presumed that MMP-7 participates in pro-MMP-2 and pro-MMP-9 activation. MMP-13, who has a wide range of substrates and is detected at early stages of tumor, activates MMP-2, MMP-3, and MMP-14. And as positive regulation, MMP-14 activates MMP-2 and newly synthetized MMP-13. Interestingly, MMP-13 expression pattern is elevated in the tumor edges promoting the infiltration of immune cells and the exit of cancer cells which in turn allows microinvasion.

Role of MMPs in different stages of cancer

Throughout the progression of cancer, different metalloproteinases are expressed in the microenvironment in order to obtain benefits such as opening space for tumor growth, bringing the bloodstream closer and thus obtaining nutrients or releasing cancer cells to establish secondary tumors in other tissues.

(a) Tumor growth

There is strong evidence indicating that MMPs generate growth-promoting signals; MMP-$9^{-/-}$-deficient mice showed a decreased in cancer cell proliferation. There are several ways which some MMPs can regulate growth of cancer cells (i) by releasing precursors bound to cell membrane of some growth factors, such as TGF-α, which is released by MMPs or ADAMs; (ii) modulating bioavailability of growth factors sequestered by ECM proteins after degradation by MMPs; or (iii) regulating proliferative signals by integrins. On the contrary, MMPs can inhibit tumor growth activating TGF-β or stimulating the production of pro-apoptotic molecules such as TNF-α or Fas ligand [43]. When MMPs degrade ECM, it reveals hidden sites in the ECM where different receptors can be bound.

(b) Apoptosis

Importantly, cancer cell survival depends on the evasion of apoptosis in the presence of genetic instability; low levels of oxygen and nutrients; immune surveillance and detachment from the ECM. MMPs have both apoptotic and anti-apoptotic functions: MMP-3 induces apoptosis when overexpressed in mammary epithelial cells, degrading laminin; MMP-7 releases FASL (a transmembrane stimulator of the death receptor FAS) and cleaves pro-heparin-binding epidermal growth factor to generate the mature form promoting cell survival via ERBB4 receptor; and MMP-11 inhibits cancer cell apoptosis, since overexpression decreases spontaneous apoptosis of tumor xenografts, but conversely, cancer cells injected on $MMP^{-/-}$ null mice have a higher rate of spontaneous apoptosis than wild-type host. Also, MMPs or ADAMs cleave VE-cadherin, PECAM-1, and E-cadherin during the apoptosis of endothelial or epithelial cells.

(c) Angiogenesis

Angiogenesis is a complex mechanism, essential for tumor growth and development, where new blood vessels are formed from existing vessels. Various studies had previously reported the high expression of MMP-2 and MMP-9, being MMP-2 predominant in early phases of tumor growth, while MMP-9 during late ones. MMP-2 expression in cancer cells decreases angiogenesis in a chicken model, and in the MMP-$2^{-/-}$-deficient mice; MMP-2 cleaves type IV collagen, exposing cryptic integrin-binding sites that are involved in migration and in vitro angiogenesis. MMP-9 promotes angiogenesis by different ways: one is driving the vessel sprout by release of vascular endothelial growth factor (VEGF) and another is by proteolysis of angiogenesis inhibitors like angiostatin, endostatin, tumstatin, etc. [44]. MMP-14 promotes tumor angiogenesis since antibodies directed against the catalytic domain or the MMP-14 null mice inhibit invasion and capillary tube formation in vitro; MMP-14 can degrade the fibrin matrix surrounding newly formed vessels, thereby potentially allowing the endothelial cells to invade further [45].

(d) Invasion and metastasis

Once the tumor has grown, the next step, the metastasis is a multistage mechanism, starting from losing the intercellular connections and releasing of single cells of the tumor through anoikis prevention. Then, cancer cells migrate and cross through several ECM barriers. First, they cross the epithelial basement membrane and invade the surrounding stroma, next enter the blood or lymphatic vessels and extravasate, and finally establish a secondary growth in the new location. Tumor cells are insensible to anoikis may be due to the loss of cell connections; the process where there is a transition in the phenotype from epithelial to mesenchymal is called EMT, where an increase in migration, invasion, and metastasis is observed. Major inducer of EMT is TGF-β, activated by MMP-28 [46].

There is extensive experimental evidence for the role of MMPs in metastasis and in vitro invasion assays. For instance, cleavage of E-cadherin and CD44 results in the release of fragments of the extracellular domains and an increased invasive behavior [47]. In the course of invasion, MMPs relocalize to the invadopodia (special actin-rich surface protrusions), on the cell membrane where the increased ability to promote invasion is associated with ECM degradation. The MMPs localized in the invadopodia are MMP-2, MMP-9, and MMP-14. MMP-14 is recruited to invadopodia through its transmembrane and cytoplasmic domains [48].

MMPs as targets in cancer therapy

As it has been shown, some MMPs are overexpressed in order to facilitate the matrix remodeling for gaining space, breaking adhesive interactions, promoting angiogenesis, and facilitating cancer cells scape to other tissues. These enzymes are conveniently used to facilitate the intake of nutrients and oxygen in blood vessels as well as growth factors. However, it should be emphasized that not all MMPs have the same effects. Although some MMPs share some substrates, not all of them have the same behavior and regulation in the pathology.

Possible MMPs targets

Here, we present some MMPs which role in one or more stages in cancer makes them an attractive target to cancer therapy (in Table 3, there is the summary). MMP-1 promotes the degradation of the interstitial collagen and is regulated by TIMP-1. The overexpression of both MMP-1 and TIMP-1 is associated with an elevated invasive and migratory capacity in hepatocellular carcinoma, most likely by ECM degradation in the process of EMT.

In breast cancer, it is established that MMP-8 expression not only suppresses tumor progression and metastasis but also lengthens relapse-free survival of patients [44]. As mentioned earlier, MMP-8 is produced by neutrophils when they are activated with inflammatory stimuli and in cancer cells, it increases adhesion from tumor cells to ECM [49]. Consequently, the invasion is reduced and neutrophils could attack. So, the presence of MMP-8 is the result of an inflammatory response, which is necessary to fight cancer. Conversely, when the tumor microenvironment turns the immune response to be regulatory, inflammation cells that attack the tumor are no longer recruited.

TABLE 3 Role of MMPs in cancer.

Family	Enzyme	Function in cancer
Collagenases	MMP-1 MMP-8 MMP-13	Invasion, promotes metastasis Angiogenesis, migration Growth, invasion, and angiogenesis
Enamelysin	MMP-20	Synthesized in odontogenic tumors
Gelatinases	MMP-2 MMP-9	Proteolytic degradation of extracellular proteins in tumor invasion, collagenolytic pathway driver for lymphatic vessel formation, tumor angiogenesis Angiogenesis, proteolytic degradation of extracellular proteins during tumor invasion
Matrilysins	MMP-7 MMP-26	Contributes to invasive potential, proliferation, anti-apoptotic, and immune surveillance Activates MMP-9 in prostate cancer, role in early skin carcinogenesis
Membrane-type	MMP-14 MMP-15 MMP-16 MMP-17 MMP-24 MMP-25	Cleaves other pro-MMPs (mainly MMP-2) to activate them, role in invasive blood vessel growth, and promoting metastasis. In vitro has been shown to promote invasion In vitro shown to play role in epithelial to mesenchymal transition, promotes angiogenesis In vitro promotes invasion and metastasis Induces angiogenesis promote growth and metastasis Progression in brain tumors, aides in migration and metastasis In vitro tumor growth promoter
Metalloelastase	MMP-12	Protective inhibition of tumor growth, anti-angiogenic
Stromelysins	MMP-3 MMP-10 MMP-11	Invasion, metastasis, and epithelial to mesenchymal transition Invasion, migration, and growth; prevents tumor cell apoptosis; produces angiogenic and metastatic factors Produced by peritumoral stromal fibroblasts; regulates early tumor invasion, implantation, and expansion; prevents apoptosis of early cancer cells
Others	MMP-19 MMP-21 MMP-23A MMP-28	In vitro modulates proliferation, adhesion, and metastasis Expression changes associated with cancer prognosis Expression levels altered in multiple cancers Urinary levels decreased in renal cell carcinoma Promotes epithelial to mesenchymal transition, promotes invasion and metastasis

MMP-14 is a potential target for cancer treatments because it is associated with bad prognostics of breast cancer. MMP-14 is membrane-bound matrix metalloproteinase (MT1-MMP) with a collagenase activity that permits to remodel tissue and as long as it is membrane bound, enables cells to get through the extracellular matrix. Besides MMP-14 activates MMP-2 by cleaving its pro-domain, therefore, the expression of MMP-2 is highly linked to MMP-14. Studies and clinical reports correlate MMP-14 expression with tumor progression, metastasis, angiogenesis, invasion, and growth.

Unfortunately, some cancer patients come up with another tumor after being treated, and this is caused mainly because some cancer cells scape from the initial tumor by metastasis and rise in silence in other tissues like bones. During this time, it is difficult to detect new tumors;

TABLE 4 Potential role of MMP as biomarkers, alone or in combination (&).

Cancer type	Modulation	MMP involved
Lung squamous	Upregulation	MMP-11 & MMP-12
Colon cancer	Upregulation	MMP-11, MMP-7 & MMP-13 & MMP-28
Esophageal cancer	Upregulation	MMP-11, MMP-12
Breast cancer	Upregulation	MMP-13, MMP-1, MMP-2
Renal cell carcinoma	Upregulation	MMP-14, MMP-17, MMP-23B
Pancreatic carcinoma	Upregulation	MMP-7
Liver cancer	Upregulation	MMP-1, MMP-3, MMP-7 & MMP-2, MMP-9
Thyroid cancer	Upregulation	MMP-11 & MMP-19
Renal cell carcinoma	Downregulation	MMP-15, MMP-20, MMP-24

therefore, the patient stops treatment after removing the original. This time is named as the relapse-free survival and is different per patient; however, epidemiological studies can be performed. A study in Spain helped to make a prognosis of patients' relapse-free survival depending on the MMPs' expression. Patients who had had breast cancer were followed up for new sprouts. New tumors had in average high expression of MMP-9 and MMP-11 but also TIMP-1 and TIMP-2. Specifically, MMP-9 and TIMP-2 were expressed principally by tumor cells; fibroblast around the tumor expressed MMP-1, MMP-7, MMP-9, MMP-11, MMP-13, TIMP-2, and TIMP-3; and mononuclear inflammatory cells expressed MMP-7, MMP-9, MMP-11, MMP-13, MMP-14, TIMP-1, and TIMP-2 [50]. The richness of this study is that reveals that the neighboring cells contribute to generating a favorable microenvironment for tumor growth. It is important to note that the attempt to regulate the high levels of MMPs' activity with TIMPs is not lost at all, but overall imbalance worsens the pathology of the malady. In Table 4, we show some MMPs whose expression changes during the tumor formation and growth and would be important to evaluate the changes as biomarkers for cancer progression.

Strategies for anticancer therapy

Based on the important role of several MMPs in cancer progression and diseases, it is logical to assume the efforts to develop specific inhibitors avoiding unwanted side effects. However, after many years to develop synthetic MMP inhibitors, only a single inhibitor, Periostat, a tetracycline derivative, is used in periodontal disease [51]. More than 50 clinical trials conducted with active-site MMP inhibitors have failed due to the onset of significant dose-limiting musculoskeletal toxicity or lack of efficacy [52]. Most of these compounds act to perturb the critical coordinated zinc in the catalytic domain, which results in the loss of enzymatic activity; it is likely that the sequence conservation and structural similarity in the catalytic domain among MMPs contribute to the nonspecific inhibition of these compounds.

Although there have been promising results with significantly better survival rates, the toxicity must be addressed [53].

In order to produce drugs with better antitumor effect, developing inhibitors should be specific only to those members of the MMPs, which has the pro-cancerogenic properties. Also, it is very important to distinguish the stage of cancer progression in order to use the proper strategy: in early stages, the inhibition of the MMPs' activity could lead to the suppression of tumor growth, whereas in the fully formed and vascularized tumor, MMP activity is not critical for survival.

There are main strategies for MMPs' inhibition at the level of transcription, activation, and inhibition of their catalytic activity. At the transcription level, the inhibition of MMPs can be achieved by interfering with extracellular factors such as interferon; blocking signal transduction pathways, like MAPK or ERK pathway; and modulating nuclear factor of transcription, e.g., NF-κB or AP-1 [54,55]. Regarding MMPs' activity regulation, as previously described, most MMPs are secreted as inactive zymogens, the strategy proposed is considered the use of monoclonal antibodies to inhibit MMPs. The next goal is to inhibit active MMPs with exogenous or endogenous factors, such as α-2 microglobulin, or specific TIMP [56], and small molecules or peptides identify by phage display [57]. Here, we describe deeply relevant strategies and prospects:

(a) Highly selective chemical inhibitors

Among the MMP inhibitors (MMPI) are synthetic peptides, nonpeptidic molecules, chemically modified tetracyclines, and bisphosphonates. Regarding synthetic peptides, some act as competitive inhibitors that inhibit MMP activity, mainly by interacting with Zn^{2+} ions located in catalytic sites of the enzyme. Hydroxypyrones and hydroxythiopyrones are alternative zinc-binding groups (ZBGs) that, when combined with synthetic peptide backbones, comprise a novel class of MMP inhibitors. Particularly, Batimastat (BB-94)—a hydroxamate derivative with low water solubility and a broad-spectrum of inhibition, was one of the first peptides to start clinical trials. Later, Marimastat (BB-2516) with better solubility was introduced, however, showed side effects, such as musculoskeletal pain [32].

The second group of MMPIs is nonpeptidomimetic with high specificity, because these agents are based on the 3D conformation of the MMP active site. This category comprises of Prinomastat (AG-3340), Tanomastat (BAY12–9566), and MMI 270 B (CGS 27023 A). Some also exhibit musculoskeletal side effects. The third category of MMPIs are chemically modified tetracyclines, such as Metastat (COL-3, CMT-3) minocycline, and doxycycline. The modifications lower systemic toxicity and lacks antimicrobial activity. The fourth group of MMPIs are bisphosphonates, besides selected inhibition of MMPs' activity; they are involved in the inhibition of the mevalonate pathway, osteoclast activity, and bone reabsorption. Certain bisphosphonates influence the gene and protein expression of several MMPs and TIMPs, especially in breast cancer. The other agent is letrozole—a reversible nonsteroidal inhibitor of P450 aromatase, which additionally to its main function inhibits the gelatinases (MMP-2 and MMP-9) released by breast cancer cells and limits the metastatic potential of these cells.

As previously described, the main reason for the failure of some chemical inhibitors is the similarity among the conserved catalytic domain of MMPs, along with the high doses required to an adequate effect. Therefore, a different approach is needed. One strategy uses a biochemical and structural screening paradigm, JNJ0966 is a highly selective compound that inhibited the activation of MMP-9 zymogen and subsequent generation of catalytically active enzyme. JNJ0966 had no effect on MMP-1, MMP-2, MMP-3, MMP-9, or MMP-14 catalytic activity and did not inhibit activation of the highly related MMP-2 zymogen; the success in the specificity is due to the interaction of JNJ0966 with a structural pocket in proximity to the MMP-9 zymogen cleavage site near Arg-106, which is distinct from the catalytic domain. JNJ0966 was efficacious in reducing disease severity in a mouse experimental autoimmune encephalomyelitis model, demonstrating the viability of this therapeutic approach [58].

Indeed, one of the major concerns of chemical inhibitors is the toxicity of healthy tissues. With this in mind, there are novel efforts, like in vivo monitoring of the activity of specific MMPs, with fluorinated and nonfluorinated, e.g., analogues of a secondary sulfonamide-based lead structure, named compound 2, and test their efficacy as in vivo inhibitors and tracers of the gelatinases, MMP-2 and MMP-9. Using a murine neuroinflammatory model, compound 2 resulted a highly effective in vivo inhibitor of both MMP-2 and MMP-9 activity with little or no adverse effects even after long-term daily oral administration [59]. In vivo monitoring could give important information of localization of MMP, their activity, and the efficacy of new inhibitors.

(b) Phage display

As many chemical inhibitors are nonspecific for the desired MMP, none of them has been approved as treatment for cancer and therefore other strategies are being proposed to develop specific inhibitors of the enzymatic activity of each MMP. Since the 2000's enormous efforts have been done in developing targeted drugs. Phage display technology has been a power tool used to select high-affinity sequences that impede catalytic activity of certain MMPs. With this tool, it has been identified residues that inhibit the activity of the gelatinases MMP-2 and MMP-9 and the membrane-bounded type 1 MT1-MMP or MMP-14, which are metalloproteinases overexpressed in different types of cancers [60,61].

Using different selection strategies, it seems that the tertiary structure is important for enhancing high-affinity interactions with the enzymes, because cyclic sequences are being selected. The selected sequences are then the raw material for developing artificial sequences, antibodies, or macrocomplexes containing anticancer drugs, hoping they could be the next generation of cancer treatments. In prostate carcinoma, the goal was to identify the peptides able to interfere with cell adhesion, spreading, motility, and invasion [62]. Recent strategies include engineering approach combining phage display, rational design, and d-amino acid screening to obtain synthetic inhibitor to MMP-2 with high potency, target selectivity, and proteolytic stability [63].

By using phage display, it was selected a sequence with affinity to MMP-2 and MMP-9 which contained HWGF. This was used to synthetize the peptide CTTHWGFTLC who reveal to inhibit migration in an in vitro model and tumor growth, invasion, and survival in vivo models [57].

(c) Monoclonal antibodies

Monoclonal antibodies provide an exquisite specificity capable of distinguishing between closely related proteinase family members. Other important features of monoclonal antibodies are stability in serum, potential to cross blood-brain barrier, novel design as prodrugs, and improved effector functions which means significant advantages over small molecules inhibitors [64–66].

One strategy used to produce monoclonal antibodies against MMPs with the aim of inhibiting their enzymatic activity is immunizing with a synthetic molecule carrying the zinc-histidine complex present in these enzymes. On a matter of fact, this would block the ion disabling the enzyme from lowering the energy level of the chemical reaction, just as TIMPs do. With this strategy in mind, the mouse hybridomas producing SDS3 and SDS4 monoclonal antibodies were recovered, who inhibited not only MMP-9 and MMP-2 but also lower MMP-1, MMP-7, MMP-12, and MMP-14 activities [67]. These monoclonal antibodies have been criticized for being nonspecific, what would turn into a problem if they were used to target molecules to deliver chemotherapy drugs. Although all these MMPs are overexpressed at different levels in various tumoral microenvironments, other noncancer cells still produce them. So, the scientists should design a drug that also includes other darts in order to use these antibodies; otherwise, chemotherapy would spread all around the patient's body. To overcome this problem, microinjection directly in tumor site should be considered.

Monoclonal antibodies against MMP-9 (human gelatinase B) were prepared by immunizing mice with MMP-9 purified from human neutrophils. REGA-3G12 antibody had the highest capacity to inhibit the catalytic activity of MMP-9 and had no cross-reactivity with other MMPs [68]. Mouse monoclonal antibody AB0041 binds to an allosteric site in human MMP-9, inhibiting its enzymatic activity; reducing TNF-α in ulcerative colitis model, and tumor growth and metastasis in a mouse model of colorectal carcinoma. The antibody named GS-5745 or andecaliximab was then proposed to use in human trials [69], it was combined with a modified anticancer drug named FOLFOX-6 (leucovorin, fluorouracil and oxaliplatin). The clinical trial in phase III, GAMMA-1 failed to show overall improvement and the progression-free survival for untreated HER2-negative gastroesophageal adenocarcinomas; however, there was a slight higher survival median for old patients (65 years or more) compared to younger patients (less than 65 years).

One novel technique of highly efficient selection method for proteinase inhibitory mAbs was achieved by coexpressing 3 recombinant proteins in the periplasmic space of *Escherichia coli*—an antibody clone, a proteinase of interest, and a β-lactamase modified by the insertion of a proteinase-cleavable peptide sequence. During the functional selection, inhibitory antibodies prevent the proteinase from cleaving the modified β-lactamase, thereby allowing the cell to survive in the presence of ampicillin. Using this method, antibodies from synthetic human antibody libraries were identified for MMP-14 and MMP-9 important in both metastasis and tumor growth; antibody L13 inhibited MMP-9 but not MMP-2/MMP-12/MMP-14 [70].

The scFv variable library was used to select inhibition to human recombinant MMP-9 activity; the sequence was then inserted in the CDR of a human antibody, CALY-001. This antibody modulates the accumulation of collagen in intestinal grafts of fibrotic intestinal models by reducing it. To inhibit MMP-14, a peptide sequence GACFSIAHECGA, named peptide G, was selected by its ability catalytic activity but not other MMPs. This peptide reduces cell migration and invasion and diminishes tumor growth in xenografts; the fragment with sequence was inserted in the CDR of an antibody in order to gain stability, obtaining the 1F8 antibody similar to the antibody DX-2400 which blocks MMP-14's catalytic activity. Antibody 3A2 was obtained after a 27-aa peptide long discovered with phage display having a strong inhibition comparable to that of the chemical inhibitor GM6001 and TIMP-2 (tissue inhibitor of metalloproteinases); the major stability half-life was achieved when inserted in an IgG antibody [71,72].

Two scFvs antibodies were selected based on their affinity to the hemopexin domain of the MMP-14, scFv CHA, and scFv CHL. As hemopexin-like domain is an important conserved sequence between MMPs, the scFvs were tested for specific interactions among the other MMPs. The Fab 3369 is another antibody derived from a phage display library; it was selected for inhibiting the enzymatic activity of MMP-14 [73]. All these antibodies showed reduce extracellular matrix degradation, tumor growth, invasion, metastasis, and tube formation.

DX-2400 antibody was the first to target MMP-14 enzymatic activity, with promising results. Initially, it revealed to inhibit tube formation and cleave pro-MMP-2 and decreased generalized MMPs' activities. Next as an approach, tumors in animals treated with the antibody and a chemotherapy drug (either paclitaxel or bevacizumab) delayed tumor growth and metastasis [74]. Also, in murine breast cancer, animals treated with DX-2400 antibody develop a less immunosuppressive response compared to the controls, which leads to a better response to radiotherapy. There was less secretion of TGF-β and macrophages produced nitric oxide by the inducible nitric oxide synthase, enhancing an inflammatory response [75]. Immediately, toxicity test was done, founding no histological abnormalities, so the antibody could be proposed as a candidate for clinical trials. By 2012, two companies signed a contract to commercialize it and clinical trials are being developed.

More monoclonal antibodies have been designed, but they do not inhibit completely the catalytic activity of MMP-14. An important characteristic that was kept in mind when developing the antibody 9E8 was to increase specificity to molecules related to tumors. 9E8 antibody was derived from immunization and was found to recognize MMP-14 in the collagen-binding site, resulting in avoiding the breakdown of the pro-MMP-2 in its active form. The MMP-14 did not lose all the enzymatic activity but the main discovery was that the activation of MMP-2 was enough to reduce lymph vessel formation and metastasis [76]. There is a high correlation between worsen tumor progression and the expression of MMP-14 and active MMP-2, so this antibody is a promising solution for cancer and other disorders where there is neovascularization, because MMP-14 can still be functional without neovascularization induction.

On the contrary, the LEM-2/15 antibody specifically inhibits the collagenolytic activity of the MMP-14, inhibiting the catalytic activity of type 1 collagen and gelatin. This antibody was generated by immunizing an animal with a cyclic peptide containing a sequence far away from the catalytic site actually where part of TIMPs bound [77]. Just as TIMPs, LEM-2/15 causes a conformational change that narrows the substrate-binding cleft and, as a consequence, the enzymatic activity decreases.

Another strategy is the use of self-assembling peptides that are complexes that carry doxorubicin or paclitaxel, or other chemotherapy drugs, with a fragment of MMP-2's substrate. By this, the complex is targeted to cells expressing that enzyme, so the substrate fragment could be cleaved and the drug delivered. The mechanism discriminated between cancer cells overexpressing MMP-2 and cancer cells without the enzyme, increasing mortality in the first group of cells. Presumably, the results in vivo were congruent; mice with human tumor xenografts were treated with a protein complex carrying paclitaxel and not only reduced the tumor volume but, more important, increased mice survival [78].

(d) Drug delivery system

One of the main setbacks of chemotherapy is unwanted side effects, because of the highly cytotoxic compounds and doses needed to treat cancer. Strong efforts to develop formulations activated specifically by tumor cells would allow a reduction of systemic exposure while maintaining therapeutic concentrations in the tumor. The drug delivery system is one approach to solve this problem; the goal is to have a very specific molecule either peptide or antibody binding only to tumor cells and releasing a highly toxic compound within the cancer cells to specifically destroy them; the release can be regulated by enzymatic cleavage of proteinases. It has been extensively described previously the upregulation of MMPs during early and late stages of cancer, which makes MMPs a suitable candidate for this system. Here, we present some examples of these approaches.

Nanoformulations are a pharmaceutical packaging system designed for delivering the payload directly to the pharmacological target [79]. Organic or inorganic compounds can be used to prepare nanoformulation and generate nanoparticles. Some systems are (i) organic self-assembling phospholipids to formulate liposomal drug delivery systems (doxorubicin-calyx), successfully used in the treatment of breast cancer [80]; (ii) silica mesoporous-based nanoparticles (large porous structures and are able to encapsulate a variety of therapeutics) [81]; and (iii) block copolymers with enhanced biocompatibility and self-assembling features to generate nanosized drug delivery systems such as poly(dimethylsiloxane)-poly-b-(methyloxazoline) known as PDMS-PMOXA [82].

As mentioned above, MMP-9 is one MMP involved in tumor progression and enhanced in the tumor environment. One delivery system was developed with surface-modified PDMS-PMOXA polymersomes designed to release their cytotoxic payload upon digestion by MMP-9. Paclitaxel-loaded particles were tested on breast cancer cells; after treatment with the polymersomes, MCF7 cells significantly reduced cell viability, and this effect was abolished after the addition of MMP-inhibitors, suggesting proteolytic activation. Similar effect was observed in zebra fish embryos, xenografted with mKate2-expressing MCF7 cells; the amount of tumor cells decreased after treatment with PDMS-PMOXA-SRL-paclitaxel polymersomes detected by the copies of the heterologous expressed fluorescent protein, suggesting that polymersomes modified with an MMP-9 labile peptide and loaded with paclitaxel can be formulated and that these particles exert pharmacological activity upon enzymatic digestion [83].

Antibody-drug conjugates (ADCs) are a relatively new class of biological drugs—created by attaching a therapeutic agent to an antibody via a linker; this system provides several beneficial features: (i) the payloads used are highly toxic and therefore could not be used alone as single agents at an efficacious dose without severe off-target side effects; (ii) the monoclonal antibody endows exquisite specificity to a given antigen, thereby providing targeted delivery of the payload to the tumor cell, and also contributes to longer half-life which in some cases can allow weekly dosing, which has a positive impact on patient; and (iii) developments in linker and conjugation technology are designed to impair the release of payload before reaching its target, avoiding off-target toxicity [84].

By 2019, there are currently seven ADCs approved by the FDA on the market (all for oncology indications) and over eighty ADCs currently under clinical development. MMPs seem good targets for ADCs since they are overexpressed on the tumor/target cell surface, with lower expression in normal tissue. Ideally, linker-release technologies should demonstrate good internalization properties, as intracellular tracking to the lysosomal and endosomal compartments is key for many to work well. However, if the complex is recycled to the cell surface and payload is released into the tumor microenvironment, there can be bystander effects upon surrounding tumor cells which may increase the cytotoxic effect. One initial development of a noninternalizing ADC was employing a broad-spectrum of nonselective MMP inhibitor for conjugation and linked this to an MMP-9-targeting antibody, and the resulting ADC fully inhibits MMP-9 but not MMP-2, suggesting high specificity in the inhibition [85].

Conclusions

Although the results of some targeted drug experiments are encouraging in terms of tumor growth, they are criticized for being toxic for other noncancer cells, which also express the targeted molecule. Furthermore, even more supporting experiments reveal that the reason why a percentage of chemotherapy drugs arrive to the tumor is not because by affinity but by an increase in permeability and irrigation to the tissue [86,87]. It is important to remember that the bloodstream travels along the body at different velocities but on average 2 km/h and everything in it moves at a similar velocity. For this reason, it is being said that the mechanism that slows down the drug's impulse is the same physiology of tumors, but this does not imply that the drug can be released in other capillaries. So, the drugs should be designed to target specific molecules found only in tumors. So, the development of drugs has to be designed considering not only for the efficacy and affinity to a tumor but also for the physiology of the organism. That is why many attempts to develop specific treatments for cancer cells pass animal tests, but not in human trials, as variability between individual augments because of genetics, environment, nutrition, etc.

Here, we present options to take the advantage of the physiology and functions of some of the overexpressed molecules in tumors, the matrix metalloproteinases, because although they are ubiquitously present in several normal tissues, most of them are highly expressed in some tumors. Initial attempts to inhibit tumor growth by inhibiting MMP, failed because inhibitors were directed to the active site which is very conserved among family members and the amounts required to reach a therapeutic result, were excessive and caused musculoskeletal syndrome among several side effects; findings that MMP activity can be inhibited specifically by targeting molecular structures outside of the catalytic domain (named exosites) helped to overcome this problem [88].

Identification of the different exosites present in different MMPs allows targeting with synthetic, low molecular weight compounds or antibodies, specific function of a single MMP. This approach has led to the generation of highly selective mAbs to MMP-9, (AB0041 and AB0046, GS-5745), effective against colorectal carcinoma; mAbs to MMP-14 (LEM-2/15, LEM-2/63, and LEM-1/58), LEM-2/15 specifically inhibits MMP-14 degradation of gelatin and type I collagen without affecting its capacity to activate pro-MMP-2, an important function of MMP-14. Conversely, another antibody to MMP-14 (9E8), which is also highly selective for this MMP, has no effect on MMP-14 proteolytic activity, but inhibits pro-MMP-2 activation, showing that specific MMP functions can be selectively inhibited.

MMPs are secreted as inactive proenzymes, using endogenous or intrinsic MMP inhibitors; the enzymatic activity is inhibited by the intramolecular interaction of the catalytic domain with the N-terminal "pro" domain. Since the pro-domains differ from one MMP to another, the differences can be exploited to generate specific protein inhibitors. The exogenous addition of the pro-domains of ADAM10 and ADAM17 results in selective inhibition of the respective enzyme, without cross-reactivity in spite of the high similarity of the two ADAMs.

Protein engineering improves the inhibitory activity of antibodies and TIMPs. Anti-MMP-14 antibodies that effectively reduce tumor growth and metastasis in preclinical models have been generated by the selection of a phage display library of single-chain variable fragment (scFv), followed by protein engineering to increase their affinity and inhibitory activity. Point mutations in the sequence of TIMP-2 increase binding to and inhibition of MMP-14 by 9–14-folds; TIMP-2 mutants generated selectively block MMP-14 activity with an inhibition constant of 0.9 pmol/L, the strongest inhibitor of this MMP thus far generated.

In order to improve the affinity and inhibitory activity, the selection of a phage display library of single-chain variable fragment (scFv) followed by protein engineering specificity has been used in mouse models with certain success. For instance, mAb DX-400 is a high-affinity, highly selective inhibitor of MMP-14 that retards tumor growth and metastasis in several in vivo mouse models of breast cancer and melanoma, both as a single agent and in combination with paclitaxel or bevacizumab; mAb fragments (scFv) have also been developed to MMP-1, MMP-2, and MMP-3 by a combination of phage display library screening and combinatorial mutagenesis [89].

Taking advantage of high affinity, specificity, and selectivity of the MMPIs or antibodies, the drug delivery system can direct highly toxic compounds to be internalized in the cancer cells and after being hydrolyzed by the metalloproteinase become active and become a useful tool to break resistance to cancer therapies. Novel diagnostic techniques can analyze the MMP expression in very small biological samples or even single tumor cells; tumor cell DNA can also be detected in the circulation, providing a potential surrogate of metastasis. These techniques can allow the identification of the MMPs produced by a tumor, rapid assessment of the treatment efficacy, and therefore a precision medicine approach to anti-MMP treatment. As mentioned earlier, the failure of early clinical trials could be due to the lack of sensitive techniques to detect an improvement in the therapy and also the early diagnosis may allow the implementation of therapies at the beginning of the cancer where the effect of inhibition of MMPs' activity, involved in the tumor growth and metastasis, becomes essential.

Acknowledgments

The authors would like to thank Claudia A. Morales-Garay and Alejandra A. Morales-Garay for help with the design of some figures and tables.

References

[1] Bissell MJ, Radisky D. Putting tumours in context. Nat Rev Cancer 2001;1(1):46–54.
[2] Egeblad M, Werb Z. New functions for the matrix metalloproteinases in cancer progression. Nat Rev Cancer 2002;2(3):161–74.

[3] Scheau C, et al. The role of matrix metalloproteinases in the epithelial-mesenchymal transition of hepatocellular carcinoma. Anal Cell Pathol (Amst) 2019;2019:9423907.
[4] Cui N, Hu M, Khalil RA. Biochemical and biological attributes of matrix metalloproteinases. Prog Mol Biol Transl Sci 2017;147:1–73.
[5] Page-McCaw A, Ewald AJ, Werb Z. Matrix metalloproteinases and the regulation of tissue remodelling. Nat Rev Mol Cell Biol 2007;8(3):221–33.
[6] John A, Tuszynski G. The role of matrix metalloproteinases in tumor angiogenesis and tumor metastasis. Pathol Oncol Res 2001;7(1):14–23.
[7] Chen Q, et al. Matrix metalloproteinases: inflammatory regulators of cell behaviors in vascular formation and remodeling. Mediators Inflamm 2013;2013:928315.
[8] Tokuhara CK, et al. Updating the role of matrix metalloproteinases in mineralized tissue and related diseases. J Appl Oral Sci 2019;27:e20180596.
[9] Garay-Canales CA, Licona-Limon I, Ortega E. Distinct epitopes on CD13 mediate opposite consequences for cell adhesion. Biomed Res Int 2018;2018:4093435.
[10] Licona-Limon I, et al. CD13 mediates phagocytosis in human monocytic cells. J Leukoc Biol 2015;98(1):85–98.
[11] Gross J, Lapiere CM. Collagenolytic activity in amphibian tissues: a tissue culture assay. Proc Natl Acad Sci U S A 1962;48:1014–22.
[12] Pulkoski-Gross AE. Historical perspective of matrix metalloproteases. Front Biosci (Schol Ed) 2015;7:125–49.
[13] Balitskii KP, Pridato OE. Qualitative changes in collagen and its role in the process of carcinogenesis. Ukr Biokhim Zh 1967;39(2):220–2.
[14] Brkic M, et al. Friends or foes: matrix metalloproteinases and their multifaceted roles in neurodegenerative diseases. Mediators Inflamm 2015;2015:620581.
[15] Kapoor C, et al. Seesaw of matrix metalloproteinases (MMPs). J Cancer Res Ther 2016;12(1):28–35.
[16] Du J, et al. KEGG-PATH: Kyoto encyclopedia of genes and genomes-based pathway analysis using a path analysis model. Mol Biosyst 2014;10(9):2441–7.
[17] Kanehisa M, Goto S. KEGG: Kyoto encyclopedia of genes and genomes. Nucleic Acids Res 2000;28(1):27–30.
[18] Ala-aho R, Kahari VM. Collagenases in cancer. Biochimie 2005;87(3–4):273–86.
[19] Lo Presti R, Hopps E, Caimi G. Gelatinases and physical exercise: a systematic review of evidence from human studies. Medicine (Baltimore) 2017;96(37):e8072.
[20] Lech AM, Wiera G, Mozrzymas JW. Matrix metalloproteinase-3 in brain physiology and neurodegeneration. Adv Clin Exp Med 2019;28(12):1717–22.
[21] Remy L, Trespeuch C. Matrilysin-1 and cancer pathology. Med Sci (Paris) 2005;21(5):498–502.
[22] Guo JG, et al. High MMP-26 expression in glioma is correlated with poor clinical outcome of patients. Oncol Lett 2018;16(2):2237–42.
[23] Itoh Y. Membrane-type matrix metalloproteinases: their functions and regulations. Matrix Biol 2015;44-46:207–23.
[24] Molet S, et al. Increase in macrophage elastase (MMP-12) in lungs from patients with chronic obstructive pulmonary disease. Inflamm Res 2005;54(1):31–6.
[25] Gutierrez-Cantu FJ, et al. Amelogenin and enamelysin localization in human dental germs. In Vitro Cell Dev Biol Anim 2011;47(5–6):355–60.
[26] Brooks PC, et al. Localization of matrix metalloproteinase MMP-2 to the surface of invasive cells by interaction with integrin alpha v beta 3. Cell 1996;85(5):683–93.
[27] Yu WH, et al. CD44 anchors the assembly of matrilysin/MMP-7 with heparin-binding epidermal growth factor precursor and ErbB4 and regulates female reproductive organ remodeling. Genes Dev 2002;16(3):307–23.
[28] Sternlicht MD, Werb Z. How matrix metalloproteinases regulate cell behavior. Annu Rev Cell Dev Biol 2001;17:463–516.
[29] An HJ, et al. The prognostic role of tissue and serum MMP-1 and TIMP-1 expression in patients with non-small cell lung cancer. Pathol Res Pract 2016;212(5):357–64.
[30] Limb GA, et al. Matrix metalloproteinase-1 associates with intracellular organelles and confers resistance to lamin A/C degradation during apoptosis. Am J Pathol 2005;166(5):1555–63.
[31] Nawaz M, et al. Extracellular vesicles and matrix remodeling enzymes: the emerging roles in extracellular matrix remodeling, progression of diseases and tissue repair. Cell 2018;7(10):167.

[32] Jablonska-Trypuc A, Matejczyk M, Rosochacki S. Matrix metalloproteinases (MMPs), the main extracellular matrix (ECM) enzymes in collagen degradation, as a target for anticancer drugs. J Enzyme Inhib Med Chem 2016;31(sup1):177–83.
[33] Liu J, Khalil RA. Matrix metalloproteinase inhibitors as investigational and therapeutic tools in unrestrained tissue remodeling and pathological disorders. Prog Mol Biol Transl Sci 2017;148:355–420.
[34] Zurac S, et al. Variations in the expression of TIMP1, TIMP2 and TIMP3 in cutaneous melanoma with regression and their possible function as prognostic predictors. Oncol Lett 2016;11(5):3354–60.
[35] Duarte S, et al. Matrix metalloproteinases in liver injury, repair and fibrosis. Matrix Biol 2015;44-46:147–56.
[36] Inagaki Y, et al. Novel aminopeptidase N (APN/CD13) inhibitor 24F can suppress invasion of hepatocellular carcinoma cells as well as angiogenesis. Biosci Trends 2010;4(2):56–60.
[37] Kessenbrock K, Plaks V, Werb Z. Matrix metalloproteinases: regulators of the tumor microenvironment. Cell 2010;141(1):52–67.
[38] Chambers AF, Matrisian LM. Changing views of the role of matrix metalloproteinases in metastasis. J Natl Cancer Inst 1997;89(17):1260–70.
[39] Gobin E, et al. A pan-cancer perspective of matrix metalloproteases (MMP) gene expression profile and their diagnostic/prognostic potential. BMC Cancer 2019;19(1):581.
[40] Giunciuglio D, et al. Invasive phenotype of MCF10A cells overexpressing c-Ha-ras and c-erbB-2 oncogenes. Int J Cancer 1995;63(6):815–22.
[41] Kaya M, et al. A single ets-related transcription factor, E1AF, confers invasive phenotype on human cancer cells. Oncogene 1996;12(2):221–7.
[42] Shindoh M, Higashino F, Kohgo T. E1AF, an ets-oncogene family transcription factor. Cancer Lett 2004;216 (1):1–8.
[43] Maretzky T, et al. ADAM10 mediates E-cadherin shedding and regulates epithelial cell-cell adhesion, migration, and beta-catenin translocation. Proc Natl Acad Sci U S A 2005;102(26):9182–7.
[44] Radisky ES, Raeeszadeh-Sarmazdeh M, Radisky DC. Therapeutic potential of matrix metalloproteinase inhibition in breast cancer. J Cell Biochem 2017;118(11):3531–48.
[45] Galvez BG, et al. Membrane type 1-matrix metalloproteinase is activated during migration of human endothelial cells and modulates endothelial motility and matrix remodeling. J Biol Chem 2001;276(40):37491–500.
[46] Illman SA, et al. Epilysin (MMP-28) induces TGF-beta mediated epithelial to mesenchymal transition in lung carcinoma cells. J Cell Sci 2006;119(Pt 18):3856–65.
[47] Noe V, et al. Release of an invasion promoter E-cadherin fragment by matrilysin and stromelysin-1. J Cell Sci 2001;114(Pt 1):111–8.
[48] Nakahara H, et al. Transmembrane/cytoplasmic domain-mediated membrane type 1-matrix metalloprotease docking to invadopodia is required for cell invasion. Proc Natl Acad Sci U S A 1997;94(15):7959–64.
[49] Gutierrez-Fernandez A, et al. Matrix metalloproteinase-8 functions as a metastasis suppressor through modulation of tumor cell adhesion and invasion. Cancer Res 2008;68(8):2755–63.
[50] Vizoso FJ, et al. Study of matrix metalloproteinases and their inhibitors in breast cancer. Br J Cancer 2007;96 (6):903–11.
[51] Golub LM, et al. Doxycycline inhibits neutrophil (PMN)-type matrix metalloproteinases in human adult periodontitis gingiva. J Clin Periodontol 1995;22(2):100–9.
[52] Skiles JW, Gonnella NC, Jeng AY. The design, structure, and clinical update of small molecular weight matrix metalloproteinase inhibitors. Curr Med Chem 2004;11(22):2911–77.
[53] King J, et al. Randomised double blind placebo control study of adjuvant treatment with the metalloproteinase inhibitor, Marimastat in patients with inoperable colorectal hepatic metastases: significant survival advantage in patients with musculoskeletal side-effects. Anticancer Res 2003;23(1B):639–45.
[54] Folgueras AR, et al. Matrix metalloproteinases in cancer: from new functions to improved inhibition strategies. Int J Dev Biol 2004;48(5–6):411–24.
[55] Karin M, Chang L. AP-1–glucocorticoid receptor crosstalk taken to a higher level. J Endocrinol 2001;169 (3):447–51.
[56] Vihinen P, Kahari VM. Matrix metalloproteinases in cancer: prognostic markers and therapeutic targets. Int J Cancer 2002;99(2):157–66.
[57] Koivunen E, et al. Tumor targeting with a selective gelatinase inhibitor. Nat Biotechnol 1999;17(8):768–74.

[58] Scannevin RH, et al. Discovery of a highly selective chemical inhibitor of matrix metalloproteinase-9 (MMP-9) that allosterically inhibits zymogen activation. J Biol Chem 2017;292(43):17963–74.
[59] Beutel B, et al. New in vivo compatible matrix metalloproteinase (MMP)-2 and MMP-9 inhibitors. Bioconjug Chem 2018;29(11):3715–25.
[60] Kridel SJ, et al. Substrate hydrolysis by matrix metalloproteinase-9. J Biol Chem 2001;276(23):20572–8.
[61] Lu G, et al. Selection of peptide inhibitor to matrix metalloproteinase-2 using phage display and its effects on pancreatic cancer cell lines PANC-1 and CFPAC-1. Int J Biol Sci 2012;8(5):650–62.
[62] Romanov VI, Durand DB, Petrenko VA. Phage display selection of peptides that affect prostate carcinoma cells attachment and invasion. Prostate 2001;47(4):239–51.
[63] Maola K, et al. Engineered peptide macrocycles can inhibit matrix metalloproteinases with high selectivity. Angew Chem Int Ed Engl 2019;58(34):11801–5.
[64] Yu YJ, et al. Therapeutic bispecific antibodies cross the blood-brain barrier in nonhuman primates. Sci Transl Med 2014;6(261)261ra154.
[65] Bagshawe KD, Sharma SK, Begent RH. Antibody-directed enzyme prodrug therapy (ADEPT) for cancer. Expert Opin Biol Ther 2004;4(11):1777–89.
[66] Wang X, Mathieu M, Brezski RJ. IgG Fc engineering to modulate antibody effector functions. Protein Cell 2018;9 (1):63–73.
[67] Sela-Passwell N, et al. Antibodies targeting the catalytic zinc complex of activated matrix metalloproteinases show therapeutic potential. Nat Med 2011;18(1):143–7.
[68] Paemen L, et al. Monoclonal antibodies specific for natural human neutrophil gelatinase B used for affinity purification, quantitation by two-site ELISA and inhibition of enzymatic activity. Eur J Biochem 1995;234(3):759–65.
[69] Marshall DC, et al. Selective allosteric inhibition of MMP9 is efficacious in preclinical models of ulcerative colitis and colorectal Cancer. PLoS One 2015;10(5)e0127063.
[70] Lopez T, et al. Functional selection of protease inhibitory antibodies. Proc Natl Acad Sci U S A 2019;116 (33):16314–9.
[71] Fischer T, Riedl R. Inhibitory antibodies designed for matrix metalloproteinase modulation. Molecules 2019;24 (12):2265.
[72] Remacle AG, et al. Selective function-blocking monoclonal human antibody highlights the important role of membrane type-1 matrix metalloproteinase (MT1-MMP) in metastasis. Oncotarget 2017;8(2):2781–99.
[73] Ling B, et al. A novel immunotherapy targeting MMP-14 limits hypoxia, immune suppression and metastasis in triple-negative breast cancer models. Oncotarget 2017;8(35):58372–85.
[74] Devy L, et al. Selective inhibition of matrix metalloproteinase-14 blocks tumor growth, invasion, and angiogenesis. Cancer Res 2009;69(4):1517–26.
[75] Ager EI, et al. Blockade of MMP14 activity in murine breast carcinomas: implications for macrophages, vessels, and radiotherapy. J Natl Cancer Inst 2015;107(4)djv017.
[76] Ingvarsen S, et al. Targeting a single function of the multifunctional matrix metalloprotease MT1-MMP: impact on lymphangiogenesis. J Biol Chem 2013;288(15):10195–204.
[77] Udi Y, et al. Inhibition mechanism of membrane metalloprotease by an exosite-swiveling conformational antibody. Structure 2015;23(1):104–15.
[78] Hua D, et al. Potent tumor targeting drug release system comprising MMP-2 specific peptide fragment with self-assembling characteristics. Drug Des Devel Ther 2014;8:1839–49.
[79] Kamaly N, et al. Degradable controlled-release polymers and polymeric nanoparticles: mechanisms of controlling drug release. Chem Rev 2016;116(4):2602–63.
[80] O'Brien ME, et al. Reduced cardiotoxicity and comparable efficacy in a phase III trial of pegylated liposomal doxorubicin HCl (CAELYX/Doxil) versus conventional doxorubicin for first-line treatment of metastatic breast cancer. Ann Oncol 2004;15(3):440–9.
[81] Wang Y, et al. Mesoporous silica nanoparticles in drug delivery and biomedical applications. Nanomedicine 2015;11(2):313–27.
[82] Kaditi E, et al. Block copolymers for drug delivery nano systems (DDnSs). Curr Med Chem 2012;19(29):5088–100.
[83] Porta F, et al. Synthesis and characterization of PDMS-PMOXA-based polymersomes sensitive to MMP-9 for application in breast cancer. Mol Pharm 2018;15(11):4884–97.
[84] Beck A, et al. Strategies and challenges for the next generation of antibody-drug conjugates. Nat Rev Drug Discov 2017;16(5):315–37.

[85] Love EA, et al. Developing an antibody-drug conjugate approach to selective inhibition of an extracellular protein. Chembiochem 2019;20(6):754–8.
[86] Chau Y, et al. Investigation of targeting mechanism of new dextran-peptide-methotrexate conjugates using biodistribution study in matrix-metalloproteinase-overexpressing tumor xenograft model. J Pharm Sci 2006;95(3):542–51.
[87] Chau Y, et al. Antitumor efficacy of a novel polymer-peptide-drug conjugate in human tumor xenograft models. Int J Cancer 2006;118(6):1519–26.
[88] Levin M, et al. Next generation matrix metalloproteinase inhibitors—novel strategies bring new prospects. Biochim Biophys Acta, Mol Cell Res 2017;1864(11 Pt A):1927–39.
[89] Winer A, Adams S, Mignatti P. Matrix metalloproteinase inhibitors in cancer therapy: turning past failures into future successes. Mol Cancer Ther 2018;17(6):1147–55.

Index

Note: Page numbers followed by *f* indicate figures, and *t* indicate tables.

K

L

M

N

O

Kolibri 360 GmbH
Ferdinand-Jühlke-Straße 7
99095 Erfurt, Deutschland
produktsicherheit@kolibri360.de